Lecture Notes in Computer Science   1372

Edited by G. Goos, J. Hartmanis and J. van Leeuwen

Springer-Verlag Berlin Heidelberg GmbH

Serge Vaudenay (Ed.)

Fast Software Encryption

5th International Workshop, FSE '98
Paris, France, March 23-25, 1998
Proceedings

 Springer

Series Editors

Gerhard Goos, Karlsruhe University, Germany
Juris Hartmanis, Cornell University, NY, USA
Jan van Leeuwen, Utrecht University, The Netherlands

Volume Editor

Serge Vaudenay
Ecole Normale Supérieure, DMI
45, rue d'Ulm, F-75230 Paris Cedex 05, France
E-mail: serge.vaudenay@ens.fr

Cataloging-in-Publication data applied for

Die Deutsche Bibliothek - CIP-Einheitsaufnahme

Fast software encryption : 5th international workshop ;
proceedings / FSE '98, Paris, France, March 23 - 25, 1998. Serge
Vaudenay (ed.).

(Lecture notes in computer science ; Vol. 1372)
ISBN 978-3-540-64265-7

CR Subject Classification (1991): E.3, E.4, F.2.1, G.4

ISSN 0302-9743
ISBN 978-3-540-64265-7   ISBN 978-3-540-69710-7 (eBook)
DOI 10.1007/978-3-540-69710-7

Typesetting: Camera-ready by author
SPIN 10631900   06/3142 – 5 4 3 2 1 0   Printed on acid-free paper

Preface

Fast Software Encryption (FSE) is an annual research workshop devoted to the promotion of research on classical encryption algorithms and related cryptographic primitives such as hash functions.

When public key cryptography started to receive wide attention in the 1980s, the much older and more basic art of secret key cryptography was sidelined at many research conferences. This motivated Ross Anderson to organise the first FSE in Cambridge, England in December 1993; subsequent workshops followed at Leuven, Belgium (December 1994), Cambridge again (February 1996), and Haifa, Israel (January 1997).

These proceedings contain the papers due to be presented at the fifth FSE workshop in March 1998 at the Hotel du Louvre in Paris. This event is organized by the Ecole Normale Supérieure and the Centre National pour la Recherche Scientifique (CNRS) in cooperation with the International Association for Cryptologic Research (IACR), and has attracted the kind support of Gemplus and Microsoft, the world leaders in smart cards and software – two domains very closely connected to our research concerns.

As in previous years, the FSE workshop presents new advances in the design and analysis of (non-public-key) cryptographic primitives: block ciphers, stream ciphers, message authentication codes, hash functions, and pseudorandom generators. It is the tradition that new algorithm designs are presented with concrete examples; other researchers then try to break them and give results at the next workshop. This continuing competition between cryptographer and cryptanalyst has forced the evolutionary pace. It has been highly effective both at advancing the state of the art, and at enabling the community to develop simultaneously its practical technology and its theoretical insight into the security of cryptographic primitives.

It has been a great pleasure to organize the fifth FSE workshop in Paris and to chair the programme committee. The committee consists of Ross Anderson (Cambridge University), Eli Biham (Technion), Don Coppersmith (IBM Research), Cunsheng Ding (National University of Singapore), Dieter Gollmann (Microsoft Research), Lars Knudsen (University of Bergen), Jim Massey (ETH Zurich), Mitsuru Matsui (Mitsubishi), and Bart Preneel (Katolieke Universiteit Leuven).

We received 39 submissions. Each was carefully reviewed by at least three committee members, the authors were sent substantial reports, and 20 papers were selected for presentation at the workshop. These papers are presented in this book.

This year, our FSE proceedings consist of:

- six papers which break previously proposed algorithms,
- two papers on the cryptanalysis of modes of operation of block ciphers,
- one paper on the analysis of pseudorandom generators,

– three new dedicated stream cipher proposals,
– three new dedicated block cipher proposals,
– two papers on the design and analysis of hash functions, and
– three papers on the analysis of dedicated constructions with formal security arguments.

As with previous FSE workshops, the proceedings are published in Springer-Verlag's Lecture Notes in Computer Science series. (Previous proceedings were volumes 809, 1008, 1039, and 1267.) As with previous workshops, the entire review and selection process was done electronically using email and the LaTeX text processing language.

To conclude, I would like to thank all the authors who submitted papers and all members of the programme committee, as well as Alex Biryukov and Stafford Tavares, who acted as external referees. I also wish to thank the IACR, Springer-Verlag, the Hotel du Louvre, the Ecole Normale Supérieure, and particularly Jacques Stern, Brigitte van Elsen, and Dominique Ho Tin Noe. Finally, we are grateful to our generous sponsors Gemplus and Microsoft, and particularly, David Naccache, Gideon Yuval, and Ramarathnam Venkatesan.

January 1998   **Serge Vaudenay**

Table of Contents

New Results in Linear Cryptanalysis of RC5

Ali Aydın Selçuk *

Department of Computer Science and Electrical Engineering
University of Maryland Baltimore County
Baltimore, MD, 21250, USA
aselcu1@cs.umbc.edu

Abstract. We show that the linear cryptanalytic attack on RC5 that was published by Kaliski and Yin at Crypto'95 does not work as expected due to the failure of some hidden assumptions involved. Then we present new linear attacks on RC5. Our attacks use the same linear approximation as the one used by Kaliski and Yin. Therefore, the plaintext requirement of our attack is around $4w^{2r-2}$ which is impractically high for reasonably high values of w and r. These new attacks has also significances beyond the linear cryptanalysis of RC5 to show how linear cryptanalysis can carry on when the approximation used has a non-zero bias for the wrong key values. We also discuss certain issues about linear cryptanalysis of RC5 that need to be resolved for a better linear attack.

Keywords: Cryptology, cryptanalysis, block ciphers, RC5, linear cryptanalysis.

1 Introduction

RC5 is a secret key block cipher designed by Rivest [5]. Kaliski and Yin [1] published a linear cryptanalytic attack on RC5 at Crypto'95, which still remains as the only general linear attack on RC5 that has been published in the open literature. We are going to refer to this attack as the *K-Y Attack*. In this paper, we show that the K-Y Attack does not work as expected due to the failure of some hidden assumptions involved. Then we present some new attacks. Our attacks are based on the same linear approximation used in the K-Y Attack, but they are different from that attack in the way they use the approximation to recover the secret key.

We first briefly review RC5 and linear cryptanalysis. RC5 has a variable block size, a variable number of rounds, and a variable length secret key. A particular RC5 algorithm is defined by these three parameters and denoted as RC5-$w/r/b$: w, the word size in bits (half of a block is called a *word*); b, the key size in bytes; r, the number of rounds. For the encryption algorithm, we adopt the notation used in [1]. The algorithm is as follows:

---

* This research was done while the author was visiting RSA Laboratories.

S. Vaudenay (Ed.): Fast Software Encryption – FSE'98, LNCS 1372, pp. 1–16, 1998.

$$L_1 = L_0 + S_0$$
$$R_1 = R_0 + S_1$$
$$\textbf{for } i = 2 \textbf{ to } 2r + 1 \textbf{ do}$$
$$L_i = R_{i-1}$$
$$R_i = ((L_{i-1} \oplus R_{i-1}) \lll R_{i-1}) + S_i$$

In the algorithm, "$+$" denotes addition modulo 2^w, "$\oplus$" denotes bitwise xor, "$\lll$" denotes left rotation. The ith iteration of the for loop is referred as the ith *half-round*. The *first* half-round refers to the two initial equations. L_{i-1}, R_{i-1} denote the left and the right halves of the input and S_i denotes the subkey at the ith half-round. (L_0, R_0) is the plaintext, (L_{2r+1}, R_{2r+1}) is the ciphertext.

Linear cryptanalysis is a kind of statistical correlation attack for block ciphers which was developed by Matsui [4] in 1993. The basic idea of linear cryptanalysis is to find a linear relation, which is called an *approximation*, among the plaintext, ciphertext and key bits, such that the probability of the approximation is different from $1/2$. But, if wrong values are substituted for the key bits in the approximation, the approximation will behave randomly (i.e. its probability will be $1/2$). The attacker collects plaintext/ciphertext pairs which are encrypted under the same key. Then he tries all possible combinations for the key bits involved in the approximation with all the plaintext/ciphertext pairs he has collected. The correct key bit combination is distinguished by its non-random behavior. Matsui [3] showed that the success probability of the attack is proportional to $N|p - 1/2|^2$ where N is the number of plaintext/ciphertext pairs collected.

Some specific notation used in this paper is as follows. RC5-w/r denotes the RC5 scheme with w bit words and r rounds where each round key is generated independently. $x[i]$ denotes the ith bit of a binary string x, and $x[i \ldots j]$ denotes the ith through j^{th} bits of x; n denotes $2r + 1$.

The remainder of the paper is organized as follows. In §2 we discuss the hidden assumptions in the K-Y Attack and explain why they do not hold. In §3–7 we present our attacks and discuss their success rates. In §8 we conclude with some open research problems regarding linear cryptanalysis of RC5 and discuss briefly the factors that make linear cryptanalysis of RC5 harder than linear cryptanalysis of DES-like ciphers.

2   Hidden Assumptions in the K-Y Attack

In this section, we briefly discuss the K-Y Attack and show the hidden assumptions that cause the attack to fail.

2.1   The Attack

Let T denote $S_1[0] \oplus S_3[0] \oplus \cdots \oplus S_{2r-1}[0]$. The approximation used in the K-Y Attack is

$$R_0[0] \oplus L_{2r}[0] = T \tag{1}$$

which is obtained by combining the half-round approximation

$$R_1[0] = R_0[0] \oplus S_1[0]$$

and the half-round approximation

$$R_i[0] = L_{i-1}[0] \oplus S_i[0]$$

for $i = 3, 5, \ldots, 2r - 1$. The probability of this approximation, which we denote by p, is $\frac{1}{2} + \frac{1}{2w^{r-1}}$.

The K-Y Attack was different from all the previously published linear cryptanalytic attacks in the way that it was composed of multiple steps where each step aimed to recover one bit of the round key. The outline of the attack algorithm is as follows:

Step 1: Guess $S_n[0]$ by using the data with $L_n \bmod w = 1$.
Step 2: Guess T by using the data with $L_n \bmod w = 0$.
Step 3: For $i = 1, \ldots, w - 1$, guess $S_n[i]$ by using the data with $L_n \bmod w = i$.

The point that is important for us in this algorithm is that, at each step, $L_n \bmod w$ is *fixed* to a certain value.

2.2   Hidden Assumptions

We implemented this attack on RC5-16/2 with $2|p - 1/2|^{-2}$ plaintext/ciphertext pairs for each different value of $L_n \bmod w$ (i.e. $w \times 2|p - 1/2|^{-2}$ texts in total). The success rate we observed for recovering S_n was around 11–15% as opposed to the 95–99% that was expected by Kaliski and Yin. Even more surprisingly, the success rate did not improve as we increased the amount of data used. These results led us to the following observations.

Let AH_i, for $i = 3, 5, \ldots, 2r - 1$, denote the event that the ith half-round approximation $R_i[0] = L_{i-1}[0] \oplus S_i[0]$ holds. The probability of the approximation $P(\text{AH}_i)$ can be calculated as

$$P(\text{AH}_i) = P(\text{AH}_i \mid R_{i-1} \bmod w = 0) \cdot P(R_{i-1} \bmod w = 0)$$
$$+ P(\text{AH}_i \mid R_{i-1} \bmod w \neq 0) \cdot P(R_{i-1} \bmod w \neq 0).$$

$P(\text{AH}_i \mid R_{i-1} \bmod w = 0)$ is always equal to 1. $P(R_{i-1} \bmod w = 0)$ is equal to $1/w$ and $P(\text{AH}_i \mid R_{i-1} \bmod w \neq 0)$ is equal to $1/2$, hence $P(\text{AH}_i)$ is equal to $1/2 + 1/2w$, *given that the input (or the output) of the ith half-round is uniformly random.*

An important point in the K-Y Attack is that at each step the value of $L_n \bmod w$ is *fixed* to a certain value and it is implicitly assumed that the probability of Approximation (1) does not depend on $L_n \bmod w$. This assumption is based on two other assumptions:

1. The probability $P(R_{i-1} \bmod w = 0)$ does not depend on $L_n \bmod w$;
2. The probability $P(\text{AH}_i \mid R_{i-1} \bmod w \neq 0)$ does not depend on $L_n \bmod w$.

We will refer to these two assumptions as *Assumption 1* and *Assumption 2*, respectively.

2.3   On Assumption 1

We observe that the probability of a zero rotation in the $n - 2^{nd}$ half-round, i.e. $P(R_{n-3} \bmod w = 0)$, depends on $L_n \bmod w$, hence Assumption 1 does not hold:

$$
\begin{aligned}
R_{n-3} &= L_{n-2} \\
&= ((R_{n-1} - S_{n-1}) \ggg R_{n-2}) \oplus R_{n-2} \\
&= ((L_n - S_{n-1}) \ggg R_{n-2}) \oplus R_{n-2}.
\end{aligned}
$$

Therefore, when $L_n - S_{n-1}$ is fixed, the distribution of R_{n-3} is *not* uniform. Hence, the probability $P(R_{n-3} \bmod w = 0)$, and therefore the probability of Approximation (1), is *not* independent of $L_n \bmod w$.

As an example, let δ denote the difference $L_n - S_{n-1} \bmod w$ and ρ' denote $R_{n-3} \bmod w$. The probability $P(\rho' = 0 \,|\, \delta = 0)$ for $w = 16$ is $1.56/w$ as opposed to the expected probability $1/w$.

2.4   On Assumption 2

We also observe that Assumption 2, just like Assumption 1, does not hold for the $n - 2^{nd}$ half-round; i.e. $P(\mathrm{AH}_{n-2} \,|\, \rho' \neq 0)$ is *not* independent of $L_n \bmod w$:

First, we observe that the half-round approximation

$$
R_{n-2}[0] = L_{n-3}[0] \oplus S_{n-2}[0] \tag{2}
$$

can be expressed in terms of *only* R_{n-2}, S_{n-2}, and ρ'. The approximation is

$$
\begin{aligned}
R_{n-2}[0] &= L_{n-3}[0] \oplus S_{n-2}[0] \\
&= (R_{n-2} - S_{n-2})[\rho'] \oplus R_{n-3}[0] \oplus S_{n-2}[0] \\
&= (R_{n-2} - S_{n-2})[\rho'] \oplus \rho'[0] \oplus S_{n-2}[0].
\end{aligned}
$$

Second, we know from Section 2.3 that ρ' (i.e. $R_{n-3} \bmod w$) is equal to $(((L_n - S_{n-1}) \ggg R_{n-2}) \oplus R_{n-2}) \bmod w$. Therefore, we observe that conditions on $L_n - S_{n-1}$ and ρ' together give information about $R_{n-2} \bmod w$.

These two observations imply that when $L_n - S_{n-1}$ and S_{n-2} are fixed, the condition $\rho' \neq 0$ gives information about Approximation (2) and possibly causes the probability $P(\mathrm{AH}_{n-2} \,|\, \rho' \neq 0)$ to be different from $1/2$. For example, for $w = 16$, $L_n - S_{n-1} = 1$, $S_{n-2} = 0$, the probability $P(\mathrm{AH}_{n-2} \,|\, \rho' \neq 0)$ is equal to 0.494 as opposed to 0.5.

At this point we should remark that the probability of the approximation does not depend on the top $w - \lg w$ bits of S_{n-2}. This fact is because ρ' and $L_n - S_{n-1}$ give information about only the last $\lg w$ bits of R_{n-2}; therefore $(R_{n-2} - S_{n-2})[\rho']$ has a uniform distribution when $\rho' \geq \lg w$, regardless of S_{n-2}.

2.5   Overall Impact of the Assumptions

In every step of the attack the difference $L_n - S_{n-1}$ is fixed. When $L_n - S_{n-1}$ is fixed, R_{n-3} and R_{n-2} have a non-uniform distribution. This fact has two effects on the probability of the $n - 2^{nd}$ half-round approximation: First, the probability $P(R_{n-3} \bmod w = 0)$ may be different from $1/w$. Second, the probability $P(AH_{n-2} \mid R_{n-3} \bmod w \neq 0)$ may be different from $1/2$.

Table 4 in Appendix A lists the bias of Approximation (2) for different values of $L_n - S_{n-1} \bmod w$ and $S_{n-2} \bmod w$ for $w = 16$. What is particularly important in Table 4 regarding the attack of Kaliski and Yin is the negative entries which correspond to the case $p < 1/2$. Success of Steps 2 and 3 of the attack depends on the assumption that $p > 1/2$ for every single value of $L_n \bmod w$. If the $(i, j)^{th}$ entry of Table 4 is negative, then for $S_{n-2} \bmod w = j$ the attack fails with very high probability at the step where $L_n - S_{n-1} \bmod w$ is fixed to i and the failure probability goes to one as the amount of data used goes to infinity. With respect to the numbers in Table 4, we calculated that the average success rate of the attack for recovering the last round key S_n in RC5-16/2 goes to 9.375% as the amount of data goes to infinity. We also calculated the success rate with $2|p - 1/2|^{-2}$ texts as 13.9%. These results matched our experimental results in Section 2.2 very well.

3 New Attacks

We developed a number of new linear cryptanalytic attacks on RC5. They all use Approximation (1), but they are different from the attack of Kaliski and Yin in the way they use the approximation to recover the round key S_n. Our attacks are similar to "Algorithm 2" of Matsui [3] which is sometimes referred as the *1R-method*. We unroll the last round and substitute the actual value of $L_{n-1}[0]$ in Approximation (1), which is $(R_n - S_n)[\rho] \oplus L_n[0]$, where ρ denotes $L_n \bmod w$ (i.e. the rotation amount in the last half-round). So, the approximation becomes

$$R_0[0] \oplus (R_n - S_n)[\rho] \oplus L_n[0] = T. \tag{3}$$

An important difference of our attacks from the 1R-method of Matsui is that, when we substitute a wrong value s for S_n in Approximation (3), the bias of the approximation is *not* zero. Moreover, the bias can be expressed in terms of s, S_n, ρ and the probability of the approximation, as will be shown in Section 4.

Our attacks can be classified into two types: In the first one, we fix ρ (i.e. $L_n \bmod w$) to a certain value at each step and we aim to recover one key bit at a time. We will refer to the attacks of this type as the *1-bit attacks*. In the second type of attack, we aim to recover a group of consecutive key bits at the same time. We will refer to the attacks of this type as the *multi-bit attacks*. We will describe the attacks in more detail in Sections 5 and 6.

An important issue regarding the experimental comparison of the attacks presented in the following sections is that they are all run on relatively small versions of RC5 such as $r = 2, 4$. The reason for this choice of small parameters is just to make the experiments computationally feasible. Our attack techniques all use the same approximation (i.e. Approximation (3)), and they differ only in the way they use this approximation to recover the secret key. Therefore, increasing the number of rounds does not have much effect on the relative performance of

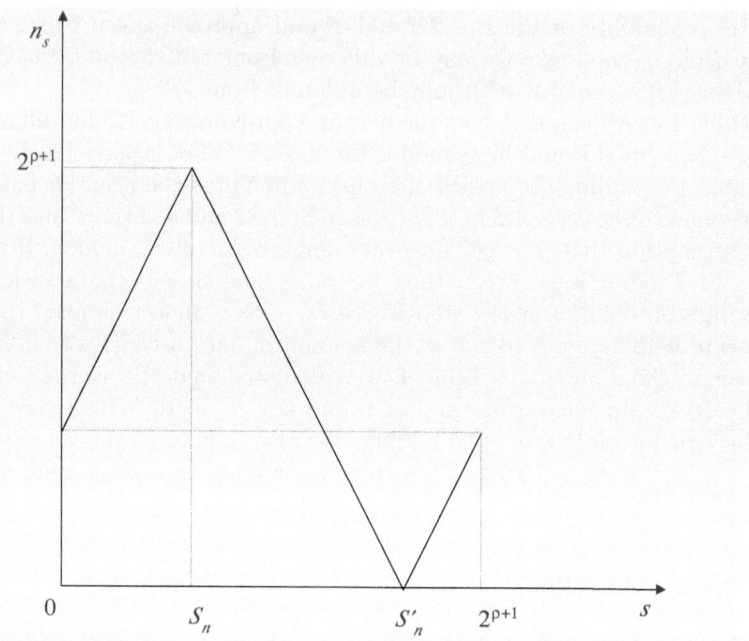

Fig. 1. The variable s denotes the possible key values that can be tried for $S_n[0\ldots\rho]$. n_s denotes the number of different values of $R_n[0\ldots\rho]$ such that $(R_n - s)[\rho] = (R_n - S_n)[\rho]$. S'_n is the key that differs from S_n only at the ρth bit.

these attack techniques and the experiments with relatively small values of r give a general comparison of the attacks.

4   Bias for a Wrong Key

As mentioned in Section 3, something special with Approximation (3) is that, when a wrong key value s is substituted for S_n, the bias of the approximation is not zero. An important observation to understand the behavior of the approximation is the following. When s is substituted for S_n in Approximation (3), the result is the same as the result for S_n if and only if $(R_n - s)[\rho]$ is the same as $(R_n - S_n)[\rho]$. These two bits agree when one of the following two conditions is satisfied:

Let S_{min} denote $\min\{S_n[0\ldots\rho - 1], s[0\ldots\rho - 1]\}$ and similarly, let S_{max} denote $\max\{S_n[0\ldots\rho - 1], s[0\ldots\rho - 1]\}$:

1. $s[\rho] = S_n[\rho]$, and $R_n[0\ldots\rho - 1] < S_{min}$ or $R_n[0\ldots\rho - 1] \geq S_{max}$.
2. $s[\rho] \neq S_n[\rho]$ and $S_{min} \leq R_n[0\ldots\rho - 1] < S_{max}$.

Let n_s denote the number of different values of $R_n[0\ldots\rho]$ such that we have $(R_n - s)[\rho] = (R_n - S_n)[\rho]$. Figure 1 illustrates the value of n_s for $0 \leq s < 2^{\rho+1}$.

More specifically, n_s is $2^{\rho+1}$ for the correct key S_n; it decreases by two as s gets further from S_n in either direction; and it is zero at S'_n, which denotes the key that differs from S_n only at the ρth bit.

Assuming that the probability that the approximation holds and the probability that the result of the approximation is the same for both s and S_n are independent (where both probabilities are taken over the plaintext), we obtain a similar figure for the bias of the approximation with s substituted for S_n. Figure 2 shows the expected bias of Approximation (3) for different values of s. Let N denote the number of plaintext/ciphertext pairs satisfying $L_n \bmod w = \rho$ for some fixed ρ. Let U_s denote the number of those texts such that the left side of Approximation (3) is 1 when we substitute s for S_n, and let B_s denote the bias $U_s - N/2$. Figure 2 illustrates the expected bias $E[B_s]$ for $0 \le s < 2^{\rho+1}$, assuming $E[B_{S_n}] > 0$.

The significance of Figure 2 is that it shows what the expected bias of Approximation (3) will be when $L_n \bmod w$ is fixed and a wrong value s is substituted for the round key S_n. This behavior of the bias has a crucial role in the attacks we develop in this paper, especially in the 1-bit attacks (see Section 5).

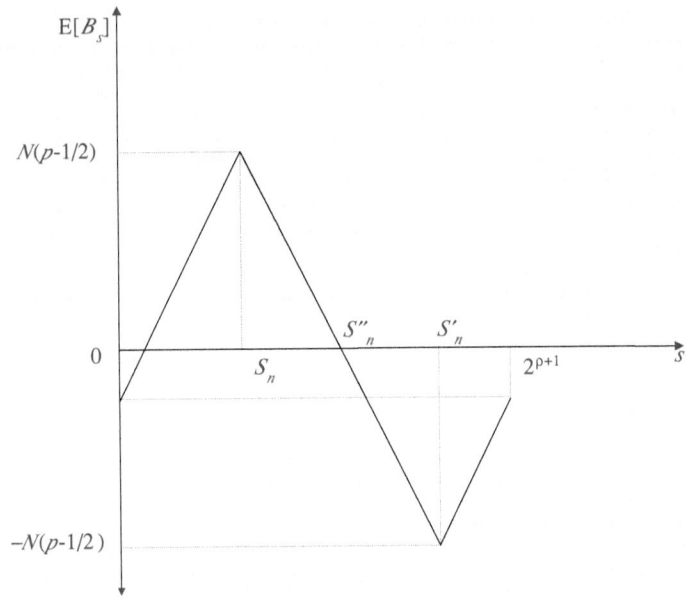

Fig. 2. Expected bias for different values of s for $L_n \bmod w = \rho$. S''_n is the key that differs from the correct key S_n only at the $\rho - 1^{st}$ bit. p denotes the probability of the approximation given $L_n \bmod w = \rho$.

5   1-Bit Attacks

In this section we discuss the attacks that recover the round key S_n in a bitwise fashion (i.e. recovering one bit at a time). The aim of these attacks is to recover the key bit $S_n[\rho - 1]$ by using the data with $L_n \bmod w = \rho$ and given that the key bits $S_n[0 \ldots \rho - 2]$ are already recovered. The idea of attacking the $\rho - 1^{st}$ bit instead of the ρth one is inspired by the fact that p may be either less or greater than $1/2$ depending on the value of ρ, S_{n-1} and S_{n-2} (see Appendix A). Moreover, S_n', the number that differs from S_n only at the ρth bit, has the exact inverse bias of the correct key S_n, (i.e. $B_{S_n'} = -B_{S_n}$) over the data with $L_n \bmod w = \rho$ (Figure 2). Therefore, we cannot distinguish between S_n and S_n' by using the data with $L_n \bmod w = \rho$ since we do not know if $p > 1/2$ or not; and we cannot know if $S_n[\rho]$ is 0 or 1. But this is not the case for $S_n[\rho - 1]$. S_n'', the number that differs from S_n only at the $\rho - 1^{st}$ bit, has a zero expected bias over the data with $L_n \bmod w = \rho$ regardless of p (Figure 2). Therefore, we can distinguish between S_n and S_n'', and hence find out $S_n[\rho - 1]$ by using the data with $L_n \bmod w = \rho$, even if $p < 1/2$.

All of our 1-bit attacks are based on a generic attack algorithm. Assume we have recovered the key bits $S_n[0 \ldots \rho - 2]$ and let s_0 and s_1 denote the two candidates $0 \| S_n[0 \ldots \rho - 2]$ and $1 \| S_n[0 \ldots \rho - 2]$ respectively, where $\|$ denotes string concatenation. A_{s_i}, for $i = 0, 1$, is a statistical variable which is supposed to be large for the correct key and small for a wrong key. The generic attack algorithm is as follows:

Generic 1-Bit Attack
Step 1: Compute A_{s_i} for $i = 0, 1$.
Step 2: If $A_{s_0} \geq A_{s_1}$ guess $S_n[\rho - 1] = 0$; otherwise guess $S_n[\rho - 1] = 1$.

Our 1-bit attacks are defined by their definition of the variable A_{s_i}. Let $\mathcal{S}_{s_i}$ denote the set of points in the $2^{\rho-2}$ neighborhood of s_i; i.e. the set defined by $\mathcal{S}_{s_i} = \{s : |s - s_i| \leq 2^{\rho-2}\}$:

Attack 1: $A_{s_i} = |B_{s_i}|$.
Attack 2: $A_{s_i} = |\sum_{s \in \mathcal{S}_{s_i}} B_s|$.
Attack 3: $A_{s_i} = \sum_{s \in \mathcal{S}_{s_i}} |B_s|$.
Attack 4: $A_{s_i} = \max_{s \in \mathcal{S}_{s_i}} \{|B_s|\}$.

Intuitively, Attack 1 simply compares the bias of s_0 and s_1. The other three attacks on the other hand, also use the biases of the points in the $2^{\rho-2}$ neighborhood of s_0 and s_1 (the choice of $2^{\rho-2}$ is because $s_0 + 2^{\rho-2}$ is the mid-point of s_0 and s_1). As will be discussed shortly, our experiments have shown that Attack 1 has the best success rate among the four.

We calculate the success rate of Attack 1 as

$$\int_{-\infty}^{\infty} \int_{-|x+2\sqrt{N}|p-1/2|\,|}^{|x+2\sqrt{N}|p-1/2|\,|} \frac{1}{\sqrt{2\pi}} e^{-y^2/2} dy \frac{1}{\sqrt{2\pi}} e^{-x^2/2} dx, \qquad (4)$$

and the success rate of Attack 2 as

$$\int_{-\infty}^{\infty} \int_{-|x+1.83\sqrt{N}|p-1/2||}^{|x+1.83\sqrt{N}|p-1/2||} \frac{1}{\sqrt{2\pi}} e^{-y^2/2} dy \frac{1}{\sqrt{2\pi}} e^{-x^2/2} dx. \tag{5}$$

It is not straightforward to obtain a closed-form theoretical result for the success rate of Attacks 3 and 4. Therefore, we compared the attacks experimentally on RC5-16/2 on a sample of 10,000 different cases and for different values of ρ. The experimental results indicated that Attack 1 is the best among the four. For Attacks 1 and 2, the experimental results matched the theoretical results given in (4) and (5) very well.

5.1   A Generalization

Attacks 1 and 2 are special cases of a more general attack, which we call *Attack G*:

Let $\mathcal{S}_{s_i,d}$ denote the d neighborhood of s_i; i.e. $\mathcal{S}_{s_i,d} = \{s : |s - s_i| \le d\}$. Attack G uses the generic attack algorithm with $A_{s_i} = |\sum_{s \in \mathcal{S}_{s_i,d}}(B_s)|$. Notice that this is the same as Attack 1 for $d = 0$ and the same as Attack 2 for $d = 2^{\rho-2}$.

We calculate the success rate of Attack G approximately as

$$\int_{-\infty}^{\infty} \int_{-|x+2(1-\frac{d}{3\cdot 2^\rho-2d})\sqrt{N}|p-1/2||}^{|x+2(1-\frac{d}{3\cdot 2^\rho-2d})\sqrt{N}|p-1/2||} \frac{1}{\sqrt{2\pi}} e^{-y^2/2} dy \frac{1}{\sqrt{2\pi}} e^{-x^2/2} dx,$$

which is maximized at $d = 0$. This result implies that Attack 1 has the highest success rate among all versions of Attack G.

5.2   An Improvement

A limitation of Attack 1, and also other 1-bit attacks discussed so far, is a zero bias (i.e. $p = 1/2$) which occurs for certain values of ρ, S_{n-1} and S_{n-2} (see Appendix A). In such cases, these attacks are no better than random guessing. One way to overcome this problem is to use the data with $L_n \bmod w > \rho$ as well as those with $L_n \bmod w = \rho$ to recover $S_n[\rho - 1]$.

We present such a modification of Attack 1, which we call *Attack 1'*. It uses the data with $L_n \bmod w = \rho+1$ as well as the data with $L_n \bmod w = \rho$ to guess $S_n[\rho - 1]$:

As in Attack 1, let s_0 and s_1 denote the two candidates $0\|S_n[0 \ldots \rho - 2]$ and $1\|S_n[0 \ldots \rho - 2]$ respectively, and similarly let s_i denote $i\|S_n[0 \ldots \rho - 2]$ for $i = 00, 01, 10, 11$. The idea of Attack 1' is to compare the bias for four possible key candidates $s_{00}, s_{01}, s_{10}, s_{11}$. If the bias is maximized for s_{00} or s_{10}, we guess $S_n[\rho - 1]$ as 0, otherwise we guess it as 1. The calculation of the bias for these four points is as follows. As in Attack 1, B_{s_0} and B_{s_1} denote the bias for s_0 and s_1 taken over the data with $L_n \bmod w = \rho$. Similarly, B_{s_i} denotes the bias for s_i for $i = 00, 01, 10, 11$, *but taken over the data with $L_n \bmod w =$*

$\rho + 1$. We guess $S_n[\rho - 1] = 0$ if $|B_{s_0}| + \max\{|B_{s_{00}}|, |B_{s_{10}}|\}$ is greater than $|B_{s_1}| + \max\{|B_{s_{01}}|, |B_{s_{11}}|\}$. Otherwise, we guess $S_n[\rho-1] = 1$. (This is the generic 1-bit attack algorithm where A_{s_i} is defined as $|B_{s_i}| + \max\{|B_{s_{0i}}|, |B_{s_{1i}}|\}$.)

We experimentally compared Attack 1 and Attack 1' on RC5-16/2. The experimental success rates are given in Table 1. N denotes the available number of texts for each particular value of $L_n \bmod w$. The results show that Attack 1' is significantly better than Attack 1.

N	1,000	4,000	10,000	40,000	100,000
Attack 1	79.8%	91.9%	96.2%	98.4%	99.1%
Attack 1'	86.8%	96.4%	98.0%	99.5%	99.9%

Table 1. Success rate of Attack 1 and 1' on RC5-16/2 for recovering one bit of the last round key S_n. The experimental results show that Attack 1' is significantly better than Attack 1.

6   Multi-bit Attack

The idea of the multi-bit attack is quite straightforward: Instead of fixing $L_n \bmod w$ at each step, calculate the bias over the data with many different values of $L_n \bmod w$. We know that when $L_n \bmod w$ is fixed, the behavior of the bias of a wrong key is not random (see Section 4). By taking the bias over many different values of $L_n \bmod w$, we hope that the bias will behave more "normally" (i.e. zero expected bias for a wrong key, positive expected bias for the correct key).

6.1   The Attack

Although the formal description of the multi-bit attack may appear complicated. In fact, it is really intuitive. Suppose we have already recovered the key bits $S_n[0 \ldots k]$ and we are going to recover the next ℓ bits $S_n[k + 1 \ldots k + \ell]$. The bias for each key candidate is computed over the data with $k+1 \leq L_n \bmod w \leq k + \ell$. The one with the highest bias is accepted. The formal description of the algorithm is as follows:

U_s denotes the number of the texts such that the left side of Approximation (3) is 1 when we substitute s for S_n. The expression $S_n[0 \ldots k]$ denotes the part of the round key that has been recovered so far. ℓ denotes the number of the key bits that is attacked at one iteration of the algorithm. Once these ℓ bits are recovered, the algorithm is repeated for the next ℓ bits of S_n.

Attack M (k, ℓ)

Step 1: For $0 \leq i < 2^\ell$, compute $U_{i \| S_n[0...k]}$ over the data with $k + 1 \leq L_n \bmod w \leq k + \ell$.

Step 2: Accept i that maximizes the bias $|U_{i\|S_n[0...k]} - N/2|$, where N is the number of data with $k + 1 \leq L_n \bmod w \leq k + \ell$.

The choice of the parameter ℓ is a matter of trade-off. The computational complexity of the attack increases as ℓ gets larger. More specifically, the number of active text bits at an iteration of Attack M is $k + 1 + \ell$, and the number of active key bits is ℓ. Hence, the computational complexity of an iteration of Attack M is $2^{2\ell+k+1}$ (see Matsui [3]). Therefore, the computational complexity of recovering a round key of w bits is $2^{2\ell} \cdot \frac{2^w-1}{2^\ell-1}$, for ℓ dividing w. On the other hand, the reliability of the guesses also increases as ℓ gets larger, especially those of the low order bits, as will be shown in Section 6.2. Therefore, the value of ℓ should be decided with respect to the constraints of the available computational power, time and the desired success rate.

But Attack M has some limitations. For example, suppose we are trying to recover the key bits $S_n[k + 1 \ldots k + \ell]$, and let S'_n denote the key that is the same as S_n in every bit except for the $k + \ell^{th}$ one. The bias for S_n and S'_n taken over the data with $k + 1 \leq L_n \bmod w < k + \ell$ will be exactly the same, since they are exactly the same at bits $0, 1, \ldots, k + \ell - 1$. Taken over the data with $L_n \bmod w = k + \ell$, the bias for S'_n will be the inverse of the bias for S_n (see Section 4). Therefore, when the bias for $L_n \bmod w = k + \ell$ is negative (i.e. $p < 1/2$), we incorrectly deduce that S'_n is the correct key with very high probability! Similar arguments apply to the lower order bits as well, but their effect is less significant. This fact implies that the guesses for the higher order bits will not be very reliable, as illustrated by the experimental results in Section 6.2.

6.2   Experimental Results

We tested Attack M on RC5-16/2 for $\ell = 6, 8, 10$ on a sample of 10,000 different cases. Our results are given in Table 2. The entries in the tables are given as a percentage of the 10,000 trials. The ith column of the tables denotes the percentage of guesses that are correct at the bits lower than i, but wrong at the ith bit. Another important point about the tables is that, the data amount N denotes the available number of texts for *each* different value of $L_n \bmod w$ (e.g. for $\ell = 10$, the total number of texts used is $10 \times N$). We chose this way of presentation to make the comparison between the tables easier.

The experimental results show that the success rate for the lower order bits improves as ℓ increases. But this improvement becomes less significant for higher data amounts. Another important point is that increasing the data amount does not help beyond a certain point and the failure rates at the low order bits are almost stabilized around 0.8–0.9%.

The high failure rates at higher order bits are due to the effect discussed at the end of Section 6.1, and the success rate for these bits cannot be improved beyond a certain point, even with unlimited amount of data. Therefore, we suggest discarding the top two bits guessed, and starting the next iteration of the attack to include these bits as well. The size of the discarded part may be different for

$\ell = 6$	failure at bit					
N	0	1	2	3	4	5
1,000	8.4%	5.5%	5.0%	4.4%	5.4%	8.1%
10,000	1.6%	1.9%	2.4%	1.9%	4.1%	7.1%
100,000	1.2%	1.4%	1.6%	1.5%	3.4%	7.2%
1,000,000	0.9%	1.5%	1.7%	1.6%	3.6%	7.2%

$\ell = 8$	failure at bit							
N	0	1	2	3	4	5	6	7
1,000	7.1%	4.7%	3.9%	3.7%	3.8%	4.9%	6.0%	7.7%
10,000	1.4%	1.3%	1.4%	1.4%	1.9%	2.2%	4.1%	6.8%
100,000	0.9%	0.9%	0.8%	1.3%	1.5%	1.5%	3.4%	7.5%
1,000,000	0.9%	0.9%	0.9%	1.2%	1.1%	1.3%	3.4%	7.3%

$\ell = 10$	failure at bit									
N	0	1	2	3	4	5	6	7	8	9
1,000	5.6%	4.7%	4.3%	3.6%	3.8%	4.0%	3.8%	4.8%	5.6%	7.6%
10,000	1.1%	1.2%	1.3%	1.2%	1.4%	1.4%	1.6%	2.1%	3.7%	7.2%
100,000	0.9%	0.9%	1.0%	0.9%	0.9%	1.1%	1.4%	1.3%	3.6%	6.9%
1,000,000	0.8%	0.7%	0.9%	0.9%	0.9%	1.2%	1.1%	1.4%	3.2%	7.2%

Table 2. Failure rates of Attack M on RC5-16/2 for $\ell = 6, 8, 10$. The ith column represents the percentage of guesses that are correct at the bits lower than i, but wrong at the ith bit. The results show that the attack gets better as ℓ increases; but this improvement is less significant when the amount of data used is higher.

different word sizes, and should be determined experimentally (or theoretically if possible). We denote the size of the discarded part by j and add the following step to the algorithm Attack M:

Step 3: Discard the top j bits of the key estimated.

7 Comparison of 1-Bit and Multi-Bit Attacks

We compared the two attack strategies experimentally on RC5-16/r for $r = 2, 4$. Table 3 lists the success rates of Attack 1' and Attack M for guessing the first eight bits of S_n. Attack 1' represents the most successful 1-bit attack. N denotes the number of plaintexts available for each different value of $L_n \bmod w$. As discussed in Section 3, even though the experiments are run for relatively small values of r, they give a general comparison of the attacks, mainly because the relative performance of the attacks will not be affected much by an increase in r since they all use the same approximation.

The results suggest that Attack M has a better success rate for smaller amounts of data but Attack 1' becomes better as the amount of available data

r	N	Attack 1'	Attack M
2	1,000	32.1%	65.5%
	10,000	88.2%	88.7%
	100,000	98.3%	91.7%
	1,000,000	99.5%	92.1%
4	1,000	0.6%	0.7%
	10,000	0.5%	0.8%
	100,000	2.3%	6.5%
	1,000,000	20.3%	22.0%

Table 3. Success rates of Attack 1' and Attack M on RC5-16/r for recovering the first eight bits of S_n. The results show that Attack M is better for smaller amounts of available data. The success rates fall sharply as r increases.

increases. An advantage of Attack 1' over Attack M is that the success rate of Attack M does not improve much beyond 92% regardless of the increase in the data amount. But there is no such limit on the success rate of Attack 1'. Besides, the 1-bit attacks have two other advantages over the multi-bit attacks. First, they are computationally less expensive. Second, a wrong guess in a 1-bit attack can be detected earlier and can be corrected more easily since the biases after a wrong bit guess will be significantly smaller than what is expected.

The dramatic decrease in the success rates as the number of rounds r increases suggests that our attacks are not practical enough to break RC5 for larger values of r and w. This thought is also supported by the fact that all of our attacks are based on Approximation (1) which has a quite low bias for larger values of r and w. At this point, it is not possible to calculate the exact success rates for a given amount of data. This fact is due to the lack of a concrete formula for the relation between the probability of Approximation (1) and $L_n \bmod w$. However, we conjecture that the data requirement for a significant success rate will be comparable to $|p - 1/2|^{-2}$, that is $4w^{2r-2}$ which is impractically high for reasonably high values of w and r (i.e. $w \geq 32$, $r \geq 6$).

8   Conclusions

We presented some new results about linear cryptanalysis of RC5. First, we showed that the attack of Kaliski and Yin [1] does not work as expected due to some unexpected consequences of fixing $L_n \bmod w$. We studied the statistical behavior of Approximation (1). Then, we presented some new techniques of using this approximation to recover the last round key S_n.

Our results on the attack of Kaliski and Yin has significances beyond the linear cryptanalysis of RC5: It is significant to emphasize that hidden assumptions may have extremely serious consequences. It is also significant to show that extreme care has to be taken when applying a method developed for a specific cipher to a cipher of different type.

The attacks we presented in this paper are examples of how linear cryptanalysis can carry on when the bias is different from zero for a wrong key substituted in the approximation. At this point, it is not possible to calculate the exact success rate of our attacks due to the lack of a concrete formula for the relation between the probability of Approximation (1) and L_n mod w. However, we conjecture that the data requirement for a significant success rate will be comparable to $|p - 1/2|^{-2}$, which is impractically high for our approximation. Therefore, we believe that RC5 still remains secure against linear cryptanalysis.

There are many open research problems that are to be solved about the linear cryptanalysis of RC5. An important one is to obtain a theoretical result for the relation between the probability of Approximation (1), and $L_n - S_{n-1}$ and S_{n-2}. In this way, it will be possible to obtain theoretical results for the success rate of the attacks that are based on Approximation (1), including the ones presented in this paper. Moreover, it should be possible to use such a relation in an attack which obtains further information about the round keys S_{n-1} and S_{n-2}.

Another significant improvement will be to develop better linear cryptanalytic attacks than the ones presented here. However, any attack based on Approximation (1) will be limited by the low bias of that approximation. Therefore, finding a better approximation is essential to improving the linear cryptanalysis of RC5 significantly. But any researcher trying to find a better linear approximation should be aware of a proposition of Kaliski and Yin [1] that states a limitation of linear approximations of RC5.

A way to circumvent this limitation may be to use non-linear Approximations [2]; not just at the end rounds, but at the intermediate rounds as well. The main reason for using linear approximations in DES-like ciphers is that it is easy to find approximations of S-boxes since they are relatively small. Moreover, if an approximation of an S-box is linear it can be distributed to and stated in terms of the input, output and key bits of that round. But this argument is not true for RC5 since it does not have small sub-blocks like S-boxes. Moreover, using linear approximations does not have the advantage of being easily distributed to input, output and key bits as it is in DES. Therefore we believe that, at least theoretically, finding a non-linear approximation of RC5 is not substantially more difficult than finding a linear approximation. But it should be noted that finding an approximation for RC5 is not an easy task in general since there are no small sub-blocks like S-boxes.

As a last minute note, we have recently found out that the probability p of Approximation (1) is *not* equal to $\frac{1}{2} + \frac{1}{2w^{r-1}}$ which was calculated by Kaliski and Yin. The reason of this unexpected result is that the two consecutive half-round approximations $R_i[0] = L_{i-1}[0] \oplus S_i[0]$ and $R_{i+2}[0] = L_{i+1}[0] \oplus S_{i+2}[0]$ are not independent, and therefore, the piling-up lemma cannot be used to calculate the probability of Approximation (1). We experimentally found out that the probability p is extremely key dependent and it can be a lot different from $\frac{1}{2} + \frac{1}{2w^{r-1}}$ depending on the key. Even when averaged over the keys, the probability of Approximation (1) is a lot different from $\frac{1}{2} + \frac{1}{2w^{r-1}}$. Now, this new finding leaves the whole issue of linear cryptanalysis of RC5 as an open question.

Acknowledgments

I gratefully acknowledge the invaluable discussions with Matt Robshaw and Lisa Yin. I would like to thank to Alan Sherman for his comments on the paper and I would like to thank to Leo Reyzin for his help with the implementation of the attacks. I am thankful to all my friends at the RSA Labs for their support and friendship during my visit there.

References

1. B. S. Kaliski and Y. L. Yin. On differential and linear cryptanalysis of the RC5 encryption algorithm. In D. Coppersmith, editor, *Advances in Cryptology — Crypto'95*, pages 171–184. Springer Verlag, New York, 1995.
2. L. Knudsen and M. Robshaw. Non-linear approximations in linear cryptanalysis. In U. Maurer, editor, *Advances in Cryptology — Eurocrypt'96*, pages 224–236. Springer-Verlag, New York, 1996.
3. M. Matsui. Linear cryptanalysis of DES cipher (I). *Journal of Cryptology*, to appear.
4. M. Matsui. Linear cryptanalysis method for DES cipher. In T. Helleseth, editor, *Advances in Cryptology — Eurocrypt'93*, pages 386–397. Springer-Verlag, Berlin, 1994.
5. R. Rivest. The RC5 encryption algorithm. In *Fast Software Encryption, Second International Workshop*, pages 86–96. Springer-Verlag, New York, 1995.

A   Bias of the Approximation

The probability of the half-round approximation

$$R_{n-2}[0] = L_{n-3}[0] \oplus S_{n-2}[0] \tag{6}$$

depends on the value of $(L_n - S_{n-1}) \bmod w$ and $S_{n-2} \bmod w$. This fact implies that when L_n, S_{n-1} and S_{n-2} are fixed, the bias of the approximation may be different from its average bias $1/2w$. Table 4 lists the bias of Approximation (6) for different values of $(L_n - S_{n-1}) \bmod w$ and $S_{n-2} \bmod w$ for $w = 16$. These values are computed by exhaustively going through all possible values of (L_n, R_n). The parameter δ denotes the difference $(L_n - S_{n-1}) \bmod w$. The entries of the tables are the actual biases as a proportion of the average bias $1/2w$ (i.e. $(p - 1/2)/(1/2w)$).

δ	0	1	2	3	4	5	6	7
				$S_{n-2} \bmod w$				
0	2.25	0.75	1.75	2.13	1.63	1.50	0.75	1.88
1	0.50	-1.25	1.50	2.63	-0.38	-0.25	1.00	1.88
2	2.25	1.75	0.75	2.13	-0.38	0.50	1.75	-0.13
3	1.00	0.25	2.50	0.13	2.63	0.75	2.50	0.38
4	0.75	0.25	0.25	0.63	0.13	1.00	0.25	0.88
5	1.00	0.25	0.00	1.13	0.13	1.25	0.50	0.88
6	-0.25	1.25	1.25	-0.38	0.13	1.00	-0.75	1.88
7	0.50	2.75	2.00	-0.38	2.13	1.25	2.00	1.38
8	0.75	1.00	1.00	1.13	1.13	0.75	1.00	0.88
9	1.00	1.00	0.75	1.63	1.13	1.00	1.25	0.88
10	0.75	1.00	1.00	1.13	1.13	0.75	1.00	0.88
11	1.50	1.50	0.75	1.13	2.13	1.00	1.75	1.38
12	0.75	1.00	0.50	0.63	1.13	0.75	0.50	0.38
13	1.00	1.00	0.25	1.13	1.13	1.00	0.75	0.38
14	0.75	1.00	0.50	0.63	0.13	1.75	-0.50	1.38
15	1.50	2.50	1.25	0.63	2.13	2.00	2.25	0.88

δ	8	9	10	11	12	13	14	15
				$S_{n-2} \bmod w$				
0	1.88	0.88	1.38	2.25	1.75	1.38	1.13	1.75
1	0.38	-1.38	1.38	2.50	0.00	-0.63	1.13	2.00
2	1.88	1.88	0.38	2.25	-0.25	0.38	2.13	-0.25
3	2.38	2.63	-0.13	2.50	0.50	2.88	0.13	2.00
4	-0.13	0.88	1.38	0.25	0.75	0.38	1.13	0.25
5	0.38	0.63	1.38	0.50	1.00	0.38	1.13	0.50
6	0.88	-0.13	2.38	-0.75	0.75	0.38	0.13	1.25
7	0.38	2.63	1.88	0.50	1.50	1.88	1.13	1.50
8	0.63	0.88	0.88	1.00	1.00	0.88	1.13	1.00
9	1.13	0.63	0.88	1.25	1.25	0.88	1.13	1.25
10	0.63	0.88	0.88	1.00	1.00	0.88	1.13	1.00
11	1.13	1.63	0.38	1.25	1.75	1.38	1.13	1.25
12	0.63	0.88	0.38	0.50	1.00	0.88	0.63	0.50
13	1.13	0.63	0.38	0.75	1.25	0.88	0.63	0.75
14	1.63	-0.13	1.38	-0.50	1.00	0.88	0.63	0.50
15	1.13	2.63	0.88	0.75	1.75	2.38	1.63	0.75

Table 4. Bias of the $n - 2^{nd}$ half-round approximation as a proportion of the expected bias $1/2w$ when $L_n \bmod w$ is fixed, for $w = 16$. The variable δ denotes the difference $(L_n - S_{n-1}) \bmod w$. The numbers in the table show that how big an impact the hidden assumptions had on the bias of the $n - 2^{nd}$ half-round approximation.

Higher Order Differential Attack
of a CAST Cipher

Shiho Moriai[1], Takeshi Shimoyama[1], and Toshinobu Kaneko[1,2]

[1] TAO (Telecommunications Advancement Organization of Japan)
{shiho,shimo}@yokohama.tao.or.jp
[2] Science University of Tokyo
kaneko@ee.noda.sut.ac.jp

Abstract. This paper proposes a new higher order differential attack. The higher order differential attack proposed at FSE'97 by Jakobsen and Knudsen used exhaustive search for recovering the last round key. Our new attack improves the complexity to the cost of solving a linear system of equations. As an example we show the higher order differential attack of a CAST cipher with 5 rounds. The required number of chosen plaintexts is 2^{17} and the required complexity is less than 2^{25} times the computation of the round function. Our experimental results show that the last round key of the CAST cipher with 5 rounds can be recovered in less than 15 seconds on an UltraSPARC station.

1 Introduction

Higher order differential attack is one of the powerful algebraic cryptanalyses. It is useful for attacking ciphers which can be represented as Boolean polynomials with low degrees. After Lai mentioned cryptographic significance of derivatives of Boolean functions in [12], Knudsen used this notion to attack ciphers which were secure against conventional differential attacks[11]. At FSE'97 Jakobsen and Knudsen[7] gave an extension of Knudsen's attacks and broke the cipher with quadratic functions such as the cipher $\mathcal{KN}$[14] and the scheme by Kiefer[10]. These were provably secure ciphers against differential and linear cryptanalysis. Furthermore, at ISW'97, Shimoyama, Moriai, and Kaneko[18] essentially reduced the complexity and the number of chosen plaintexts required for the higher order differential attack of the cipher $\mathcal{KN}$. In this paper we generalize the higher order differential attack described in [18] and apply it to CAST ciphers.

CAST ciphers are a family of symmetric ciphers constructed using the CAST design procedure[1] proposed by Adams and Tavares. The CAST design procedure describes that they appear to have good resistance to differential cryptanalysis[5], linear cryptanalysis[15], and related-key cryptanalysis[4]. A known attack on CAST ciphers is the attack which uses weaknesses of non-surjective round functions and it requires 2^{32} known texts for a CAST cipher with 6 rounds[16].

In this paper we demonstrate that some of symmetric ciphers constructed using the CAST design procedure can be broken by our higher order differential

S. Vaudenay (Ed.): Fast Software Encryption – FSE'98, LNCS 1372, pp. 17–31, 1998.
© Springer-Verlag Berlin Heidelberg 1998

attack, if the number of rounds is small. CAST-128 is a famous example CAST cipher used in several commercial applications, but this is not our target. CAST-128 seems resistant to our attack.

CAST ciphers use the Feistel structure used in DES. The CAST design procedure allows a wide variety of round functions. It has substitution boxes (S-boxes) with fewer input bits than output bits (e.g. 8×32). There are several proposals for S-boxes. For example, [3] suggested constructing the S-boxes from bent functions. Later on [6] CAST ciphers with random S-boxes were proposed. In our attack, we use the S-boxes proposed for CAST-128[1,2] based on bent functions. As for operations used for combining input and subkey or output results of S-boxes, the CAST design procedure describes that a simple way is to specify that all operations are XORs. Although other operations (addition and subtraction modulo 2^{32}, multiplication modulo ($2^{32} \pm 1$), etc.) may be used instead, we assume that the CAST cipher of our target uses XORs for all operations.

We explain the higher order differential attack of this CAST cipher with 5 rounds. We begin by finding the Boolean polynomials of all output bits of S-boxes. The polynomials show that all degrees are 4. When all operations in the round function are XORs, the degree of the round function is at most 4. If the right half of plaintext is fixed at any value, the degree of the right half of the 4-th round is at most 16, and the 16-th order differential becomes constant. Thus we can construct the attack equations for recovering the last (i.e. 5-th) round key.

In [7], exhaustive search was used for finding the true key. If their attack were applied to this CAST cipher with 5 rounds, the required complexity would be 2^{48} times the computation of the round function using 2^{17} chosen plaintexts. Our new attack can recover the last round key by solving the linear system of equations. As a result, the required number of chosen plaintexts is 2^{17} and the required complexity is reduced to less than 2^{25} times the computation of the round function. Our experimental results show that all last round key bits of the CAST cipher can be recovered in less than 15 seconds on a Sun Ultra 2 workstation (UltraSPARC, 200MHz).

2   Higher Order Differential Attack

2.1   Preliminaries

Definition 1. *Let $GF(2)^n$ be the n-dimensional vector space over $GF(2)$. We denote addition on $GF(2)^n$ by $+$. We define $V^{(l)}[a_l, \ldots, a_1]$ as the l-dimensional subspace of $GF(2)^n$ which is the set of all 2^l possible linear combinations of $a_l, \ldots, a_1$, where each a_i is in $GF(2)^n$ and linearly independent. We often use $V^{(l)}$ for $V^{(l)}[a_l, \ldots, a_1]$ when $\{a_l, \ldots, a_1\}$ is understood.*

Definition 2. *Let $GF(2)[X]$ be the polynomial ring of $X = \{x_{n-1}, \ldots, x_0\}$ over $GF(2)$. Let Id be the ideal of $GF(2)[X]$ generated by $x_{n-1}^2 + x_{n-1}, \ldots, x_0^2 + x_0 \in GF(2)[X]$. We define $R[X]$ as the quotient ring of $GF(2)[X]$ modulo Id, i.e.*

$GF(2)[X]/Id$. We call $R[X]$ the Boolean polynomial ring of X, and call each element of it a Boolean polynomial. Since an element in $R[X]$ is regarded as a function: $GF(2)^n \rightarrow GF(2)$, it is also called a Boolean polynomial function or a Boolean function.

Definition 3. Let X and K be sets of variables. We define $R[X, K]$ as a Boolean polynomial ring of $X \cup K$, i.e. $R[X \cup K]$. For element $f(X, K) \in R[X, K]$, we define $\deg_X(f)$ as the total degree of f with respective to X whose coefficients are in $R[K]$.

Definition 4. We call a vector of n Boolean functions a vector Boolean function. For example $f = (f_{n-1}, \ldots, f_0)$ is a vector Boolean function where f_i is a Boolean function. Each element of a vector Boolean function is called a coordinate Boolean function.

For an iterated block cipher with block size $2n$ bits and key size s bits, we denote a plaintext by $x = (x_{2n-1}, \ldots, x_0) \in GF(2)^{2n}$, a key by $k = (k_{s-1}, \ldots, k_0) \in GF(2)^s$, and a ciphertext by $y = (y_{2n-1}, \ldots, y_0) \in GF(2)^{2n}$. The ciphertext y is represented by a vector Boolean function $y = G(x, k) = (g_{2n-1}(x, k), \ldots, g_0(x, k)) \in R[X, K]^{2n}$, where X and K are sets of variables: $X = \{x_{2n-1}, \ldots, x_0\}$, $K = \{k_{s-1}, \ldots, k_0\}$. A coordinate Boolean function of $G(x, k)$ is a Boolean function $g[k](x)$ on X when k is fixed. In general $g[k](x)$ is represented as follows,

$$g[k](x) = \sum c_{i_{2n-1}, \ldots, i_0}(k) \cdot x_{2n-1}^{i_{2n-1}} \cdots x_0^{i_0},$$

where i_* is 0 or 1.

Definition 5. We define the i-th order differential of $g[k](x)$ with respect to X, denoted by $\Delta_{(a_i, \ldots, a_1)}^{(i)} g[k](x)$, as follows,

$$\Delta_{(a)}^{(1)} g[k](x) = g[k](x) + g[k](x + a)$$
$$\Delta_{(a_i, \ldots, a_1)}^{(i)} g[k](x) = \Delta_{(a_i)}^{(1)}(\Delta_{(a_{i-1}, \ldots, a_1)}^{(i-1)} g[k](x))$$

where $\{a_i, \ldots, a_1\} \subseteq GF(2)^{2n}$ are linearly independent and any $a \in GF(2)^{2n}$. In this paper, since we consider only the higher order differential with respect to X, we omit "with respect to X."

Definition 6. We define the i-th order differential of vector Boolean function $G = (g_{2n-1}, \ldots, g_0)$ as follows,

$$\Delta_{(a_i, \ldots, a_1)}^{(i)} G = (\Delta_{(a_i, \ldots, a_1)}^{(i)} g_{2n-1}, \ldots, \Delta_{(a_i, \ldots, a_1)}^{(i)} g_0).$$

The following propositions are known on the higher order differential of Boolean functions.

Proposition 7. [12] *The following equation holds for any $a \in GF(2)^{2n}$ and $\{a_i, \ldots, a_1\} \subseteq GF(2)^{2n}$.*

$$\Delta^{(i)}_{(a_i,\ldots,a_1)} g[k](a) = \sum_{x \in V^{(i)}[a_i,\ldots,a_1]} g[k](x + a)$$

Proposition 8. [18] *Let $\{a_{d+1}, \ldots, a_1\} \subseteq GF(2)^{2n}$ be linearly independent. If $\deg_X(g[k](x)) = d$, then we have the following equations.*

$$\Delta^{(d)}_{(a_d,\ldots,a_1)} g[k](x) \in R[K],$$
$$\Delta^{(d+1)}_{(a_{d+1},\ldots,a_1)} g[k](x) = 0$$

2.2   Attack Procedure

The following is the attack procedure of our higher order differential attack of an iterated block cipher with block size $2n$ and R rounds. Let $k^{(i)}$ be the i-th round subkey, and each subkey be m bits, i.e. $k^{(i)} = (k^{(i)}_{m-1}, \ldots, k^{(i)}_0)$. Let $K^{(i)}$ be a set of variables of $k^{(i)}$, i.e. $K^{(i)} = \{k^{(i)}_{m-1}, \ldots, k^{(i)}_0\}$.

Here we describe "$(R-1)$-round attack", where we find a certain constant value which is independent of the key (e.g. the higher order differential of the output of the $(R-1)$-th round), and construct the attack equations for recovering the last round key. Of course a "$(R-2)$-round attack" is possible though solving of the attack equations becomes rather difficult.

We assume that the attacker has or can compute all chosen plaintexts in $V^{(d)}[a_d, \ldots, a_1] + a$ and the corresponding ciphertexts, where d is the total degree of the output of the $(R-1)$-th round, and a is any value in $GF(2)^{2n}$. For some block ciphers, the total degree of the output of the $(R-1)$-th round may be difficult with choices of $\{a_d, \ldots, a_1\}$ and a. However we don't consider it in this paper.

Even if the algorithm of the cipher is not open (i.e. if it is a blackbox), our attack is applicable when we know the total degree d of the output of the $(R-1)$-th round by some ways. In this case, we start from step 2.

1. Find the degree of round function.   In attacking iterated ciphers by higher order differential attacks, it is useful to represent the round function by Boolean polynomials. We can get the degree over $GF(2)$ of each output bit of the round function from these polynomials. The information on which terms are included in the polynomials is also helpful in step 3.

We begin by representing S-boxes by Boolean polynomial functions. When the description of the S-boxes is not given as some algebraic expressions, we construct Boolean polynomial functions from the description tables (see Section 4.1).

2. Compute the higher order differential of output of the $(R-1)$-th round. Our higher order differential attack is possible, for an integer $1 \leq d \leq 2n$, when the d-th order differential of the output of the $(R-1)$-th round is a certain constant value which is independent of the key. When is this condition true? One is when the degree of the output of the $(R-1)$-th round is $d-1$. Another is when the input and subkeys are combined with XORs simply, and the degree of the output of the $(R-1)$-th round is d. In this case, the total degree of the output of the $(R-1)$-th round with respect to X is equal to the total degree with respect to X and $K^{(i)}$ $(i = 1, \cdots, R-1)$ before the degree reaches $2n$ (see [18, Proposition 1]).

In these cases, the d-th order differential of the output of the $(R-1)$-th round can be computed by using Proposition 1 without knowing the true key.

3. Construct attack equations for recovering the last round key. We give the details in the case of a Feistel cipher. Let $x = (x_L, x_R)$ and $y = (y_L(x), y_R(x))$, where x_L denotes the left half of plaintext, x_R denotes the right half, y_L denotes the Boolean polynomial function of left half of ciphertext, and y_R denotes the vector Boolean polynomial function of right half. Let $\tilde{y}_R(x)$ be the vector Boolean polynomial function of the right half of the output of the $(R-1)$-th round. Then we have

$$F[k^{(R)}](y_R(x)) + y_L(x) = \tilde{y}_R(x).$$

If the d-th order differential of $\tilde{y}_R(x)$ is constant, we have the following equation for linearly independent $\{a_i, \ldots, a_1\} \subseteq GF(2)^{2n}$ and any $a \in GF(2)^{2n}$.

$$\Delta^{(d)}_{(a_d,\ldots,a_1)} F[k^{(R)}](y_R(a)) + \Delta^{(d)}_{(a_d,\ldots,a_1)} y_L(a) = \Delta^{(d)}_{(a_d,\ldots,a_1)} \tilde{y}_R(a) \quad (= const.)$$

If we have all plaintexts in $V^{(d)}[a_d, \ldots, a_1] + a$ and corresponding ciphertexts, we obtain the following equation by computing each term using Proposition 1.

$$\sum_{x \in V^{(d)}[a_d,\ldots,a_1]+a} \left(F[k^{(R)}](y_R(x)) + y_L(x) \right) = \sum_{x \in V^{(d)}[a_d,\ldots,a_1]+a} \tilde{y}_R(x) \quad (1)$$

If the total degree of F is D (≥ 1), equation (1) has degree $D-1$ with respect to $k^{(R)}$. This is because we can rewrite the first term of equation (1) as follows. (The first order differential of a function of degree D has degree $D-1$.)

$$\sum_{x \in V^{(d)}+a} F[k^{(R)}](y_R(x))$$

$$= \sum_{x \in V^{(d)}+a\setminus\{a\}} F[k^{(R)}](y_R(x)) + F[k^{(R)}](y_R(a))$$

$$= \sum_{x \in V^{(d)}+a\setminus\{a\}} \left(F[k^{(R)}](y_R(x)) + F[k^{(R)}](y_R(a)) \right)$$

$$= \sum_{x \in V^{(d)}+a\setminus\{a\}} \Delta_{y'} F[k^{(R)}](y_R(x)) \quad (y' = y_R(x) + y_R(a))$$

Since F is a vector Boolean function composed of n coordinate Boolean functions, equation (1) forms the system of algebraic equations of degree $D-1$ with m unknowns. (Note that m is the number of bits of the last round key $k^{(R)}$.) We have some ways to solve the system of algebraic equations, and in this paper we take a similar way as one described in [8]. That is, we transform it to the system of linear equations where we regard all monomials on $k^{(R)}$ in equation (1) as independent unknown variables. Hereafter M denotes the number of the unknown variables. When $D = 2$, the unknown variables are $\{k_{m-1}^{(R)}, \ldots, k_0^{(R)}\}$ and $M = m$. When $D = 3$, the unknown variables are $\{k_{m-1}^{(R)}, \ldots, k_0^{(R)}, k_{m-1}^{(R)} k_{m-2}^{(R)}, \ldots, k_0^{(R)} k_1^{(R)}\}$ and $M = m + {}_mC_2$. Similarly, when $D = 4$, $M = m + {}_mC_2 + {}_mC_3$. When the total degree of F is D, M is at most $\sum_{i=1}^{D-1} {}_mC_i$. Actually, M is much smaller than this upper bound because coefficients of some of the unknown variables can cancel each other out, or because some of these unknown variables don't exist for some F. Finding a small M is important for reducing the complexity. General theory on a tighter upper bound of M will appear in another paper.

If the number of unknown variables of the linear equations $(= M)$ is larger than n, we have to set up equations (1) using plaintexts in different d-dimensional spaces $V^{(d)}[a_d', \ldots, a_1'] + a'$ to determine M unknowns. However, this does not increase the required number of chosen plaintexts by $\lceil \frac{M}{n} \rceil$ times because some plaintexts can be used repeatedly. That is, for an integer $\delta > d$, we can obtain ${}_\delta C_d$ different d-dimensional vector spaces from δ-dimensional vector space. Therefore, if we let δ_{min} be the smallest δ s.t. $\lceil \frac{M}{n} \rceil \leq {}_\delta C_d$, then the required number of the chosen plaintexts is at most $2^{\delta_{min}}$.

2.3   Comparison with Jakobsen-Knudsen[7]

In this section we compare the complexity of our higher order different attack with Jakobsen and Knudsen's attack[7]. The dominant complexity is setting up the system of linear equations, i.e. computing the coefficients (see also Section 4.3). For the second and third terms of equation (1), $R \times 2^{\delta_{min}}$ times the computation of the round function is needed. For the first term of equation (1), at most $(M + 1) \times 2^{\delta_{min}}$ times the computation of the round function[1] is required. Therefore, the required complexity is at most $(M + R + 1) \times 2^{\delta_{min}}$. On the other hand, the required complexity for Jakobsen and Knudsen's attack[7] was 2^{m+d}[7, Theorem 1]. Since $d \approx \delta_{min}$ and $M + R + 1 \ll 2^m$, the complexity is reduced.

3   CAST

The family of the ciphers constructed using the CAST design procedure[1] are known as CAST ciphers, and [1] describes that they appear to have good resistance to differential cryptanalysis[5], linear cryptanalysis[15], and related-key cryptanalysis[4].

---

[1] There is another way of computing the coefficients of the terms of degree $D-1$ with less complexity. See Appendix B.

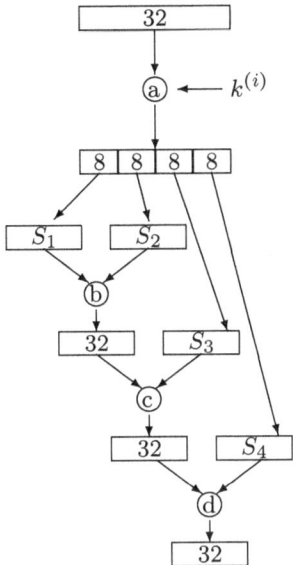

Fig. 1. CAST round function

CAST ciphers are based on the framework of the Feistel cipher. The round function is specified as follows (see also Fig.1.). A 32-bit data half is input to the function along with a subkey $k^{(i)}$. These two quantities are combined using operation "a" and the 32-bit result is split into four 8-bit pieces. Each piece is input to a different 8×32 S-box (S_1, S_2, S_3, and S_4). S-boxes S_1 and S_2 are combined using operation "b"; the result is combined with S_3 using operation "c"; this second result is combined with S_4 using operation "d". The final 32-bit result is the output of the round function.

The CAST design procedure allows a wide variety of possible round functions: 4 S-boxes and 4 operations (a,b,c, and d). As for S-boxes, [3] suggested constructing the S-boxes from bent functions. Later on [6] CAST with random S-boxes was proposed. In our attack, we use the S-boxes based on bent functions proposed for CAST-128. As for operations, a simple way to define the round function is to specify that all operations are XORs, which is addition on $GF(2)$, although other operations may be used instead. Actually, according to [1], some care in the choice of operation "a" can conceivably give intrinsic immunity to differential and linear cryptanalysis. The immunity to higher order differential for choices of operations (a,b,c, and d) will be discussed in Section 5.

As for the number of rounds, it seems that the CAST design procedure doesn't specify a concrete number. However, in [1] it is described that CAST ciphers possess a number of improvements compared to DES in both the round function and the key schedule which provide good cryptographic properties in fewer rounds[2] than DES. There are also several key schedules for CAST ciphers, but for the purpose of our attack the key schedule makes no difference.

4 Higher Order Differential Attack of a CAST Cipher

4.1 Boolean Polynomials of S-boxes

We begin by representing S-boxes by Boolean polynomial functions. We use the S-boxes proposed for CAST-128. The description of the S-boxes is given by tables. One way to construct them can be seen in [19]. Another more efficient method using a matrix transformation is also known. The obtained Boolean polynomials of S-boxes occupy a lot of space, and we show those of only some bits of S_1 in Appendix A.

From the obtained Boolean polynomials, it is confirmed that all the degrees of all output bits of all S-boxes are 4, which doesn't contradict the property of bent functions: the degree of a bent function: $GF(2)^{2n} \to GF(2)$ is at most n. When the operations a,b,c, and d are XORs, all the degrees of all output bits of the round function are at most 4. We discuss the higher order differential attack of this CAST cipher with 5 rounds.

4.2 Linear Equations for Recovering the Last Round Key

If the right half of plaintext is fixed at any value, the degree of the right half of the 4-th round, $\tilde{y}_R(x_L)$, is at most 16, and the 16-th order differential of $\tilde{y}_R(x_L)$ becomes constant. Therefore, we can compute it without knowing the true key, and we have the following attack equations for recovering the last round key, $k^{(5)}$.

$$\sum_{x_L \in V^{(16)}+a} F[k^{(5)}](y_R(x_L)) + \sum_{x_L \in V^{(16)}+a} y_L(x_L) = \sum_{x_L \in V^{(16)}+a} \tilde{y}_R(x_L), \quad (2)$$

where $y_R, y_L, \tilde{y}_R : GF(2)^{32} \to GF(2)^{32}$ and $a \in GF(2)^{32}$.

As we described in Section 2.2, since the total degree of F is 4, equation (2) has degree 3 with respect to $\{k_{31}^{(5)}, k_{30}^{(5)}, \ldots, k_0^{(5)}\}$. It follows that equation (2) forms a system of equations of degree 3 with 32 unknowns. Hereafter, we write $\{k_{31}, k_{30}, \ldots, k_0\}$ for $\{k_{31}^{(5)}, k_{30}^{(5)}, \ldots, k_0^{(5)}\}$ for simplicity.

Here we transform the system of equations of degree 3 to a system of linear equations with M unknowns. For decreasing the complexity, it is important to find as small M as possible. In this paper we find a small M by considering the structure of the round function of CAST ciphers. The output of round function F is the sum (XOR) of the outputs of S_1, S_2, S_3, and S_4, whose sets of input variables are disjoint: i.e. the set of input variables of S_1 is $\{k_{31}, k_{30}, \ldots, k_{24}\}$, that of S_2 is $\{k_{23}, k_{22}, \ldots, k_{16}\}$, that of S_3 is $\{k_{15}, k_{14}, \ldots, k_8\}$, and that of S_4 is

---

[2] For example, CAST-128 is a 12 or 16 round Feistel cipher[RFC2144].

$\{k_7, k_6, \ldots, k_0\}$. Consequently, all the terms included in equation (2) are products of variables from one of the sets above. Therefore, equation (2) is transformed to the system of linear equations below with the following M unknown variables, where $M = 32 + (4 \times {}_8C_2) + (4 \times {}_8C_3) = 368$.

$$\underbrace{\{k_0, k_1, \ldots, k_{31}}_{\text{degree-1 (32)}}, \underbrace{k_0 k_1, k_0 k_2, \ldots, k_{30} k_{31}}_{\text{degree-2 } (4\times{}_8C_2)}, \underbrace{k_0 k_1 k_2, k_0 k_1 k_3, \ldots, k_{29} k_{30} k_{31}\}}_{\text{degree-3 } (4\times{}_8C_3)}$$

$$\begin{pmatrix} a_{0,0} & a_{0,1} & \cdots & a_{0,M-1} \\ a_{1,0} & a_{1,1} & \cdots & a_{1,M-1} \\ \vdots & \vdots & \ddots & \vdots \\ a_{31,0} & a_{31,1} & \cdots & a_{31,M-1} \end{pmatrix} \begin{pmatrix} k_0 \\ k_1 \\ \vdots \\ k_{31} \\ \hline k_0 k_1 \\ k_0 k_2 \\ \vdots \\ k_{30} k_{31} \\ \hline k_0 k_1 k_2 \\ k_0 k_1 k_3 \\ \vdots \\ k_{29} k_{30} k_{31} \end{pmatrix} = \begin{pmatrix} b_0 \\ b_1 \\ \vdots \\ b_{31} \end{pmatrix}$$

We need M equations to determine the M unknown variables. However, since F, y_L, and $\tilde{y}_R$ are vector functions composed of $n\,(=32)$ functions, only n equations are obtained from equation (2). Therefore, we have to compute equation (2) for $\lceil M/n \rceil \,(= \lceil 368/32 \rceil = 12)$ different $V^{(16)}[a'_{16}, \ldots, a'_1]$. This does not increase the required number of chosen plaintexts by as many as $\lceil M/n \rceil \,(= 12)$. Because we can take ${}_{17}C_{16}\,(= 17)$ different $V^{(16)}$ from $V^{(17)}[a_{17}, \ldots, a_1]$, it only doubles the required number of chosen plaintexts.

In order to set up the system of linear equations above, we compute $M' \times M$ coefficient matrix described below, where $M' > M$. We prepare $M' \times M$ coefficient matrix because $M \times M$ matrix is not always normal. Our experimental results show that $M' = 32 \times 12$ is enough to determine the key.

How to compute coefficients $a_{i,j}$ and $b_i \in GF(2)$ in the matrices is as follows. Here we describe the computation of only the coefficients of upper 32 rows. The remaining coefficients can be computed using 11 different $V^{(16)}[a'_{16}, \ldots, a'_1]$ in the same way.

$$
\begin{pmatrix}
a_{0,0} & a_{0,1} & \cdots\cdots & a_{0,M-1} \\
a_{1,0} & a_{1,1} & \cdots\cdots & a_{1,M-1} \\
\vdots & \vdots & \ddots & \vdots \\
a_{31,0} & a_{31,1} & \cdots\cdots & a_{31,M-1} \\
\vdots & \vdots & & \vdots \\
\vdots & \vdots & & \vdots \\
\vdots & \vdots & \ddots & \vdots \\
\vdots & \vdots & & \vdots \\
\vdots & \vdots & & \vdots \\
a_{M-1,0} & a_{M-1,1} & \cdots\cdots & a_{M-1,M-1} \\
\vdots & \vdots & \ddots & \vdots \\
a_{M'-1,0} & a_{M'-1,1} & \cdots\cdots & a_{M'-1,M-1}
\end{pmatrix}
\begin{pmatrix}
k_0 \\ k_1 \\ \vdots \\ k_{31} \\ k_0 k_1 \\ k_0 k_2 \\ \vdots \\ k_{30} k_{31} \\ k_0 k_1 k_2 \\ k_0 k_1 k_3 \\ \vdots \\ k_{29} k_{30} k_{31}
\end{pmatrix}
=
\begin{pmatrix}
b_0 \\ b_1 \\ \vdots \\ b_{31} \\ \vdots \\ \vdots \\ \vdots \\ \vdots \\ \vdots \\ b_{M-1} \\ \vdots \\ b_{M'-1}
\end{pmatrix}
$$

All coefficients $a_{i,j}$ and b_i can be computed by using

$$
\mathcal{F}_j = \sum_{x_L \in V^{(16)}+a} F[e_j](y_R(x_L)) \tag{3}
$$

where e_j $(0 \le j \le M)$ is as follows:

$$
\begin{aligned}
e_j &= \bar{e}_{i_1} & (0 \le j < 32) \\
e_j &= \bar{e}_{i_1} + \bar{e}_{i_2} & (32 \le j < 144) \\
e_j &= \bar{e}_{i_1} + \bar{e}_{i_2} + \bar{e}_{i_3} & (144 \le j < M) \\
e_j &= (0,\dots,0) \in GF(2)^{32} & (j = M),
\end{aligned}
$$

where $\bar{e}_i = (0,\dots,\overset{i}{1},\dots,0) \in GF(2)^{32}$, and $0 \le i_1 < i_2 < i_3 \le 31$.

Let $B = {}^t(b_0, b_1, \dots, b_{31})$. B is computed as follows:

$$
B = \mathcal{F}_M + \sum_{x_L \in V^{(16)}+a} y_L(x_L) + \sum_{x_L \in V^{(16)}+a} \tilde{y}_R(r_L) \tag{4}
$$

Let $A_j = {}^t(a_{0,j}, a_{1,j}, \dots, a_{31,j})$ $(0 \le j < M)$. Elements of A_j are coefficients of the unknown variable located at the j-th row. When $0 \le j < 32$, A_j is a column vector of coefficients of k_j. Therefore, A_j is computed as follows:

$$
A_j = \mathcal{F}_j + \mathcal{F}_M.
$$

When A_j is a column vector of coefficients of $k_{i_1} \cdot k_{i_2}$, i.e. when $32 \le j < 144$, A_j is computed as follows:

$$
A_j = \mathcal{F}_j + \mathcal{F}_{i_1} + \mathcal{F}_{i_2} + \mathcal{F}_M.
$$

When A_j is a column vector of coefficients of $k_{i_1} \cdot k_{i_2} \cdot k_{i_3}$, i.e. when $144 \le j < M$, A_j can be computed similarly. We have another method with less complexity in Appendix B.

4.3   Complexity

This section discusses the required complexity for our higher order differential attack. Most of the execution time is spend in the following procedures.

– computing ciphertexts & higher order differentials
– computing all coefficients in the system of linear equations
– solving the linear equations

Computing ciphertexts & higher order differential

In order to compute equation (4), we have to prepare 12 sums of 2^{16} ciphertexts (=output of 5-th round) and output of the 4-th round. This can be done with 2^{17} ciphertexts and output of the 4-th round as explained in the previous section. Therefore, the required complexity is 5×2^{17} times the computation of the round function. Note that we assume that working out the sum (i.e. XOR) is negligible compared with the computation of the round function.

Computing all coefficients in the system of linear equations

All coefficients in the system of linear equations can be computed by computing equation (3) for e_j ($0 \leq j \leq 368$). Therefore, the required complexity is $(368 + 1) \times 2^{17}$ times the computation of the round function. This is the dominant part of the higher order differential attack. Since in [7] Jakobsen and Knudsen described that the average complexity was $2^{31} \times 2^{17}$, our attack has achieved speedup by 2^{23} times.

Solving the linear equations

We used Gauss-Jordan's elimination method for solving the linear equations. The size of matrix is $M' \times M$, where $M' = 384$ and $M = 368$. The required complexity is negligible compared with the computations above.

Consequently, the total complexity is $(5 + 368 + 1) \times 2^{17} < 2^{26}$ times the computation of the round function. The way to reduce the complexity by half is in Appendix B.

4.4   Experimental Results

Our experimental results showed that the all last round key bits of the CAST cipher with 5 rounds could be recovered in 13.79 seconds (average time of 100 trials) on a SunUltra2 workstation (UltraSPARC, 200MHz). Table 1 shows an execution profile of the program produced by `gprof`, which is a GNU command to display call-graph profile data.

5   Discussion

In this Section, the immunity to higher order differential attack for choices of S-boxes and operations (a,b,c, and d) is discussed.

Section 4 showed that a CAST cipher with 5 rounds which uses S-boxes proposed for CAST-128 and XORs for all operations (a,b,c, and d) can be broken

procedures	CPU time	ratio
computing ciphertexts & higher order differentials	0.83 sec.	6.0 %
computing all coefficients in the linear equations	12.92 sec.	93.7 %
solving the linear equations	0.04 sec.	0.3 %
total	13.79 sec.	100%

Table 1. Execution profile of the program

by our higher order differential attack. Since the degrees of S-boxes for CAST-128 are 4, the CAST cipher can be broken up to only 5-round. However, if the degree of the round function is lower, the CAST cipher could be broken up to more number of rounds. On [6] CAST ciphers with random S-boxes are proposed, and we must be careful of the degrees of the S-boxes in such cases. Note that it is shown that when randomly generated S-boxes are used, the resulting cipher is resistant to both differential and linear attack in [13].

Let's discuss for other choices of operations (a,b,c, and d). Some modifications of operation "a" are proposed in [1]. One example is the insert of key-dependent rotation, which is used in CAST-128, i.e. $a(x,k) = a(x, k_1, k_2) = ((x + k_1) \lll k_2)$, where k_1 is a 32-bit key, k_2 is a 5-bit key, and $\lll$ is the rotation specified by k_2. If only operation "a" is extended to XOR and rotation and "b","c", and "d" are still XOR, the CAST cipher with 5 rounds of our target can be broken by our higher order differential attack, though the complexity increases (rough estimate is 2^{40}).

There are some ways to strengthen CAST-like ciphers against the higher order differential attack. One is the increase of the number of rounds. Another is the mixture of using operations on different groups (e.g. XOR, and addition, (or subtraction) modulo 2^{32}) for "b","c", and "d". This makes the degree higher so sharply that it seems difficult to cryptanalyze by the higher order differential attack at this stage. Actually this idea is used in CAST-128 and Blowfish[17]. Moreover, Blowfish uses key-dependent S-boxes. However, note that these ways are not sufficient conditions to immune to the higher order differential attacks. How to prove the security against higher order differential attacks is open.

Acknowledgments

We would like to thank the referees for many comments. We also thank Serge Vaudenay for essential advice which can improve our attack, Bruce Schneier and Kazumaro Aoki for helpful suggestions for improving the paper.

References

1. C.M.Adams, "Constructing Symmetric Ciphers Using the CAST Design Procedure," Designs, Codes and Cryptography, Vol.12, No.3, Nov., pp.283–316, Kluwer Academic Publishers, 1997.

2. C.M.Adams, "The CAST-128 Encryption Algorithm," Request for Comments (RFC) 2144, Network Working Group, Internet Engineering Task Force, May, 1997.
3. C.M.Adams and S.E.Tavares, "Designing S-boxes for ciphers resistant to differential cryptanalysis," In Proceedings of the 3rd symposium on State and Progress of Research in Cryptography, pp.181–190, 1993.
4. E.Biham, "New Types of Cryptanalytic Attacks Using Related Keys," Advances in Cryptology–EUROCRYPT'93, Lecture Notes in Computer Science 765, pp.398–409, Springer-Verlag, 1994.
5. E.Biham and A.Shamir, "Differential Cryptanalysis of DES-like Cryptosystems," Journal of Cryptology, Vol.4, No.1, pp.3–72, Springer-Verlag, 1991.
6. H.M.Heys and S.E.Tavares, "On the security of the CAST encryption algorithm," Canadian Conference on Electrical and Computer Engineering, pp.332–335, 1994.
7. T.Jakobsen and L.R.Knudsen, "The Interpolation Attack on Block Ciphers," In Preproceedings of Fast Software Encryption Workshop'97, pp.28–40, 1997.
8. T.Kaneko, "A known-plaintext attack of FEAL-4 based on the system of linear equations on difference (Extended Abstract) ," Advances in Cryptology–ASIACRYPT'91, Lecture Notes in Computer Science 739, pp.485–488, Springer-Verlag, 1993.
9. T.Kaneko, "A Known Plaintext Cryptanalytic Attack of FEAL-4," (in Japanese), IEICE Trans. Vol.76-A, No.5, May, pp.781–786, 1993.
10. K.Kiefer, "A New Design Concept for Building Secure Block Ciphers," In Proceedings of PRAGOCRYPT'96, pp.30–41, CTU Publishing House, 1996.
11. L.R.Knudsen, "Truncated and Higher Order Differentials," Fast Software Encryption–Second International Workshop, Lecture Note in Computer Science 1008, pp.196–211, Springer-Verlag, 1995.
12. X.Lai, "Higher Order Derivatives and Differential Cryptanalysis," Communications and Cryptography, pp.227–233, Kluwer Academic Publishers, 1994.
13. J.Lee, H.M.Heys, S.E.Tavares, "Resistance of a CAST-Like Encryption Algorithm to Linear and Differential Cryptanalysis," Designs, Codes and Cryptography, Vol.12, No.3, Nov., pp.267–282, Kluwer Academic Publishers, 1997.
14. K.Nyberg and L.R.Knudsen, "Provable Security Against a Differential Attack," Journal of Cryptology, Vol.8, No.1, pp.27–37, Springer-Verlag, 1995.
15. M.Matsui, "Linear Cryptanalysis Method for DES Cipher," Advances in Cryptology–EUROCRYPT'93, Lecture Notes in Computer Science 765, pp.386–397, Springer-Verlag, 1994.
16. V.Rijmen, B.Preneel, and E.De Win "On Weaknesses of Non-surjective Round Functions," Designs, Codes and Cryptography, Vol.12, No.3, Nov., pp.253–266, Kluwer Academic Publishers, 1997.
17. B.Schneier, "Description of a New Variable-Length Key, 64-Bit Block Cipher (Blowfish)," Fast Software Encryption–Cambridge Security Workshop, Lecture Note in Computer Science 809, pp.191–204, Springer-Verlag, 1994.
18. T.Shimoyama, S.Moriai, and T.Kaneko, "Improving the Higher Order Differential Attack and Cryptanalysis of the $\mathcal{KN}$ Cipher," In Pre-Proceedings of 1997 Information Security Workshop, pp.1–8, 1997. (to appear in Lecture Notes in Computer Science, Springer-Verlag)
19. T.Shimoyama, S.Amada, and S.Moriai, "Improved Fast Software Implementation of Block Ciphers (Extended Abstract)," ICICS'97, Beijing, Nov. 1997, Lecture Notes in Computer Science 1334, pp.269–273, Springer-Verlag, 1997.

A   Boolean Polynomials of S1 for CAST-128

Due to limitations of space, we show the Boolean polynomials of only 4 bits from the least significant bit of S-box S_1 of CAST-128. Those of all S-boxes of CAST-128 can be downloaded from http://www.yokohama.tao.or.jp/shiho/. We used a computer algebra system Risa/Asir to find them.

$$
\begin{aligned}
y_0 =\ & x_4x_3x_2x_1 + x_5x_4x_2x_1 + x_5x_4x_3x_1 + x_6x_3x_2x_0 + x_6x_4x_3x_0 + x_6x_4x_3x_1 \\
& + x_6x_5x_2x_1 + x_6x_5x_3x_0 + x_6x_5x_3x_1 + x_7x_3x_2x_0 + x_7x_3x_2x_1 + x_7x_4x_2x_1 \\
& + x_7x_4x_3x_1 + x_7x_4x_3x_2 + x_7x_5x_2x_0 + x_7x_5x_3x_2 + x_7x_5x_4x_3 + x_7x_6x_2x_1 \\
& + x_7x_6x_4x_1 + x_7x_6x_4x_2 + x_7x_6x_5x_2 + x_7x_6x_5x_3 + x_7x_6x_5x_4 + x_4x_2x_0 + x_4x_2x_1 \\
& + x_5x_2x_1 + x_5x_4x_0 + x_5x_4x_2 + x_5x_4x_3 + x_6x_2x_1 + x_6x_3x_1 + x_6x_4x_0 + x_6x_4x_2 \\
& + x_6x_5x_0 + x_7x_2x_0 + x_7x_2x_1 + x_7x_3x_1 + x_7x_3x_2 + x_7x_5x_0 + x_7x_5x_1 + x_7x_5x_4 \\
& + x_7x_6x_0 + x_7x_6x_1 + x_7x_6x_4 + x_7x_6x_5 + x_1x_0 + x_4x_1 + x_5x_2 + x_5x_3 + x_7x_0 \\
& + x_7x_1 + x_7x_3 + x_7x_4 + x_7x_5 + x_3 + x_4 + x_5 + x_6 + x_7.
\end{aligned}
$$

$$
\begin{aligned}
y_1 =\ & x_5x_3x_2x_0 + x_5x_4x_2x_0 + x_5x_4x_3x_0 + x_5x_4x_3x_2 + x_6x_4x_2x_0 + x_6x_4x_3x_0 \\
& + x_6x_4x_3x_2 + x_6x_5x_2x_0 + x_6x_5x_3x_0 + x_6x_5x_3x_1 + x_6x_5x_4x_0 + x_6x_5x_4x_1 \\
& + x_6x_5x_4x_3 + x_7x_3x_2x_0 + x_7x_3x_2x_1 + x_7x_4x_3x_0 + x_7x_4x_3x_2 + x_7x_5x_3x_1 \\
& + x_7x_5x_4x_0 + x_7x_5x_4x_1 + x_7x_6x_3x_1 + x_7x_6x_4x_2 + x_7x_6x_4x_3 + x_4x_2x_0 + x_4x_2x_1 \\
& + x_4x_3x_2 + x_5x_2x_0 + x_5x_2x_1 + x_5x_4x_1 + x_5x_4x_2 + x_5x_4x_3 + x_6x_2x_0 + x_6x_2x_1 \\
& + x_6x_3x_0 + x_6x_4x_0 + x_6x_4x_1 + x_6x_4x_3 + x_6x_5x_1 + x_6x_5x_3 + x_6x_5x_4 + x_7x_2x_1 \\
& + x_7x_3x_2 + x_7x_5x_1 + x_7x_5x_4 + x_7x_6x_0 + x_7x_6x_3 + x_7x_6x_5 + x_1x_0 + x_2x_1 \\
& + x_3x_1 + x_4x_1 + x_4x_3 + x_5x_1 + x_5x_2 + x_5x_3 + x_6x_1 + x_7x_0 + x_7x_5 + x_7x_6 + x_1 \\
& + x_3 + x_4 + x_5 + x_6 + x_7.
\end{aligned}
$$

$$
\begin{aligned}
y_2 =\ & x_5x_3x_2x_1 + x_5x_4x_2x_1 + x_5x_4x_3x_0 + x_5x_4x_3x_2 + x_6x_3x_2x_0 + x_6x_3x_2x_1 \\
& + x_6x_4x_2x_1 + x_6x_4x_3x_1 + x_6x_4x_3x_2 + x_6x_5x_2x_0 + x_6x_5x_3x_0 + x_6x_5x_3x_1 \\
& + x_6x_5x_4x_2 + x_7x_3x_2x_0 + x_7x_4x_2x_0 + x_7x_4x_2x_1 + x_7x_4x_3x_0 + x_7x_4x_3x_2 \\
& + x_7x_5x_4x_0 + x_7x_5x_4x_1 + x_7x_5x_4x_3 + x_7x_6x_3x_1 + x_7x_6x_4x_1 + x_7x_6x_4x_2 \\
& + x_7x_6x_4x_3 + x_7x_6x_5x_1 + x_7x_6x_5x_2 + x_3x_2x_1 + x_4x_2x_1 + x_4x_3x_1 + x_5x_2x_1 \\
& + x_5x_3x_2 + x_5x_4x_1 + x_5x_4x_2 + x_5x_4x_3 + x_6x_2x_0 + x_6x_2x_1 + x_6x_3x_0 + x_6x_3x_2 \\
& + x_6x_4x_0 + x_6x_4x_1 + x_6x_5x_0 + x_6x_5x_4 + x_7x_2x_1 + x_7x_4x_0 + x_7x_4x_2 + x_7x_4x_3 \\
& + x_7x_5x_0 + x_7x_5x_1 + x_7x_6x_0 + x_7x_6x_1 + x_7x_6x_2 + x_7x_6x_3 + x_7x_6x_4 + x_7x_6x_5 \\
& + x_1x_0 + x_2x_0 + x_2x_1 + x_3x_0 + x_3x_1 + x_3x_2 + x_4x_2 + x_5x_1 + x_5x_2 + x_5x_3 \\
& + x_5x_4 + x_6x_0 + x_6x_1 + x_6x_2 + x_6x_3 + x_6x_4 + x_6x_5 + x_7x_1 + x_7x_4 + x_0 + x_1 \\
& + x_2 + x_7 + 1.
\end{aligned}
$$

$$
\begin{aligned}
y_3 =\ & x_4x_3x_2x_0 + x_4x_3x_2x_1 + x_5x_3x_2x_1 + x_5x_4x_2x_1 + x_5x_4x_3x_1 + x_5x_4x_3x_2 \\
& + x_6x_4x_2x_0 + x_6x_4x_2x_1 + x_6x_4x_3x_1 + x_6x_4x_3x_2 + x_6x_5x_3x_1 + x_6x_5x_3x_2 \\
& + x_6x_5x_4x_1 + x_6x_5x_4x_3 + x_7x_4x_3x_0 + x_7x_4x_3x_1 + x_7x_5x_2x_0 + x_7x_5x_2x_1 \\
& + x_7x_5x_4x_0 + x_7x_5x_4x_1 + x_7x_5x_4x_2 + x_7x_6x_2x_0 + x_7x_6x_3x_0 + x_7x_6x_4x_0 \\
& + x_7x_6x_4x_1 + x_7x_6x_4x_2 + x_7x_6x_5x_0 + x_7x_6x_5x_1 + x_7x_6x_5x_2 + x_7x_6x_5x_4 \\
& + x_4x_3x_1 + x_5x_2x_1 + x_5x_3x_2 + x_5x_4x_0 + x_5x_4x_1 + x_6x_2x_1 + x_6x_3x_0 + x_6x_3x_1 \\
& + x_6x_4x_0 + x_6x_4x_2 + x_6x_4x_3 + x_6x_5x_1 + x_6x_5x_2 + x_6x_5x_3 + x_6x_5x_4 + x_7x_2x_0 \\
& + x_7x_2x_1 + x_7x_3x_0 + x_7x_3x_1 + x_7x_4x_0 + x_7x_4x_2 + x_7x_5x_1 + x_7x_6x_2 + x_7x_6x_4 \\
& + x_7x_6x_5 + x_1x_0 + x_2x_0 + x_2x_1 + x_3x_2 + x_4x_0 + x_4x_1 + x_4x_3 + x_5x_2 + x_5x_4 \\
& + x_6x_1 + x_6x_2 + x_6x_3 + x_6x_4 + x_7x_2 + x_7x_3 + x_7x_4 + x_7x_6 + x_2 + x_3 + x_5 + x_6 \\
& + x_7.
\end{aligned}
$$

B   Fast Method for Computing Coefficients of $k_{i_1} k_{i_2} k_{i_3}$

In Section 4.2, it is described that in order to compute all coefficients, the computation of

$$\forall x_L \in V^{(17)} + a, \quad F[e_j](y_R(x_L)) \tag{5}$$

for $(M+1)$ e_j is required. However, there is another method of computing the coefficients of the terms of degree 3, $k_{i_1} k_{i_2} k_{i_3}$ with less complexity. The point is the coefficients of the terms of degree 3, $k_{i_1} k_{i_2} k_{i_3}$ is linear to the input of F. Let

$$\sum_{x_L \in V^{(16)} + a} C_{i_1 i_2 i_3}(y_R(x_L)) k_{i_1} k_{i_2} k_{i_3} \tag{6}$$

be a term of degree 3 in equation (2). The degree of $C_{i_1 i_2 i_3}$ is 1 with respect to the input of F, since the degree of F is 4. Therefore, we have

$$C_{i_1 i_2 i_3}(x) = \mathcal{A}_{i_1 i_2 i_3} x + \mathcal{B}_{i_1 i_2 i_3}. \tag{7}$$

The coefficient of (6), which is what we want, is rewritten as follows.

$$\sum_{x_L \in V^{(16)} + a} C_{i_1 i_2 i_3}(y_R(x_L)) = \sum_{x_L \in V^{(16)} + a} (\mathcal{A}_{i_1 i_2 i_3} y_R(x_L) + \mathcal{B}_{i_1 i_2 i_3})$$

$$= \mathcal{A}_{i_1 i_2 i_3} \sum_{x_L \in V^{(16)} + a} y_R(x_L) \quad \left(\because \sum \mathcal{B}_{i_1 i_2 i_3} = 0 \right) \tag{8}$$

Here we define x_s as $x_s = \displaystyle\sum_{x_L \in V^{(16)} + a} y_R(x_L)$. From equation (5) we have:

$$(8) = \mathcal{A}_{i_1 i_2 i_3} x_s$$
$$= C_{i_1 i_2 i_3}(x_s) + \mathcal{B}_{i_1 i_2 i_3}$$

The first and second terms are computed as follows:

$$C_{i_1 i_2 i_3}(x_s) = F[e_{i_1 i_2 i_3}](x_s) + F[e_{i_1 i_2}](x_s) + F[e_{i_1 i_3}](x_s) + F[e_{i_2 i_3}](x_s)$$
$$+ F[e_{i_1}](x_s) + F[e_{i_2}](x_s) + F[e_{i_3}](x_s) + F[0](x_s),$$
$$\mathcal{B}_{i_1 i_2 i_3} = F[e_{i_1 i_2 i_3}](0) + F[e_{i_1 i_2}](0) + F[e_{i_1 i_3}](0) + F[e_{i_2 i_3}](0)$$
$$+ F[e_{i_1}](0) + F[e_{i_2}](0) + F[e_{i_3}](0) + F[0](0),$$

$$\text{where} \quad e_{i_1 i_2 i_3} = (0, \ldots, \overset{i_1}{1}, \ldots, \overset{i_2}{1}, \ldots, \overset{i_3}{1}, \ldots, 0) \in GF(2)^{32},$$
$$e_{i_1 i_2} = (0, \ldots, \overset{i_1}{1}, \ldots, \overset{i_2}{1}, \ldots, 0) \in GF(2)^{32},$$
$$e_{i_1} = (0, \ldots, \overset{i_1}{1}, \ldots, 0) \in GF(2)^{32},$$
$$0 = (0, \ldots, 0) \in GF(2)^{32}.$$

The complexity required for this method is $(12+1) \times (M+1)$ times the computation of the round function F. When we use this method, the total complexity is $(5 + 144 + 1) \times 2^{17} + 13 \times (368 + 1) < 2^{25}$ times the computation of the round function F.

Cryptanalysis of TWOPRIME

Don Coppersmith[1], David Wagner[2], Bruce Schneier[3], and John Kelsey[3]

[1] IBM Research, copper@watson.ibm.com
[2] U.C. Berkeley, daw@cs.berkeley.edu
[3] Counterpane Systems, {schneier,kelsey}@counterpane.com

Abstract. Ding et al [DNRS97] propose a stream generator based on several layers. We present several attacks. First, we observe that the non-surjectivity of a linear combination step allows us to recover half the key with minimal effort. Next, we show that the various bytes are insufficiently mixed by these layers, enabling an attack similar to those on two-loop Vigenere ciphers to recover the remainder of the key. Combining these techniques lets us recover the entire TWOPRIME key. We require the generator to produce 2^{33} blocks (2^{35} bytes), or 19 hours worth of output, of which we examine about one million blocks (2^{23} bytes); the computational workload can be estimated at 2^{28} operations. Another set of attacks trades off texts for time, reducing the amount of known plaintext needed to just eight blocks (64 bytes), while needing 2^{32} time and 2^{32} space. We also show how to break two variants of TWOPRIME presented in the original paper.

1 Introduction

The TWOPRIME stream cipher [DNRS97], introduced at FSE'97, uses a 128-bit key to generate 64-bit blocks of output at each time step; these output blocks are exclusive-ORed onto the plaintext to produce ciphertext. At a high level, TWOPRIME consists of a keyed (non-bijective) cryptographic function with 64-bit inputs and 64-bit outputs, which is used in a counter-like mode to generate keystream output.

The algorithm has ten layers; the first layer is driven by a counter, and the output of each layer becomes the input to the next. We exploit weaknesses of two of the layers to produce several different attacks against the scheme. Our conclusion is that there are too few layers for cryptographic strength.

One of the main contributions of the TWOPRIME work is that the algorithm was designed so that one could prove certain statements about the security of the cipher: it has high linear complexity, good cycle length, good resistance to LSFR-synthesis attacks, and so on[1]. Nonetheless, despite the proofs of various security properties, in this paper we show how to break TWOPRIME very efficiently.

---

[1] Note that it is possible to prove that using any block cipher in counter mode has good linear complexity and good cycle length—at least, in the sense that [DNRS97] proved for TWOPRIME—so in retrospect these proofs are perhaps not terribly meaningful.

S. Vaudenay (Ed.): Fast Software Encryption – FSE'98, LNCS 1372, pp. 32–48, 1998.
© Springer-Verlag Berlin Heidelberg 1998

Our attacks fall into two natural categories. The first three attacks, discussed in Sections 4–7, recover half of the key (namely, K_2, K_3). The second category (see Sections 8–9) includes two techniques which identify the remainder of the key (K_0, K_1) once we've found K_2, K_3.

The rest of the paper is organized as follows. In Section 2 we review the TWOPRIME scheme. In Section 3 we give some preliminary remarks which will be useful in the cryptanalysis. Section 4 gives a very easy attack to recover half of the key, based on the linear map of layer 7 failing to be surjective. Section 5 shows another attack that reduces the plaintext requirements; the cost for this improvement is an increase in the amount of offline computation required. Section 6 gives a more complicated attack to recover K_2, K_3 by breaking the period of $p_0 p_1$ into two periods of p_0 and p_1 respectively. The probabilistic analysis backing up this attack is mentioned in Section 7. In Section 8 and 9 we finish with two attacks which can be used to recover the remainder of the key in a more mundane manner. Section 10 discusses some of the computational requirements of each attack. Section 11 and 12 discuss variants of the original scheme, and some attacks on these variants. Conclusions are reserved for Section 14.

2   Description of TWOPRIME

The TWOPRIME scheme [DNRS97] uses a 128-bit key to generate 64-bit blocks of output at each time step; these output blocks are exclusive-ORed onto the plaintext to produce ciphertext. At a high level, TWOPRIME consists of a keyed function $F_K : \mathbf{Z}_{256}^8 \to \mathbf{Z}_{256}^8$ and a custom mode for using F to generate keystream output.

The mode is somewhat similar to counter mode: the input to F comes from two independent 32-bit counters. Each counter is initialized with a key-dependent value, and is stepped by adding a public constant and then reducing modulo a public 32-bit prime.

The key, consisting of 16 bytes $k_0, \ldots, k_{15}$, is divided into four 32-bit parts, named K_0, K_1, K_2 and K_3, with the convention

$$
\begin{aligned}
K_0 &= k_8 + k_9 2^8 + k_{10} 2^{16} + k_{11} 2^{24} \\
K_1 &= k_{12} + k_{13} 2^8 + k_{14} 2^{16} + k_{15} 2^{24} \\
K_2 &= (k_0, k_1, k_2, k_3) \\
K_3 &= (k_4, k_5, k_6, k_7).
\end{aligned}
$$

The algorithm has ten layers, which we will describe. The output of each layer becomes the input of the subsequent layer. With one exception, each output consists of eight bytes, and so is an element of $\mathbf{Z}_{256}^8$. The scheme is depicted graphically in Figure 1.

The first layer involves two primes, $p_0 = 2^{32} - 17$ and $p_1 = 2^{32} - 5$, and two fixed public integers a_0 and a_1. At time step t, the output of the first layer is the two 32-bit integers $r_0 = a_0 t + K_0 \pmod{p_0}$ and $r_1 = a_1 t + K_1 \pmod{p_1}$. Each is broken into four 8-bit bytes, yielding a total of eight bytes output.

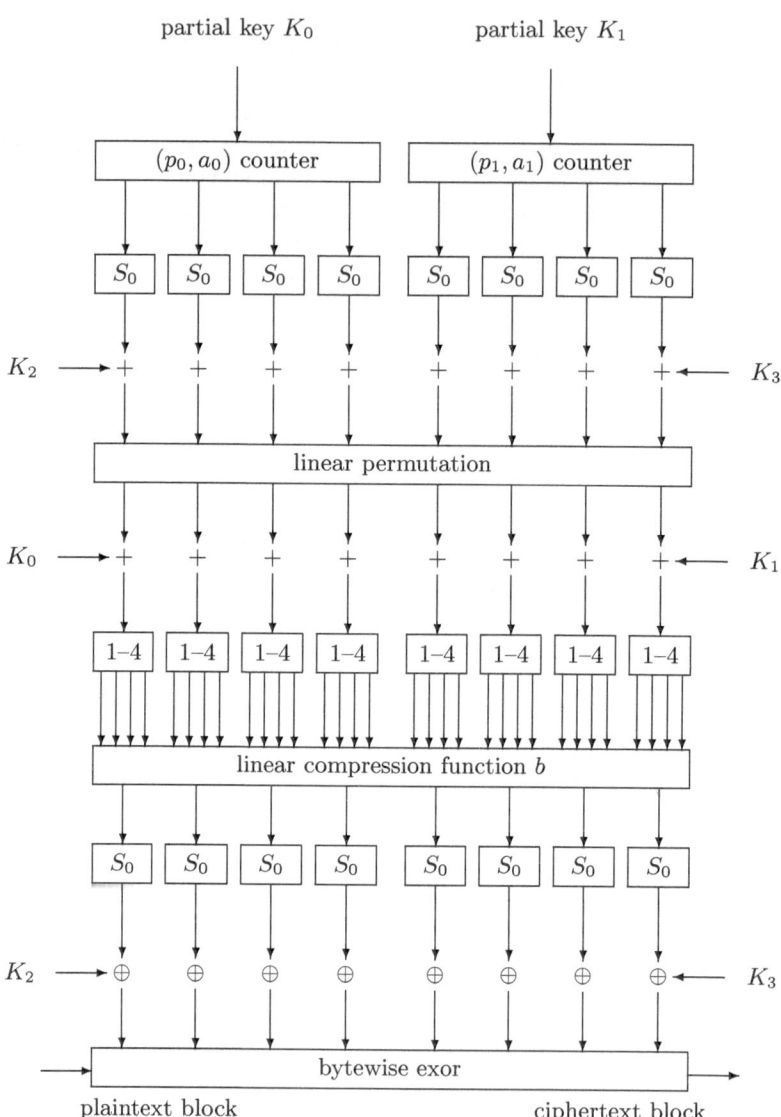

Fig. 1. Structure of the ciphering algorithm.

In the second layer, each byte x is replaced by $S_0(x) = [(x^{255} \bmod 257) \bmod 256]$. It happens that S_0 is its own inverse: $S_0(S_0(x)) = x$.

The third layer involves addition (mod 256) of the key bytes constituting K_2 and K_3.

The fourth layer is a "linear permutation": if $x_0, \ldots, x_7$ are the inputs to this layer, the outputs are

$$y_j = (\sum_{i=0}^{7} x_i) - x_j \quad (\bmod\ 256).$$

This is intended to mix the bytes; however, as we shall see, it is too weak. The only interaction between the various bytes x_i is through the single byte $\sum_i x_i$ (mod 256), and when that byte is controlled, the mixing is ineffective.

The fifth layer involves addition (mod 256) of the key bytes constituting K_0 and K_1.

The sixth layer is a non-linear expansion: each byte x is expanded to the concatenation of four bytes $S_1(x), S_2(x), S_3(x), S_4(x)$, where the S_i are various nonlinear permutations on $\mathbf{Z}_{256}$. The output of this layer is 32 bytes.

The seventh layer applies a linear compression to reduce these 32 bytes back to 8 bytes; that is, a fixed public 8×32 matrix $\{b_{ij}\}$ maps $\mathbf{Z}_{256}^{32}$ to $\mathbf{Z}_{256}^{8}$. Upon input $(X_0, \ldots, X_{31})$, the linear transform b produces the output $(Y_0, \ldots, Y_7) = b(X_0, \ldots, X_{31})$ according to the equation

$$\begin{cases} Y_0 = X_0 + X_5 + X_{10} + X_{15} + X_{16} + X_{22} + X_{24} + X_{30}, \\ Y_1 = X_1 + X_6 + X_{11} + X_{12} + X_{17} + X_{23} + X_{25} + X_{31}, \\ Y_2 = X_2 + X_7 + X_8 + X_{13} + X_{18} + X_{20} + X_{26} + X_{28}, \\ Y_3 = X_3 + X_4 + X_9 + X_{14} + X_{19} + X_{21} + X_{27} + X_{29}, \\ Y_4 = X_{16} + X_{21} + X_{26} + X_{31} + X_0 + X_6 + X_8 + X_{14}, \\ Y_5 = X_{17} + X_{22} + X_{27} + X_{28} + X_5 + X_{11} + X_{13} + X_3, \\ Y_6 = X_{18} + X_{23} + X_{24} + X_{29} + X_{10} + X_{12} + X_2 + X_4, \\ Y_7 = X_{19} + X_{20} + X_{25} + X_{30} + X_{15} + X_1 + X_7 + X_9. \end{cases} \quad (1)$$

The eighth layer applies the permutation S_0 to each byte.

In the ninth layer, bytes from K_2 and K_3 are exclusive-ORed into the bytes.

The tenth round consists of exclusive-ORing these bytes (the output of the ninth round) onto the plaintext to produce the ciphertext, or (in the case of decryption) onto the ciphertext to recover plaintext.

Let us denote by $x_i^{(j)}$ ($0 \le i \le 7, 1 \le j \le 10$) the ith byte of the output of the jth round. (For $j = 6$ we will allow $0 \le i \le 31$.) If the time step t is important we will write $x_i^{(j,t)}$. The notation $x_*^{(j)}$ will mean the whole 8-tuple of bytes $[x_i^{(j)}, 0 \le i \le 7]$.

3   Remarks on the scheme

During most of the rounds, the various bytes remain separate. During the first round, four bytes are output from one 32-bit word, and four from another. The fourth round combines bytes with a linear map, but (as has been remarked) this does a weak job of mixing them.

The seventh round combines pieces of the various bytes much more thoroughly, but only with a linear transformation. Also, the seventh round lies close to the surface, which lets us exploit the lack of diffusion in the rest of the cipher.

The designers explain that the internal structure of TWOPRIME (i.e. the function F) was chosen to resist inversion attacks (where one tries to use the output of F to work backwards). Two of our attacks succeed exactly because we can work backwards from the output of F.

In fact, we use the non-invertibility of F to our advantage in Sections 4–5. Because F is not bijective, not all intermediate values are possible. In particular, the combination of the sixth and seventh layers forms a non-surjective function, so not all 64-bit values are attainable as the output of the seventh layer. Furthermore, layers 8–10 depend only on K_2, K_3, and not on K_0, K_1. Therefore, we can isolate the effect of K_2, K_3 and attack them standing alone. Later, we can peel off layers 8–10 and use separate techniques (see Sections 8–9) to recover the remainder of the key (K_0, K_1).

4   Linear algebra

The linear recombination step (layer seven) suffers from the following regularity.
Denote by τ the 8-vector $[1, 1, 1, 1, -1, -1, -1, -1]$. The matrix b_{ij} obeys $\sum_i \tau_i b_{ij} = 0 \pmod{256}$ for all indices j. This implies that

$$\sum_{i=0}^{7} \tau_i x_i^{(7)} = 0 \pmod{256}. \tag{2}$$

We can use this information, and a few known outputs of the stream generator, to recover the half of the key (K_2, K_3).

For each byte position i we have

$$x_i^{(7)} = S_0(x_i^{(8)}) = S_0(x_i^{(9)} \oplus k_i),$$

recalling that S_0 is its own inverse. For each i this gives a fixed mapping from $x_i^{(9)}$ to $x_i^{(7)}$, independent of time and of the other bytes.

Denote by y_{ij} the unknown quantity

$$y_{ij} = S_0(j \oplus k_i)$$

which would be the value of $x_i^{(7)}$ if $x_i^{(9)} = j$. For each block of output of the stream cipher (at time t) we obtain a linear equation relating these quantities:

$$0 = \sum_{i=0}^{7} \tau_i x_i^{(7,t)} = \sum_{i=0}^{7} \tau_i y_{i, x_i^{(9,t)}} \pmod{256}.$$

After we obtain about 2,048 blocks $(16,384$ bytes) of output, we will have 2,048 linear equations in the 2,048 unknowns y_{ij}, $0 \leq i \leq 7, 0 \leq j \leq 255$. Because of homogeneity these equations will not be independent, and for fixed i we will recover y_{ij} only up to an unknown multiplicative factor and an unknown additive shift:

$$y_{ij} = \alpha_i z_{ij} + \beta_i \quad (\text{mod } 256) \tag{3}$$

with z_{ij} known but α_i, β_i unknown.

But this is clearly enough information to recover the unknown key byte k_i, using a few hundred operations of trial-and-error. For each possible value for k_i, decrypt three or four values $j = x_i^{(9)}$ into $y_{ij} = S_0(j \oplus k_i)$ and check against (3). The correct k_i will be compatible with (3), and only a few others; a few more trial decryptions should rule out the false alarms.

Having determined $(k_0, \ldots, k_7) = (K_2, K_3)$, we still[2] have to find K_0 and K_1. This seems to be more expensive (and less interesting). We see a way of finding them using about 2^{32} operations and just a few known outputs of the stream cipher. See Sections 8–9.

The present attack does require about 2048 blocks (16384 bytes) of stream output. Those known plaintext requirements are not onerous, but it is possible to reduce them even further with meet-in-the-middle techniques, which we discuss next.

5   A meet-in-the-middle attack

In this attack, we take advantage of the non-surjectivity of layer seven in a different way. It is essentially a meet-in-the-middle attack, taking advantage of unattainable values at the output of the seventh layer.

Roughly speaking, we guess (K_2, K_3) and work backwards from a block of known keystream to find the output of the seventh layer, using unattainable values to rule out incorrect guesses at (K_2, K_3). This would take 2^{64} time to implement as stated; however, we have an optimization (again based on meet-in-the-middle techniques) to reduce the complexity to 2^{32}.

As before, we rely on the crucial observation (2). If we take some keystream block $x_*^{(9)}$, then inverting layers 8–9 shows that $x_i^{(7)} = S_0(x_i^{(9)} \oplus k_i)$. Plugging into (2) gives us a relation that the correct value of the key $k_0, \ldots, k_7$ must satisfy.

So the attack proceeds as follows. We define

$$g(K_2, y_0, \ldots, y_3) = \sum_{i=0}^{3} S_0(y_i \oplus k_i) \quad (\text{mod } 256)$$

---

[2] In some situations, recovering just (K_2, K_3) might conceivably suffice. After all, this gives us enough information to predict some keystream bytes: given any seven bytes from a keystream block, we can predict the eighth unknown byte with certainty by using (2). However, we can do much better. As we shall see, recovering (K_0, K_1) in a second phase requires a bit more work, but it is still feasible.

$$h(K_3, y_4, \ldots, y_7) = \sum_{i=4}^{7} S_0(y_i \oplus k_i) \quad (\text{mod } 256).$$

We obtain eight known keystream blocks $x_*^{(9,j)}, 0 \le j \le 7$, and let

$$g'(K_2) = (g(K_2, x_0^{(9,0)}, \ldots, x_3^{(9,0)}), \ldots, g(K_2, x_0^{(9,7)}, \ldots, x_3^{(9,7)}))$$
$$h'(K_3) = (g(K_3, x_4^{(9,0)}, \ldots, x_7^{(9,0)}), \ldots, g(K_3, x_4^{(9,7)}, \ldots, x_7^{(9,7)})).$$

Note that, for the correct value of (K_2, K_3), we have $g'(K_2) = h'(K_3)$.

After all this effort to frame things in the language of meet-in-the-middle attacks, it should be clear how to recover (K_2, K_3) with standard techniques. (Here the "middle" for the meet-in-the-middle attack will be the 64-bit value $g'(K_2) = h'(K_3)$, i.e. a characteristic of the output of the seventh layer.)

First, for each guess at K_2, we compute $g'(K_2)$, and store the pair $(g'(K_2), K_2)$ in a hash table indexed on the first coordinate of the pair. After enumerating all 2^{32} possibilities for K_2, we will have constructed a hash table of size 2^{32}. Then, for each guess at K_3, we compute $h'(K_3)$ and look it up in the hash table. If we find a match $g'(K_2) = h'(K_3)$, then with high probability we will have obtained the correct values for (K_2, K_3).

We need eight keystream blocks to ensure that the test will eliminate nearly all incorrect values. One can count the number of false alarms by counting the number of solutions a, b to $g'(a) = h'(b)$. Because S_0 is highly non-linear, we are justified in expecting the functions g', h' to behave roughly like random functions of the form $\mathbf{Z}_{256}^4 \to \mathbf{Z}_{256}^8$. Combining this heuristic with the birthday paradox, we find that the probability of generating a false alarm is $1 - e^{-1} \approx 0.63$, and the expected number of false alarms is 1.

To aid the intuition, we can think of the present attack as applying a meet-in-the-middle attack *twice*, splitting the cipher first with a horizontal cut and then splitting it again with a vertical cut.

The horizontal cut is possible because layer seven fails to be surjective, and it is beneficial because layers 8–10 only depend on half of the key. (There is a slight difference, though. In a normal meet-in-the-middle attack, one computes forward part-way, backward part-way, and then meets in the middle. In our attack on TWOPRIME, because layers 6–7 fail to be surjective, we only need to compute backwards, and the forward part of the computation is substantially simplified.)

The vertical cut is made possible by the linearity of layer seven (or, more precisely, the linearity of (2)). Here the "middle" is the value $g'(K_2) = h'(K_3)$. We compute up the left half, and up the right half, and then meet in the "middle" of the output of the seventh layer. This second application of meet-in-the-middle techniques lets us isolate the effect of K_2 from that of K_3, and hence reduces the attacker's workload significantly.

In summary, we can recover (K_2, K_3) with 2^{32} offline work, 2^{32} space, and about eight blocks (64 bytes) of known keystream. As we shall see in Section 10, the computational requirements are not unreasonable.

6   Splitting the period

The previous two attacks could be avoided (in a hypothetical TWOPRIME successor) by using a different linear transformation at layer seven. So we develop here another attack against that eventuality.

This attack is similar to the attacks on two-loop Vigenere ciphers, which can be found in references [Sin68] and [Tuc70].

For an arbitrary time step t_0, let us consider the outputs at four specific time steps:

$$a = t_0$$
$$b = t_0 + p_0$$
$$c = t_0 + p_1$$
$$d = t_0 + p_0 + p_1.$$

Because the counters at layer 1 are cyclic with periods p_0 and p_1 respectively, we have

$$x_i^{(1,a)} = x_i^{(1,b)}, \quad x_i^{(1,c)} = x_i^{(1,d)}, \quad 0 \le i \le 3$$
$$x_i^{(1,a)} = x_i^{(1,c)}, \quad x_i^{(1,b)} = x_i^{(1,d)}, \quad 4 \le i \le 7,$$

and hence, because the actions of subsequent layers are time-invariant,

$$x_i^{(3,a)} = x_i^{(3,b)}, \quad x_i^{(3,c)} = x_i^{(3,d)}, \quad 0 \le i \le 3$$
$$x_i^{(3,a)} = x_i^{(3,c)}, \quad x_i^{(3,b)} = x_i^{(3,d)}, \quad 4 \le i \le 7.$$

Consider the event **E** that the following two equations both hold:

$$\sum_{i=0}^{3} x_i^{(3,a)} = \sum_{i=0}^{3} x_i^{(3,c)} \quad (\mathrm{mod}\ 256)$$

$$\sum_{i=4}^{7} x_i^{(3,a)} = \sum_{i=4}^{7} x_i^{(3,b)} \quad (\mathrm{mod}\ 256).$$

Each equation holds with probability about $1/256$ (for randomly chosen time step t_0), and the two are independent, so that event **E** holds with probability about $1/65536$. When it does hold, we have

$$\sum_{i=0}^{7} x_i^{(3,a)} = \sum_{i=0}^{7} x_i^{(3,b)} = \sum_{i=0}^{7} x_i^{(3,c)} = \sum_{i=0}^{7} x_i^{(3,d)} \quad (\mathrm{mod}\ 256).$$

This in turn implies that the outputs of layer 4 are well behaved:

$$x_i^{(4,a)} = x_i^{(4,b)}, \quad x_i^{(4,c)} = x_i^{(4,d)}, \quad 0 \le i \le 3$$
$$x_i^{(4,a)} = x_i^{(4,c)}, \quad x_i^{(4,b)} = x_i^{(4,d)}, \quad 4 \le i \le 7.$$

This can be pushed forward to give information on the outputs of layer 6:

$$x_i^{(6,a)} = x_i^{(6,b)}, \quad x_i^{(6,c)} = x_i^{(6,d)}, \quad 0 \leq i \leq 15$$
$$x_i^{(6,a)} = x_i^{(6,c)}, \quad x_i^{(6,b)} = x_i^{(6,d)}, \quad 16 \leq i \leq 31$$
$$x_i^{(6,a)} + x_i^{(6,d)} = x_i^{(6,b)} + x_i^{(6,c)}, \quad 0 \leq i \leq 31,$$

and because layer 7 is linear (mod 256), we get

$$x_i^{(7,a)} + x_i^{(7,d)} = x_i^{(7,b)} + x_i^{(7,c)} \quad (\text{mod } 256). \tag{4}$$

Suppose we know that event **E** has occurred for time step t_0, and that we have available for the output of the stream cipher $x_i^{(9,h)}$. Then from $x_i^{(7,h)} = S_0(k_i \oplus x_i^{(9,h)})$ and (4), we get a suitability test for possible values of key byte k_i. That is, for each position $0 \leq i \leq 7$, for each possible value of k_i, we test whether the values of $x_i^{(7,h)}$ obtained from $x_i^{(9,h)}$ using k_i would satisfy (4):

$$S_0(k_i \oplus x_i^{(9,a)}) + S_0(k_i \oplus x_i^{(9,d)}) \stackrel{?}{=} S_0(k_i \oplus x_i^{(9,b)}) + S_0(k_i \oplus x_i^{(9,c)}) \quad (\text{mod } 256). \tag{5}$$

Each concatenation of possible bytes $(k_0, k_1, \ldots, k_7)$ from this step represents a possible setting of (K_2, K_3) consistent with the event **E** having occurred at this time step t_0. We will call this 8-byte setting a *putative key*.

If event **E** did occur, then the correct setting of $(k_0, k_1, \ldots, k_7)$ will be represented among these possibilities. If it did not occur, we may get several false alarms.

The difficulty is that we do not know, *a priori*, whether event **E** occurred or not. We may find that for one of the byte positions i there is no possible setting of k_i satisfying (5); in this case we know that **E** did not occur at t_0 and this case can be discarded.

Our strategy will be to try about 330,000 different values of t_0, and for each one that has at least one possible setting for each of the eight bytes k_i, record the possible values of the 8-tuple $(k_0, k_1, \ldots, k_7) = (K_2, K_3)$. The correct value should show up about five times among these putative keys, and incorrect values should show up less often. Having ascertained the correct value for (K_2, K_3), we will be able to get the keys (K_0, K_1) with less difficulty in Section 8.

7   Probabilistic analysis

For our analysis it will be useful to know the following two probability distributions.

For bytes x_a, x_b, x_c, x_d, representing $x_i^{(9,a)}, \ldots, x_i^{(9,d)}$, let $N(x_a, x_b, x_c, x_d)$ be the number of key bytes k that would satisfy (5):

$$S_0(k_i \oplus x_a) + S_0(k_i \oplus x_d) \stackrel{?}{=} S_0(k_i \oplus x_b) + S_0(k_i \oplus x_c) \quad (\text{mod } 256). \tag{6}$$

We want to know the distribution $P_1(n) = \Pr(N(x_a, x_b, x_c, x_d) = n)$ when the x_h are independent random variables. We also want to know the distribution

$P_2(n) = \Pr(N(x_a, x_b, x_c, x_d) = n)$ when the x_h are known to arise from event $\mathbf{E}$, that is, when the correct key byte k_i is known to satisfy (6). The two are related by $P_2(n) = nP_1(n)$. The experimental distributions are given in the Appendix.

The first distribution is almost Poisson with mean 1: $P_1(n) = e^{-1}/n!$, with three notable exceptions.

First, $P_1(256) \approx 2/256^2 = 2^{-15}$, because with that probability we either have ($x_a = x_b$ and $x_c = x_d$), or ($x_a = x_c$ and $x_b = x_d$), and in either case all key bytes k will work.

Second, $P_1(128) \approx (1/2)/256^2 = 2^{-17}$, and similarly $P_1(64) \approx (5/4)/256^2$, $P_1(32) \approx (13/8)/256^2$, and $P_1(16), P_1(8)$ are similarly high. This happens because of idiosyncrasies of the permutation S_0. For example, in the case $n = 128$, consider the event that $x_a \oplus x_d = x_b \oplus x_c = 11111101$ in binary, and x_a and x_b agree in the second-lowest bit. This event has probability $(1/256)^2(1/2) = 2^{-17}$. When this happens, for all 128 key bytes k disagreeing with x_a in the second-lowest bit, we have $(k \oplus x_a) + (k \oplus x_d) = 257$. Then, because $S_0(x) = x^{-1}$ (mod 257) if $x \neq 0$, we have

$$S_0(k \oplus x_a) + S_0(k \oplus x_d) = S_0(k \oplus x_b) + S_0(k \oplus x_c) = 257$$

for each of these 128 values of k, so that $N(x_a, x_b, x_c, x_d) \geq 128$. This implies $P_1(128) \approx 2^{-17}$. Similar calculations obtain for $n = 64, 32, 16, 8$.

Third, it appears experimentally that $P_1(0)$ is a little higher than expected: 0.40 rather than 0.37; and $P_1(1)$ is a little lower. This may be related to the first two observations.

These deviations from the Poisson distribution, particularly the relative high values of $P_2(256)$ and $P_2(128)$, create a minor nuisance for our cryptanalysis.

When event $\mathbf{E}$ has happened, the distribution $P_2(n)$ is related to the number of trial key bytes k_i that would satisfy (6) in each byte position i. The number of 8-byte keys $(k_0, k_1, \ldots, k_7)$ is given by

$$\prod_{i=0}^{7} N(x_i^{(9,a)}, x_i^{(9,b)}, x_i^{(9,c)}, x_i^{(9,d)})$$

with expected value about $4.3^8 \approx 120,000$. This expected value is so high because of the unusually large values of $P_2(256)$ and $P_2(128)$.

When event $\mathbf{E}$ has not happened, the distribution $P_1(n)$ is relevant, and the expected number of 8-byte keys is 1. In fact with probability about $1 - (1 - 0.404)^8 \approx 0.984$ at least one of the values $N(x_i^{(9,a)}, x_i^{(9,b)}, x_i^{(9,c)}, x_i^{(9,d)})$ is zero, so that no 8-byte keys are valid; with the complementary probability 0.016, all are nonzero, and then the expected number of keys is $1/0.016 \approx 62$.

So with 330,000 experiments, the expected number of 8-byte putative keys is $5 \times 120,000 + (330,000 - 5) \times 1 = 930,000$. Among these, the correct key should appear five times, and should be easy to detect; incorrect keys should appear at most once, with possible exception of those differing from the correct key in only one or two bytes.

Remark: Although the mean number of putative keys is fairly small, the variance is huge; the standard deviation exceeds 10^{11}. This is because of the relatively high probability that, for a given time step and a given byte position, $N(x_a, x_b, x_c, x_d)$ is either 256 or 128; if several such bytes occur at the same time step, this time step will yield a huge number of putative keys. In this case an alternative data structure is called for. For example, if one time step has two or more such byte positions, declare that event **E** has probably occurred, and deduce putative values for the *remaining* six or fewer key bytes. Or we could simply list 4-byte putative keys K_2 and K_3 separately.

8   Splitting the period, again

Having determined K_2 and K_3 by the attack in Section 6, we also know the handful of positions where event **E** has occurred; we know several places where

$$\sum_{i=0}^{3} x_i^{(3,a)} = \sum_{i=0}^{3} x_i^{(3,c)} \quad (\text{mod } 256).$$

Because of the relation between $x_i^{(3)}$ and $x_i^{(2)}$ we also have

$$\sum_{i=0}^{3} x_i^{(2,a)} = \sum_{i=0}^{3} x_i^{(2,c)} \quad (\text{mod } 256),$$

whence

$$\sum_{i=0}^{3} S_0(x_i^{(1,a)}) = \sum_{i=0}^{3} S_0(x_i^{(1,c)}) \quad (\text{mod } 256). \tag{7}$$

By enumeration of 2^{32} possibilities, we can find all the possible values of the concatenation $(x_0^{(1,a)}, x_1^{(1,a)}, x_2^{(1,a)}, x_3^{(1,a)})$ and hence, by adding $p_1 a_0 \pmod{p_0}$, the concatenation $(x_0^{(1,c)}, x_1^{(1,c)}, x_2^{(1,c)}, x_3^{(1,c)})$, which satisfy (7). This whittles down the possible values of K_0 from a collection of 2^{32} to about $2^{32}/256^5 = 2^{12}$ possible values. Similar calculations reduce our choice of K_1 to about 2^{12} possible values. The correct values can be gotten by exhaustion.

9   Meet-in-the-middle, again

Another approach at recovering (K_0, K_1) is given here. We assume that we have previously identified (K_2, K_3) using any of the attacks from Sections 4–6. This attack requires only 2^{32} operations, 2^{24} space, and two known keystream blocks; therefore, it should be very fast.

Because of the form of the linear relation in layer 7, we find that the sum $x_0^{(7)} + x_2^{(7)} - x_4^{(7)} - x_6^{(7)} \pmod{256}$ depends only on the four bytes $x_i^{(5)}, i = 1, 3, 5, 7$. Use a meet-in-the-middle approach, requiring time $256^3 = 2^{24}$, to discover all the 2^{24} values of the 4-tuple $[x_i^{(5)}, i = 1, 3, 5, 7]$ that could lead to a given value for this sum. Similarly the sum $x_0^{(7)} + x_2^{(7)} - x_5^{(7)} - x_7^{(7)}$ (mod 256) depends only on the four bytes $x_i^{(5)}, i = 0, 2, 4, 6$. Combine these two lists with another meet-in-the-middle attack, and in time 2^{24} we can recover the 8-tuple $x_*^{(5)}$ from any given value of the 8-tuple $x_*^{(7)}$.

Use time 2^{24} to decrypt one ciphertext back to layer 5. For each of the 2^{32} trial subkeys K_0, compute forward to $x_i^{(3)}, 0 \leq i \leq 3$, and backward from layer 5 to $x_i^{(4)}, 0 \leq i \leq 3$. See whether there is a byte sum $\sum_{i=0}^{7} x_i^{(3)}$ which would enable the linear permutation at layer 4 to map $x_i^{(3)}, 0 \leq i \leq 3$ to $x_i^{(4)}, 0 \leq i \leq 3$. We expect 256 trial subkeys K_0 to pass this test. Similarly develop 256 trial subkeys K_1. Try each of the resulting 65,536 pairs (K_0, K_1) on another ciphertext to determine the correct pair.

10   Computational requirements

The first attack should take only a few seconds to find all of K_2 and K_3, including gathering data.

The meet-in-the-middle attack recovering (K_2, K_3) (see Section 5) requires 2^{32} hash table lookups and about 2^{33} words of memory. If we keep the entire table in memory, the 2^{32} table lookups will take only 400 seconds or so (assuming 100ns access time to main memory, which is not unreasonable).

The space requirements may be more noticeable. One simple approach is to distribute the table across a cluster of 256 workstations, each with 128 MB of memory; such a cluster would take roughly 400 seconds to find (K_2, K_3). Another simple approach, if only one workstation is available, is to trade off time for memory: by splitting the table across time, one workstation can finish in $256 \times 400 \approx 10^5$ seconds (about one month), and n workstations will finish n times as fast that. This is not out of reach, and the interested reader might be able to find better ways to reduce memory needs: for example, the parallel collision search techniques of van Oorschot and Wiener [OW96] (applied to find a "golden collision") look promising.

For the attack based on identifying occurrences of event $\mathbf{E}$ (see Sections 6–8), we need the generator to run for $p_0 + p_1 \approx 2^{33}$ time steps, generating 2^{36} bytes. At the advertised speed of 1 megabyte per second, this will take about nineteen hours. We will look at only 1,000,000 message blocks (8,000,000 bytes): 330,000 at the beginning (representing a), another 330,000 in the middle (representing both b and c, because p_0 and p_1 are so close to each other), and another 330,000 at the end. For each selection (a, b, c, d) we might need to evaluate $8 \times 256 = 2048$ trial key bytes $0 \leq k_i \leq 255, 0 \leq i \leq 7$. However, realize that much of the time we will find that, for example, key byte k_1 has no possible values, so that bytes $k_2, \ldots, k_7$ need not be examined for this case. In total about 212,000,000 key bytes need to be examined.

11 TWOPRIME-1

The same paper [DNRS97] proposes a faster version TWOPRIME-1, differing from TWOPRIME only in the seventh layer; in TWOPRIME-1, this layer preserves halves. That is, the output bytes $x_i^{(7)}, 0 \leq i \leq 3$ only depend on the input bytes $x_i^{(6)}, 0 \leq i \leq 15$, and the output bytes $x_i^{(7)}, 4 \leq i \leq 7$ only depend on the input bytes $x_i^{(6)}, 16 \leq i \leq 31$. This means that the only interaction between the left and right halves of the message occurs during the "linear permutation" in the fourth layer, and there the interaction is limited to the one byte $\sum_i x_i^{(3)}$ (mod 256). In two time steps where this sum agrees, the halves are completely separated.

So we can examine the output at time $a = t_0$ and $b = t_0 + p_0$. If $\sum_{i=4}^{7} x_i^{(3,a)} = \sum_{i=4}^{7} x_i^{(3,b)}$ (mod 256) (i.e. the second of the two conditions for event **E**), then the left-hand half of the output of each layer is the same for a as for b:

$$x_i^{(j,a)} = x_i^{(j,b)}, \quad 0 \leq i \leq 3, \quad j \neq 6$$
$$x_i^{(6,a)} = x_i^{(6,b)}, \quad 0 \leq i \leq 15.$$

In particular the left-hand halves of the outputs will agree. By identifying eight pairs (a, b) where these output halves agree, we can deduce the value of K_0 as in the TWOPRIME case. Similar computations give us K_1.

We can then use exhaustive search to compute K_2 in about 2^{32} steps. For example, if we guess the four bytes representing $(\sum_{j=0}^{7} k_j) - k_i, 0 \leq i \leq 3$, and we know the values of K_0 and K_1, we can find the left-hand half of all layers up through layer 8. We can compare the encryptions of two unrelated time steps, say a and e, to see whether

$$x_i^{(8,a)} \oplus x_i^{(8,e)} \stackrel{?}{=} x_i^{(9,a)} \oplus x_i^{(9,e)}, \quad 0 \leq i \leq 3.$$

If not, these four bytes are wrong. But if they are equal, we can use layer 8 to deduce K_2, giving us another check on our original assumptions, and furnishing us with the correct value of K_2. The calculation of K_3 is left to the reader.

We needed to run the generator for 2^{32} messages (2^{35} bytes), or ten hours, and examine about $2 \times 8 \times 256 = 4,096$ blocks (32,768 bytes). The computational requirements of 2^{32} operations are not onerous, and the interested reader might well find more efficient methods to discover K_2.

Another approach is also available. In the first phase of this attack, we recover (K_2, K_3). The key observation is that—modelling each half of layers 6–7 as a random function—only about $1 - e^{-1}$ of the 2^{32} possible values for the left half of the output of the seventh layer will actually be attainable. Therefore, in the first phase, we guess K_2, compute up the left side of the cipher to the output of the seventh layer, and discard guesses at K_2 when they produce unattainable intermediate values. Because $(1 - e^{-1})^{50} < 2^{-32}$, we see that after about 50 blocks (400 bytes) of known plaintext, there will be just one value remaining—namely, the correct value of K_2. A similar technique recovers K_3.

Now the second phase proceeds as in Section 9. For each guess at K_0, we compute forward down the left side of the cipher to the output of layer 3 and backward to the output of layer 4, checking to see whether the two are compatible. We expect 256 values of K_0 to remain, and similarly 256 values of K_1; these remaining 2^{16} possibilities can be checked by trial encryption.

In short, this second approach breaks TWOPRIME-1 with about the same time and space complexity as the corresponding attack on TWOPRIME. We require slightly more known plaintext, but 50 blocks (400 bytes) of known plaintext should be readily available in many systems.

12   ONEPRIME

The same paper [DNRS97] proposes a scheme ONEPRIME, which differs from TWOPRIME only in the first layer: instead of two primes p_0 and p_1, we have only one prime $p = 2^{64} - 59$ and fixed multiplier a. The output of the first layer at time t is

$$(x_0^{(1)}, \ldots, x_7^{(1)}) = at + (K_0, K_1) \pmod{p}.$$

A slight modification enables our attack to run against this scheme as well. Based on the value a (which was not specified in the paper), compute values Δ_0 and Δ_1 such that in the binary representation of $a\Delta_0 \pmod{p}$, the left-most 34 bits are 0 (so that the left half is 0 and the right half represents an integer smaller than 2^{30}). Similarly in the binary representation of $a\Delta_1 \pmod{p}$, the leftmost (highest order) two bits are 0, and the rightmost 32 bits are 0. Each Δ_i should be about 2^{34} and can be computed using methods from continued fractions.

Then if we select time steps

$$a = t_0$$
$$b = t_0 + \Delta_0$$
$$c = t_0 + \Delta_1$$
$$d = t_0 + \Delta_0 + \Delta_1$$

we will find, with probability exceeding $(3/4)^2 > 0.56$, that the left-hand halves of the outputs of layer 1 agree at times a and b, as well as at times c and d; and the right-hand halves agree at times a and c, as well as at times b and d. The rest of the attack proceeds as before.

We need the generator to run for somewhat longer, because $\Delta_0 > p_0$, and we need to examine someone more ciphertext, because our favorable conditions only occur with probability 0.56, but the attack is still feasible.

Another approach is also available. We can break ONEPRIME with meet-in-the-middle techniques. In fact, simply applying the attacks in Sections 5 and 9 immediately breaks ONEPRIME, without any modifications needed. This second approach requires eight blocks of known keystream as well as 2^{33} time and 2^{32} space.

13   Discussion

At a high level, the intuition behind some of our cryptanalysis is that we apply the meet-in-the-middle attack repeatedly, at two levels of abstraction. First, we divide the cipher horizontally between layers, and meet at the "middle"—the output of the seventh layer—at the highest level of abstraction. Second, we divide the cipher vertically into left and right halves, and meet in the "middle", where the "middle" is a characteristic of the output of the seventh layer.

Some of the techniques, e.g. Sections 6–8, do not fall cleanly into this model. We will ignore them for the moment.

Note that the vertical split can be viewed as decomposing the 64-bit function F into two parallel 32-bit functions G, H. In other words, splitting F vertically corresponds to writing $F(a, b) = (G(a), H(b))$. Of course, given such a parallel decomposition, we can apply a divide-and-conquer attack; since breaking a 32-bit function has complexity at most 2^{32}, such a decomposition lets us break F in at most $2 \cdot 2^{32}$ time.

So we conclude that F should be designed to resist parallel decomposition, and in particular there should be no parallel G, H that approximate F. This just comes down to ensuring there is plenty of diffusion, a well-known design principle for cipher design. This lack of diffusion helped make our attacks on TWOPRIME possible.

We can also analyze the horizontal split in terms of functional decomposition. In this case, we find that it corresponds to finding G, H such that $F = H \circ G$ (i.e. $F(a) = H(G(a))$). When we can find such G, H where G is non-surjective and H is bijective, then meet-in-the-middle attacks may allow the cryptanalyst to isolate the effect of G from the effect of H. In other words, the cryptanalyst can often analyze H without taking into account the effect of G (or the key bits that enter G); once H has been broken, the cryptanalyst can then peel off the effect of H (since it is bijective) and attack G alone. The result of such a divide-and-conquer attack would be that F is not much stronger than the strongest of G or H standing alone. TWOPRIME put some of its strength into G, and some into H, with the result that much of its strength was wasted. Far better would have been to concentrate all the strength in one of G or H and make the other as simple as possible, to avoid this potential danger.

Therefore, we suggest the following design principle, which seems broadly applicable to the construction of non-bijective cryptographic functions from a product of rounds. One should avoid introducing non-surjectivity in the middle of the function, because that may speed up meet-in-the-middle attacks and thus waste precious cryptographic strength.

Note that the latter design principle offers some intuitive justification for the structure of many of today's most successful non-bijective cryptographic functions (such as MD5, SHA, ...). The Davies-Meyer construction [Win84] builds F as $F(a) = G(a) \oplus a$. Here all the strength is concentrated in a bijective function G (usually built out of a block cipher); the non-surjectivity is introduced as late as possible, and as simply as possible. MD2 [Kal92] and Snefru [Mer90] also follow our suggested design principle: they too use a bijective function G at the core, and introduce non-surjectivity only at the endpoints (by adding simple redundancy to the input of G, and truncating its output).

This design principle is not novel. It has been discussed in more detail by Preneel in the context of the design of compression functions for hash functions; see [Pre93, e.g. Section 4.2].

14 Conclusions

Pulling it all together, we can identify three important attacks against the stream cipher TWOPRIME. First, we can break TWOPRIME with 2048 blocks of known keystream and 2^{32} work by using the techniques of Sections 4 and 9. Alternatively, we can get by with only 8 blocks of known keystream with repeated use of meet-in-the-middle attacks (Sections 5 and 9); the cost is that we need 2^{32} space as well as 2^{33} work. Finally, we can cryptanalyze TWOPRIME with 2^{33} blocks of known keystream and about 2^{28} operations by using the methods from Sections 6–8; this last attack uses no special features of the compression function in layer seven (other than its linearity). We see that, for a cipher with a 128-bit key, TWOPRIME is disappointingly weak.

We have pointed out weaknesses in two of the layers in TWOPRIME. Because TWOPRIME has only nine layers, each layer lies close to the surface, and any weakness is more easily exploited. The system needs more layers to have any serious cryptographic strength.

References

DNRS97. C. Ding, V. Niemi, A. Renvall, and A. Salomaa, "TWOPRIME: A Fast Stream Ciphering Algorithm," *Fast Software Encryption, FSE'97*, Springer LNCS volume 1267, pages 88–102, 1997.

Kal92. B.S. Kaliski, "The MD2 Message Digest Algorithm," RFC 1319, April 1992.

Mer90. R.C. Merkle, "A Fast Software One-Way hash Function," *Journal of Cryptology*, vol 3 no 1, 1990.

OW96. P.C. van Oorschot and M.J. Wiener, "Improving implementable meet-in-the-middle attacks by orders of magnitude," *CRYPTO'96*, pages 228-236, Springer-Verlag, 1996.

Pre93. B. Preneel, "Design principles for dedicated hash functions," *Fast Software Encryption*, it FSE'93, Springer LNCS volume 809, pages 71–82, 1994.

Sin68. A. Sinkov, *Elementary Cryptanalysis, A Mathematical Approach.* New York: Random House, 1968.

Tuc70. B. Tuckerman, "A study of the Vigenere-Vernam single and multiple loop enciphering systems," IBM Research Report RC2879, 14 May 1970, Yorktown Heights NY.

Win84. R. Winternitz, "Producing One-Way Hash Functions from DES," *Advances in Cryptology: Proceedings of Crypto 83*, Plenum Press, 1984, pp. 203–207.

A   Appendix

We give here the experimental distributions of $P_1(n)$ and $P_2(n)$:

n	$e^{-1}/n!$	$P_1(n)$	$P_2(n)$
0	0.3679	0.404	0
1	0.3679	0.337	0.337
2	0.1839	0.183	0.367
3	0.0613	0.062	0.185
4	0.0153	0.017	0.070
5	0.0030	0.004	0.020
6	0.0005	0.001	0.006
7	0.0001	0.0002	0.001
8	0	0.00029	0.002
16	0	0.000028	0.0004
32	0	0.000025	0.0008
64	0	0.000019	0.0012
128	0	0.000008	0.0010
256	0	0.000031	0.0078

$$\sum nP_1(n) = 1, \quad \sum nP_2(n) \approx 4.3$$

JEROBOAM

Hervé Chabanne[1] and Emmanuel Michon[2]

[1] SAGEM SA, Unité de Recherche et Développement
Systèmes et Terminaux Sécurisés, France
chabanne@urd32.sagem.fr
[2] Ecole Polytechnique, Palaiseau, France
michon@www.enst.fr[* * *]

Abstract. We introduce a new fast stream cipher, JEROBOAM, working with a key of 128 or 248 bits. JEROBOAM was designed to work with eight internal 32-bit registers called multiply-with-carry generators (mwc). These register are very easy to implement in software and produce a sequence of excellent statistical quality. Per contra, one mwc is easily cracked by a lattice reduction algorithm. Hence, we are lead to interpose a nonlinear filter between these weak registers and the pseudo-random output.

1 Introduction

The cipher JEROBOAM is designed to work efficiently on 16-bit microprocessors. The key is 128- or 248-bit long, which is quite comfortable; after a short setup equivalent in time to the encryption of a 42-byte message, JEROBOAM produces a pseudo-random stream one can use as a symmetric cipher to XOR a cleartext of any length.

JEROBOAM was designed with IDEA[8,11] as a model. It lies on a classical scheme and can be seen as a nonlinear combination of irregularily clocked pseudo-random generators. Current technical requirements bind us to use only operations directly available on all microprocessors : we use mwc as random generators and the nonlinear filter is obtained by a now classical alternance of + and ⊕[2].

The mwc, multiply-with-carry generators, are a new primitive in cryptography due to G.Marsaglia[5][1]. They allow fast computations in different prime finite fields ; their description and the way they can be cryptanalysed can be found in Sect. 2.

The complete description of JEROBOAM is given in Sect. 3. It is completed by a slow C implementation and test values in Sect. 7.

In Sect. 4, we discuss the statistical evaluation of the output stream as a pseudo-random sequence. We study in Sect. 5 the speed of JEROBOAM and give a microprocessor-independent evaluation.

[* * *] current address: Ecole Nationale Supérieure des Télécommunications, Paris, France. This work was completed during a terminal training period in SAGEM SA. The reference *Etude cryptologique du chiffreur* JEROBOAM[4] is the complete, full-documented version of this article.

[1] thanks to [10] for its pointer on the package DIEHARD due to G.Marsaglia

S. Vaudenay (Ed.): Fast Software Encryption – FSE'98, LNCS 1372, pp. 49–59, 1998.
© Springer-Verlag Berlin Heidelberg 1998

2   Specification

2.1   Multiply With Carry (mwc) Generators[5]

Definition 1. *Let $a, b \geq 1$, $0 \leq c_0 < a$, $0 \leq x_0 < b$. We call* mwc *of multiplier a and base b the sequence of integer pairs $(c_n, x_n)_{n \in \mathbb{N}}$ determined by the following relation:*

$$\begin{array}{l} c_{n+1} \text{ is the quotient} \\ x_{n+1} \text{ is the remainder} \end{array} \quad \text{in the euclidean division of } ax_n + c_n \text{ by } b. \quad (1)$$

c_n is called the carry, and we denote with $^{c_n}x_n$ the pair (c_n, x_n).

The division is just a right shift if we let b be a power of 2. In JEROBOAM, $b = 2^{16}$ and we let the carry c_n be the MSBs and x_n the LSBs of a 32-bit word w. This way, (1) becomes in ready-to-code form

$$\mathtt{w} = a(\mathtt{w} \,\&\, \mathtt{0xffff}) + \mathtt{w} \gg 16$$

where $\&$ is the logical AND and $\gg$ the right shift. We return $x_n = \mathtt{w} \,\&\, \mathtt{0xffff}$.

We can state the following result from [5]:

Proposition 2. *Let us consider the set $S = \{^c x \mid 0 \leq c < a, 0 \leq x < b\}$ and the integer $m = ab - 1$, called the modulus — remark that $\#S = m + 1$. The directed graph of the function f which transforms $^{c_n}x_n$ into $^{c_{n+1}}x_{n+1}$ in S consists of several orbits.*

The two points $^0 0$ and $^{a-1}b - 1$ are fixed points for f. If an orbit has k points, k is the order of b in a group $(\mathbb{Z}/m'\mathbb{Z}, \times)$, where m' is a divisor of m.

If we choose m to be a safe prime, ie both m and $(m-1)/2$ are prime, we get two non trivial orbits for this graph.

The following result shows how to switch from one orbit to another.

Proposition 3. *The symmetry $g : S \to S$ which transforms $^c x$ into $^{u\ 1}\,^c b - 1 - x$ commutes to f.*

Proof. Let $^{c'}x' = f(^c x)$ defined by the euclidean division by b

$$bc' + x' = ax + c, \quad \text{with } 0 \leq x' \leq b - 1$$

then

$$ab - 1 - (bc' + x') = ab - 1 - ax - c$$
$$b(a - 1 - c') + (b - 1 - x') = a(b - 1 - x) + (a - 1 - c)$$

with $0 \leq b - 1 - x' \leq b - 1$, which means $g \circ f \circ g = f$, so $f \circ g = g \circ f$.   □

This way, we can switch from a one-point orbit to the other, and, more interesting, from a $(m-1)/2$ point orbit to the other.

Choice of the Multipliers Considering only safe prime modules $m = a2^{16} - 1$ $(0 < a < 2^{16})$ leave us the choice between 392 values.

We would appreciate to use every bit in the 32-bit word w, so we impose the condition $2^{15} < a$. There are still 171 possibilities left.

To explain our final choice, let us consider the following result, which establishes a strong link between mwc and the well-known Lehmer generator $X_{n+1} = aX_n \bmod m$:

Proposition 4. *Let $X_0 = ax_0 + c_0$ if $ax_0 + c_0 < m$. In this case the multiplicative congruential generator*

$$X_{n+1} = aX_n \bmod m$$

verifies for all $n \in \mathbb{N}$:

$$c_n = X_n \bmod a, \qquad x_n = \left\lfloor \frac{X_n}{a} \right\rfloor. \tag{2}$$

The initial conditions not satisfying $ax_0 + c_0 < m$ consist of the unique pair $^{c_0}x_0 = a^{-1}b - 1$, which is a fixed point for f.

Proof. (2) is true for $n = 0$. Suppose it is true for n. Let q, r be the quotient and the remainder of the euclidean division of X_n by b, we get

$$X_n = bq + r \text{ avec } 0 \le r < b.$$

Multiply by a, it becomes

$$aX_n = abq + ar \text{ avec } 0 \le ar < ab,$$

we obtain this way the euclidean division of aX_n by ab. One can also write:

$$aX_n = (ab - 1)q + ar + q$$

is this the euclidean division of aX_n by $ab - 1$? Let us check that $0 \le ar + q < ab - 1$:

— if $0 \le r \le b - 2$, given $0 \le q \le a - 1$, one gets $ar + q = a(b - 2) + a - 1 = ab - 1 - a < ab - 1$.
— if $r = b - 1$, we know then $0 \le q \le a - 2$, il vient $ar + q = a(b - 1) + a - 2 = ab - 2 < ab - 1$.

This way, $X_{n+1} = ar + q$: it is the relation stated at the rank $n + 1$. □

D.E.Knuth defines the spectral test in [3]. This test is an evaluation of the quality of the geometric repartition of the successive t-uples $(X_n, X_{n+1}, \ldots, X_{n+t-1})$ of a Lehmer generator ; it is clear that two generators of same module m can have very different behavior with different multipliers a.

G.S.Fishman and L.R.Moore[6,9] use this test to determine the best multipliers for given modules $m = 2^{31} - 1$, 2^{32} and 2^{48}.

We choose the biggest eight a satisfying the conditions: $m = a2^{16} - 1$ safe prime, $2^{15} < a < 2^{16}$, and $\mu_3 > 0.02$ (the figure of merit μ_t and ν_t are defined in [3], p.101).

a	a(hex)	ν_2^2	ν_3^2	ν_4^2	ν_5^2	ν_6^2	μ_2	μ_3	μ_4	μ_5	μ_6
36594	8ef2	1339120837	54273	31023	3610	1148	1.75	0.0221	1.98	1.72	3.26
36804	8fc4	1354534417	65869	29375	3287	771	1.76	0.0294	1.77	1.35	0.98
37959	9447	1440885682	54529	29468	2782	667	1.82	0.0214	1.72	0.86	0.62
38568	96a8	1487490625	75165	30978	2259	1017	1.85	0.0342	1.87	0.51	2.15
40995	a023	1680590026	77923	25142	4287	1118	1.97	0.0339	1.16	2.36	2.69
42153	a4a9	1776875410	81906	53845	1827	797	2.02	0.0355	5.18	0.27	0.95
42903	a1=a797	1840667410	65865	40036	1408	1258	2.06	0.0252	2.81	0.14	3.66
43995	a3=abdb	1935560026	58906	54315	4903	930	2.11	0.0208	5.05	3.07	1.44
47529	a5=b9a9	2259005842	68806	40203	690	686	2.28	0.0243	2.56	0.02	0.54
51813	a7=ca65	2684586970	73230	39339	5164	1055	2.48	0.0244	2.25	2.97	1.79
53130	a6=cf8a	2822796901	75169	41931	4691	889	2.55	0.0248	2.49	2.28	1.04
54564	a4=d524	2977230097	83097	39742	4283	1087	2.62	0.0281	2.18	1.77	1.86
57225	a2=df89	3274700626	74934	53947	7793	1223	2.74	0.0229	3.83	7.52	2.52
61914	a0=f1da	3833343397	88869	71443	4014	750	2.97	0.0273	6.21	1.32	0.54

Remark that the whole multiplicative congruential generator is implied in the spectral test ; the bits we use in JEROBOAM are, in fact, the most significant ones.

2.2 Cryptanalysis of a mwc

We know that the sequence produced by a multiplicative congruential generator is predictible. More precisely, A.M.Frieze, J.Hastad, R.Kannan, J.C.Lagarias and A.Shamir[1] prove the following result:

Proposition 5. *Let us consider k successive terms in the sequence $X_{n+1} = aX_n \bmod m$. If we specify at least*

$$s = 1 + \log_2 \Delta_k + \frac{1}{2}\log_2 k$$

high-order bits on each term, it is possible to recover the whole sequence $(X_n)_{n\in\mathbb{N}}$ in polynomial time. Δ_k is defined below.

Proof. Consider the matrix[2]

$$A = \begin{pmatrix} m & 0 & 0 & \cdots & 0 \\ a & -1 & 0 & \cdots & 0 \\ a^2 & 0 & -1 & \cdots & 0 \\ \vdots & \vdots & \vdots & \ddots & \vdots \\ a^{k-1} & 0 & 0 & \cdots & -1 \end{pmatrix},$$

---

[2] A detailed proof of a more complete result can be found in [1].

we want to solve the system of modular equations $AX \equiv C(\mathrm{mod}\, m)$ where $X = (x_1, x_2, \ldots, x_k)$ and $C = (0, 0, \ldots, 0)$ are two column vectors.

Let us consider the lattice formed by the rows of the matrix A, and reduce it by the LLL algorithm. This way, we get a matrix $B = PA$ with a "small" norm Δ_k, defined as the biggest euclidean norm of the rows of B.

The system can then be written $BX \equiv C'(\mathrm{mod}\, m)$. If we choose the components of C' in the integer interval $[-m/2, m/2[$, and if we know in advance that $\|X\|$ is small enough, ie, $\|BX\| < m/2$, we loose the modular aspect and we are lead to solve a trivial linear system $BX = C'$ in $\mathbb{Z}$.

Knowing some high order bits, X^h, of X allows us to change the unknown and substitute X by the small unknown vector $X - X^h$, which leads us to the situation above. □

A C implementation of this proposition shows that s is roughly n/k for big m, and that a Lehmer generator can be "cracked" in a few seconds by the observation of a few MSBs. For more precise results, see [4].

3 Description of JEROBOAM

The heart of JEROBOAM consists of eight 32-bit mwc registers, a FIFO queue of two 16-bit words and a particular 16-bit word:

$\mathrm{mwc}_0, \mathrm{mwc}_1, \mathrm{mwc}_2, \mathrm{mwc}_3, \mathrm{mwc}_4, \mathrm{mwc}_5, \mathrm{mwc}_6, \mathrm{mwc}_7,\ \mathrm{queue}_1, \mathrm{queue}_2,\ \mathrm{lea}$ for <u>lea</u>der.

3.1 Setup: Key Insertion

One can choose between a 248-bit key and a 128-bit key.

248-bit key The key is given by eight 32-bit words key_0, key_1, key_2, key_3, key_4, key_5, key_6, key_7. The 32th bit of each word must be 0, and none of these words can be 0. The initial value of mwc_i is set to key_i.

queue_1 and queue_2 receive any values (we chose 0xda37, 0xc07f). lea is initially the LSBs of mwc_0.

Make 21 cycles[3] of the algorithm below, and prepare to encrypt.

128-bit key The key is given by eight 16-bit words key_0, key_1, key_2, key_3, key_4, key_5, key_6, key_7. The ith mwc receives the 32-bit word $(i + 1)2^{16} + \mathrm{key}_i$. One can choose any of the 2^{128} possible key, even the zero one! The following is identical.

---

[3] This is the first rank k beyond which the probability $(k + 1)/2^k$ of outputting one of the queue original setup values is less than the uniform probability 2^{-16}.

3.2 Elementary Cycle

1. Consider the bits of `lea` :

15	14	13	12	11	10	9	8	7	6	5	4	3	2	1	0
swi2	sup0	lea0	chop	ini1	sup2	five	fifo	ini2	lea1	ini0	sup1	lea2	swi0	cplt	swi1

2. i_0 is $4ini2 + 2ini1 + ini0$.
3. If `chop` is 1, then

$$cmb = mwc_{i_0} \oplus mwc_{i_0+1 \bmod 8} + mwc_{i_0+2 \bmod 8} \oplus mwc_{i_0+3 \bmod 8}$$

else

$$cmb = mwc_{i_0} + mwc_{i_0+1 \bmod 8} \oplus mwc_{i_0+2 \bmod 8} + mwc_{i_0+3 \bmod 8}.$$

   $+$ denotes the modular addition in $\mathbb{Z}/2^{16}\mathbb{Z}$, $\oplus$ the bit-to-bit XOR. The evaluation of these two non commutative, non associative operations is done from left to right.
4. If `five` is 1, add $mwc_{i_0+4 \bmod 8}$ to `cmb` with the appropriate alternating operation.
5. If `fifo` is 1, `cmb` enters the queue and is replaced by the output of the queue.
6. **The output 16-bit word of this cycle is `cmb`: make a XOR between `cmb` and the corresponding word in the cleartext.**
7. Advance all the `mwc`.
8. Advance once more the `mwc` indexed by $4sup2 + 2sup1 + sup0$.
9. Switch the orbit of the `mwc` indexed by $4swi2 + 2swi1 + swi0$.
10. `newlea` is the `mwc` indexed by $4lea2 + 2lea1 + lea0$.
11. If `cplt` is 1, bit-to-bit complement `newlea`.
12. The new `lea` is `newlea`.
13. Go to step 1.

4 Statistical Evaluation

We have used the statistical tests defined by D.E.Knuth[3] to check the random behavior of the output word of JEROBOAM.

These empirical tests are: frequency test, serial test, gap test, poker test, coupon collector's test, permutation test, run test, max-of-t test, collision test and serial correlation test.

We did not notice any significant bias, every bit of the output behaves as a coin-tossing experiment does, independently from his neighbours.

A full but quite boring evaluation of these statistics can be found in [4].

5 Performances

We ciphered $1, 2, \ldots, 10$ Mbyte files with a hundredth of second precise measure on different PCs.

β denotes the number of clocks necessary to the obtention of one 16-bit output word.

Microprocessor	Exploitation system	Speed (Mbyte/s)	β
Pentium 100	Windows 95	0.73	261
Pentium 120	Windows 95	0.88	260
Pentium 166	Windows 95	1.21	261
Pentium 166MMX	Windows NT 4	1.17	270
Pentium 200MMX	Windows 95	1.40	272

The C compiler is Microsoft Visual C++ version 4.2, using Pentium code generation.

Let us try to find a rapid estimation of the cost of a JEROBOAM cycle on a standard Pentium. With the slowest addressing mode on a Pentium[7]:

Operation	Operand	Size (bits)	Quantity	Cycles
unsigned multiplication	MUL	16.16→32	9	11
substraction	SUB	32	1	2
logical and	AND	16	16	2
addition	ADD	16	11.75	2
right shift	SHR	16	9	2
exclusive or	XOR	16	2.75	2
complement	NEG	16	0.5	2
Total cycle number C				181

This is 30% less of the above observation, but it is still the same ordre of magnitude.

Computing C for a N MHz frequency microprocessor, one can estimate the enciphering speed v in megabyte per second at

$$v \approx 1.9 \frac{N}{C}.$$

We finally note that the speed of JEROBOAM strongly depends on the speed of the multiplication of two unsigned 16-bit words. For instance, experiments on an ARM processor bring us to a 25 % speeding up.

6   Conclusion : On the Cryptanalysis of Two or Three mwc

We have seen that a mwc used alone is insecure.

We would now like to insist on the fact that, given the sequence formed by the sum in $\mathbb{Z}/2^{16}\mathbb{Z}$ of two mwc x_n and y_n in two different finite fields $\mathbb{Z}/p\mathbb{Z}$ and $\mathbb{Z}/q\mathbb{Z}$, we do not know how to recover the initial terms x_0 and y_0.

The output of JEROBOAM is far more tricky. In particular, this output always implies a third mwc with an XOR operation, ie another algebraic structure.

We invite the reader to determine how to recover the inital content of two mwc given the LSBs of their sum, then to incorporate a third mwc with $\oplus$, and finally crack JEROBOAM...

7   Appendix : C Source Code and Test Values

7.1   A Readable but Slow C Implementation

```
/* 16-bit stream cipher JEROBOAM 2.0
   Readable but slow C implementation */

#include <stdio.h>

typedef unsigned short w16 ;
typedef unsigned long  w32 ;

static w16 a[8]={61914,42903,57225,43995,54564,47529,53130,51813};
static w32 mwc[8];
static w16 lea,queue1,queue2 ;

#define nsetupcycle 21

void clockmwc(int i)
  { mwc[i]=(w32)(w16)mwc[i]*a[i]+(mwc[i]>>16); }
void switchmwc(int i)
  { mwc[i]=((w32)(a[i]-1)<<16)+0xffff-(w32)mwc[i]; }

w16 elemcycle()
{
int five,chop,fifo,cplt;
int lea2,lea1,lea0;
int ini2,ini1,ini0;
int sup2,sup1,sup0;
int swi2,swi1,swi0;
w16 newlea,cmb,save;
int i0,i;
swi1=(lea&0x0001)?1:0;
cplt=(lea&0x0002)?1:0;
swi0=(lea&0x0004)?1:0;
lea2=(lea&0x0008)?1:0;
sup1=(lea&0x0010)?1:0;
ini0=(lea&0x0020)?1:0;
lea1=(lea&0x0040)?1:0;
ini2=(lea&0x0080)?1:0;
fifo=(lea&0x0100)?1:0;
five=(lea&0x0200)?1:0;
sup2=(lea&0x0400)?1:0;
ini1=(lea&0x0800)?1:0;
chop=(lea&0x1000)?1:0;
lea0=(lea&0x2000)?1:0;
sup0=(lea&0x4000)?1:0;
swi2=(lea&0x8000)?1:0;
newlea=mwc[4*lea2+2*lea1+lea0];
if (cplt) newlea=~newlea;
i0=4*ini2+2*ini1+ini0;
cmb=mwc[i0];
for (i=1;i<=3+five;i++)
  {
```

```
 if (chop) cmb=cmb^mwc[(i0+i)%8]; else cmb=cmb+mwc[(i0+i)%8];
 chop=!chop;
 }
if (fifo)
 {
 save=queue1 ;
 queue1=queue2 ;
 queue2=cmb ;
 cmb=save ;
 }
for (i=0;i<8;i++) clockmwc(i);
clockmwc(4*sup2+2*sup1+sup0);
switchmwc(4*swi2+2*swi1+swi0);
lea=newlea;
return cmb;
}

void cipher(unsigned char *msg,int sizeinw16)
{
w16 *doublemsg;
int compt ;
doublemsg=(w16*)msg ; /* ! pentium is little-endian */
for (compt=0;compt<sizeinw16;compt++)
 doublemsg[compt]=doublemsg[compt]^elemcycle() ;
}

void key248(w32 key0,w32 key1,w32 key2,w32 key3,
            w32 key4,w32 key5,w32 key6,w32 key7)
{
w16 dumb[nsetupcycle];
if (!(key0&&key1&&key2&&key3&&key4&&key5&&key6&&key7))
 { printf("incorrect key: a 32-bit word is zero\n"); exit(1); }
if ((key0|key1|key2|key3|key4|key5|key6|key7)&0x80000000)
 { printf("incorrect key: a 32th bit is non-zero\n"); exit(1); }
mwc[0]=key0 ; mwc[1]=key1 ; mwc[2]=key2 ; mwc[3]=key3 ;
mwc[4]=key4 ; mwc[5]=key5 ; mwc[6]=key6 ; mwc[7]=key7 ;
lea=mwc[0] ;
queue1=0xda37 ;
queue2=0xc07f ;
cipher((unsigned char*)dumb,nsetupcycle) ;
}

void key128(w16 key0,w16 key1,w16 key2,w16 key3,
            w16 key4,w16 key5,w16 key6,w16 key7)
{
key248(0x00010000+key0,0x00020000+key1,0x00030000+key2,0x00040000+key3,
       0x00050000+key4,0x00060000+key5,0x00070000+key6,0x00080000+key7) ;
}
```

The fast but unreadable version, essentially written with C preprocessor macros will not be shown here. It can be found in [4].

7.2   Differences Between JEROBOAM 1.0 and JEROBOAM 2.0

JEROBOAM 1.0 is the cipher we study in [4].

This article presents an extended version : JEROBOAM 2.0. There are three differences:

— JEROBOAM 1.0 does not support 248-bit keys. He has just one 128-bit key-insertion function called `clef(...)`, and

$$\text{JEROBOAM 1.0:clef(\ldots)} \equiv \text{JEROBOAM 2.0:key128(\ldots)}.$$

— JEROBOAM 1.0's `chiffre(...)` function only ciphers 1024 byte blocks:

$$\text{JEROBOAM 1.0:chiffre(\ldots)} \equiv \text{JEROBOAM 2.0:cipher(\ldots,512)}.$$

— The setup of JEROBOAM 2.0 is 21 cycle long. JEROBOAM 1.0's setup is 512 cycle long.

7.3   Test Values

Here's what we get with the following 248-bit key if we encrypt the zero message:

```
key248(7323aafc,01638ef6,20903ffa,7f8750d0,2275e0d1,36da83da,4fe33cca,38743eca)
0000-000f : e4 e3 d8 12 06 36 20 73 00 2d cc 66 29 6b 4f 9d
0010-001f : ca 6e 48 a2 a2 be 9e 61 b1 f1 f9 0d 25 0b 06 66
...
```

Acknowledgements

The first author wishes to thank the CONFIDENCE's team. Both authors would like to express their gratitude to Professor Jacques Stern for fruitful discussions about JEROBOAM and the anonymous referees for their advices.

References

1. FRIEZE A.M., HASTAD J., KANNAN R., LAGARIAS J.C., SHAMIR A.,   " Reconstructing truncated integer variables satisfying linear congruences " *SIAM Jour. Comp.*, 17 (1988) 262–280.
2. ROGAWAY P.[4], COPPERSMITH D., " A software-optimized encryption algorithm " R. Anderson, editor, *Fast software encryption, Cambridge Security Workshop*, (LNCS 809) 56-63, 1994.
3. KNUTH D.E., *The Art of Computer Programming*, volume 2, Addison-Wesley, 1984.
4. MICHON E.,   Rapport de stage d'option scientifique: étude cryptologique du chiffreur JEROBOAM, Document LaTeX 2_ε, 74 ko, Ecole Polytechnique, june 1997.

---

[4] SEAL has been redesigned to SEAL 3.0. A description of this new version can been found at http://wwwcsif.cs.ucdavis.edu/rogaway/papers/list.html

5. MARSAGLIA G., Diehard, `ftp stat.fsu.edu/pub/diehard/diehard.zip`.

6. FISHMAN G.S., " Multiplicative congruential random number generators with modulus 2^β " *Math. Comp.*, 54 (1990) 331–344.

7. INTEL, *Developers' insight CDROM*, january 1997, F. Instruction format and timing, in `/design/pentium/manuals/241430_4.pdf`.

8. LAI X., MASSEY J.L., " A proposal for a new block encryption standard " I.Darmgård, editor, *EUROCRYPT 90, Advances in Cryptology*, (LNCS 473) 389–404, 1991.

9. FISHMAN G.S., MOORE L.R. " An exhaustive analysis of multiplicative congruential random number generators with modulus $2^{31} - 1$ " *SIAM Jour. Comp.*, 7 (1986), 24–45.

10. JENKINS R.J., " ISAAC " D. Gollman, editor, *Fast Software Encryption, Third International Workshop*, (LNCS 1039) 41–50, 1996.

11. LAI X., MASSEY J.L.,MURPHY S., " Markov ciphers and differential cryptanalysis " D.W.Davies, editor, *EUROCRYPT 91, Advances in Cryptology*, (LNCS 547) 17–38, 1991.

Fast Hashing and Stream Encryption with PANAMA

Joan Daemen[1] and Craig Clapp[2]

[1] Banksys, Haachtesteenweg 1442, B-1130 Brussel, Belgium
Daemen.J@banksys.be
[2] PictureTel Corporation, 100 Minuteman Rd., Andover, MA 01810, USA
craigc@pictel.com

Abstract. We present a cryptographic module that can be used both as a cryptographic hash function and as a stream cipher. High performance is achieved through a combination of low work-factor and a high degree of parallelism. Throughputs of 5.1 bits/cycle for the hashing mode and 4.7 bits/cycle for the stream cipher mode are demonstrated on a commercially available VLIW micro-processor.

1 Introduction

PANAMA is a cryptographic module that can be used both as a cryptographic hash function and a stream cipher. It is designed to be very efficient in software implementations on 32-bit architectures.

Its basic operations are on 32-bit words. The hashing state is updated by a parallel nonlinear transformation, the buffer operates as a linear feedback shift register, similar to that applied in the compression function of SHA [6]. PANAMA is largely based on the STEPRIGHTUP stream/hash module that was described in [4].

PANAMA has a low per-byte work factor while still claiming very high security. The price paid for this is a relatively high fixed computational overhead for every execution of the hash function. This makes the PANAMA hash function less suited for the hashing of messages shorter than the equivalent of a typewritten page. For the stream cipher it results in a relatively long initialization procedure. Hence, in applications where speed is critical, too frequent resynchronization should be avoided.

A typical application for PANAMA might be the encryption or decryption of video-rate data in conditional access applications (e.g. pay-TV). Set-top boxes and future digital televisions will increasingly include media processors for decoding compressed video and for performing other computationally intensive image processing tasks. This is an application space where data rates are high, high-performance processors are increasingly likely to be present, and decryption must be done yet must not unduly burden an already heavily loaded processor.

After specifying the PANAMA hash function and stream cipher, we discuss the particular design strategy and the implementation aspects. We don't attempt to give a proof of security. However, a motivation for the design choices is given.

S. Vaudenay (Ed.): Fast Software Encryption – FSE'98, LNCS 1372, pp. 60–74, 1998.

A C reference implementation of PANAMA and a PostScript and PDF versions of [4] are available from `http://www.esat.kuleuven.ac.be/~rijmen/daemen`.

2   Basic design principles

PANAMA is based on a finite state machine with a 544-bit *state* and a 8192-bit *buffer*. The state and buffer can be updated by performing an *iteration*. There are two modes for the iteration function. A *Push* mode, that allows to inject an input and generates no output, and a *Pull* mode that takes no input and generates an output. A *blank* Pull iteration is a Pull iteration in which the output is discarded.

The updating transformation of the state has high diffusion and distributed nonlinearity. Its design is aimed at providing very high nonlinearity and fast diffusion for multiple iterations. This is realised by the combination of four distinct transformations each with its specific contribution. There is one for nonlinearity, one for bit dispersion, one for inter-bit diffusion, and one for injection of buffer and input bits.

The buffer behaves as a linear feedback shift register that ensures that input bits are injected into the state over a wide interval of iterations. In the Push mode the input to the shift register is formed by the external input, in the Pull mode, by part of the state.

The PANAMA hash function is defined as performing Push iterations with message blocks as input. If all message blocks have been injected, a number of blank Pull iterations are performed to allow the last message blocks be diffused into the buffer and state. This is followed by a final Pull iteration to retrieve the hash result.

The PANAMA stream encryption scheme is initialised by doing two Push iterations to inject the key and diversification parameter followed by a number of blank Pull iterations to allow the key and parameter to be diffused into the buffer and state. After this initialisation, the scheme is ready to generate keystream bits at leisure by performing Pull iterations.

3   Specification

The state is denoted by a and consists of 17 (32-bit) words a_0 to a_{16}. The buffer b is a linear feedback shift register with 32 *stages*, each consisting of 8 words. An 8-word stage is denoted by b^j and its words by b_i^j. Both stages and words are indexed starting from 0.

The three possible *modes* for the PANAMA module are Reset, Push and Pull. In Reset mode the state and buffer are set to 0. In Push mode an 8-word input is applied and there is no output. In Pull mode there is no input and an 8-word output is delivered.

The buffer update operation is denoted by λ. We have (with $d = \lambda(b)$):

$$
\begin{aligned}
d^j &= b^{j-1} \text{ if } j \notin \{0, 25\}, \\
d^0 &= b^{31} \oplus q , \\
d_i^{25} &= b_i^{24} \oplus b_{i+2 \bmod 8}^{31} \text{ for } 0 \le i < 8 .
\end{aligned}
\tag{1}
$$

In Push mode q is the input block p, in Pull mode it is part of the state a, with its 8 component words given by

$$
q_i = a_{i+1} \text{ for } 0 \le i < 8 .
\tag{2}
$$

The state updating transformation is denoted by ρ. It is composed of a number of specific transformations:

$$
\rho = \sigma \circ \theta \circ \pi \circ \gamma .
\tag{3}
$$

Here $\circ$ denotes the (associative) composition of transformations where the rightmost transformation is executed first.

θ is an invertible linear transformation defined by:

$$
c = \theta(a) \Leftrightarrow c_i = a_i \oplus a_{i+1} \oplus a_{i+4} \text{ for } 0 \le i < 17 ,
\tag{4}
$$

with the indices taken modulo 17. The invertibility of θ follows from the fact that $1 \oplus x \oplus x^4$ is coprime to $1 \oplus x^{17}$.

γ is an invertible nonlinear transformation defined by:

$$
c = \gamma(a) \Leftrightarrow c_i = a_i \oplus (a_{i+1} \text{ OR } \overline{a_{i+2}}) \text{ for } 0 \le i < 17 ,
\tag{5}
$$

with the indices taken modulo 17. A proof for the invertibility of γ can be found in [4].

The permutation π combines cyclic word shifts and a permutation of the word positions. If we define τ_k to be a rotation over k positions from LSB to MSB, we have:

$$
c = \pi(a) \Leftrightarrow c_i = \tau_k(a_j) ,
\tag{6}
$$

with

$$
\begin{aligned}
j &= 7i \qquad \bmod 17 \quad \text{and} \\
k &= i(i+1)/2 \bmod 32 .
\end{aligned}
\tag{7}
$$

The transformation σ corresponds with bitwise addition of buffer and input words. It is given by (let $c = \sigma(a)$):

$$
\begin{aligned}
c_0 &= a_0 \quad \oplus 00000001_{\text{hex}} , \\
c_{i+1} &= a_{i+1} \oplus \ell_i \qquad\qquad \text{for } 0 \le i < 8 , \\
c_{i+9} &= a_{i+9} \oplus b_i^{16} \qquad\quad \text{for } 0 \le i < 8 .
\end{aligned}
\tag{8}
$$

In the Push mode ℓ corresponds with the input p, in the Pull mode $\ell = b^4$.

In the Pull mode the output z consists of 8 words given by

$$
z_i = a_{i+9} \text{ for } 0 \le i < 8 .
\tag{9}
$$

The transformation ρ is illustrated in Fig. 1, the Push and Pull modes of the PANAMA module are illustrated in Fig. 2.

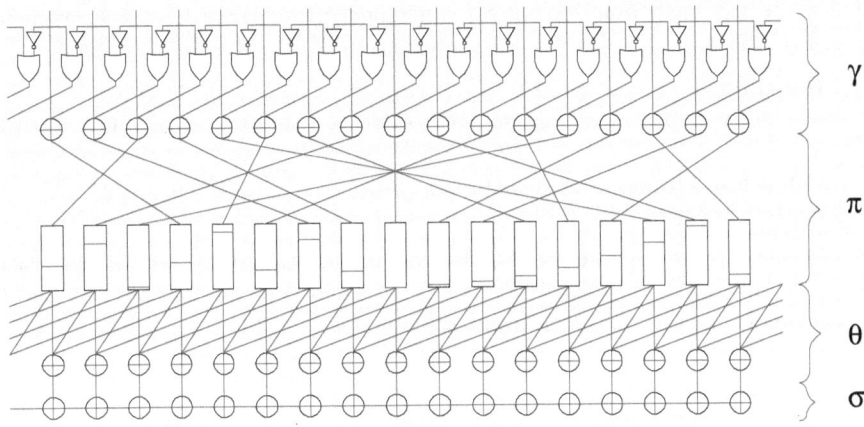

Fig. 1. The state updating transformation ρ.

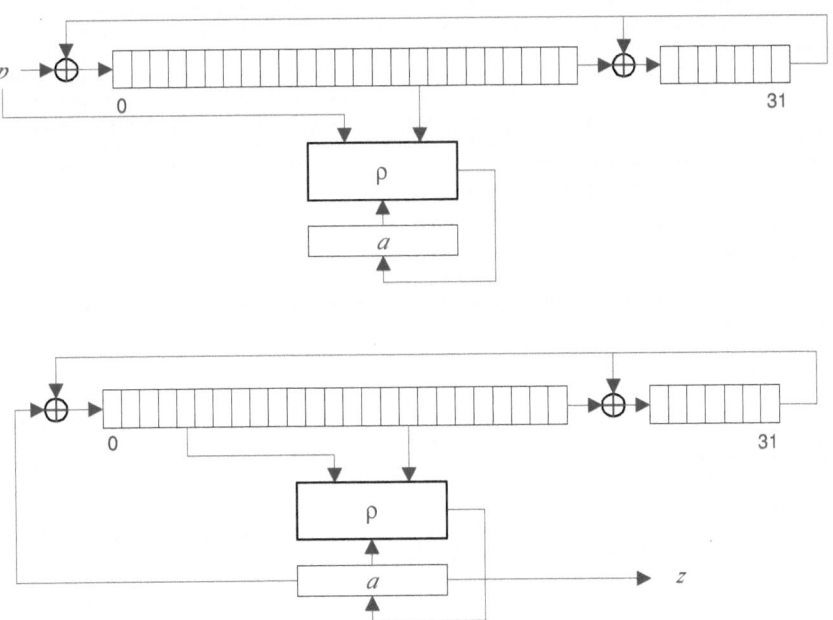

Fig. 2. Push (above) and Pull (below) modes of PANAMA.

3.1   The PANAMA hash function

The PANAMA hash function maps a message of arbitrary length M to a hash result of 256 bits. The PANAMA hash function is executed in two phases:

- **Padding** M is converted into a string M' with a length that is a multiple of 256 by appending a single 1 followed by a number d of 0-bits with $0 \leq d < 256$.
- **Iteration** The input sequence $M' = p^1 p^2 \ldots p^V$ is loaded into the PANAMA module according to Table 1.

After all input blocks have been loaded, an additional 32 blank Pull iterations are performed. Then the Hash result h is returned. The number of Push and Pull iterations to hash an V-block input sequence is $V + 33$.

Time step t	Mode	Input	Output
0	reset	–	–
$1, \ldots, V$	Push	p^t	–
$V + 1, \ldots, V + 32$	Pull	–	–
$V + 33$	Pull	–	h

Table 1. The sequence diagram of the iteration phase of the PANAMA hash function.

The design goal for the PANAMA hash function is that it should be *hermetic*. For the definition of this term we refer to [4]. In short, for a hermetic hash function, the following statements are true.

Assume we take as hash result the value of a subset of n bits of the (for PANAMA 256-bit) output:

- the expected workload of generating a collision is of the order of $2^{n/2}$ executions of the hash function,
- given an n-bit value, the expected workload of finding a message that hashes to that value is of the order of 2^n executions of the hash function,
- given a message and its n-bit hash result, the expected workload of finding a second message that hashes to the same value is of the order of 2^n executions of the hash function.

Moreover, it is infeasible to detect systematic correlations between any linear combination of input bits and any linear combination of bits of the hash result.

A hermetic hash function can be turned into a secure MAC by simply including a secret key in the message input. Its security is independent of the positions of the secret key bits in the message. It may be appended, pre-pended, inserted somewhere in the middle of the message or even be distributed over bits spread all over the message. Observe that this is not the case with hash functions of the MD4 family [8].

3.2 The PANAMA stream encryption scheme

The stream cipher is initialized by first loading the 256-bit key K, the 256-bit diversification parameter Q and performing 32 additional *blank* Pull iterations. During keystream generation an 8-word block z is delivered at the output for every iteration. In practice, the diversification parameter allows for frequent resynchronisation without the need to change the key.

The sequence diagram of the PANAMA stream encryption scheme is given in Table 2.

Time step t	Mode	Input	Output
-34	reset	–	–
-33	Push	K	–
-32	Push	Q	–
$-31, \ldots, 0$	Pull	–	–
$1, \ldots$	Pull	–	z^t

Table 2. The sequence diagram of the PANAMA stream encryption scheme.

The design goal for the PANAMA stream encryption scheme is that it should be hermetic and *K-secure* (also defined in [4]). K-security implies among other things that, given part of the keystream outputs corresponding with a given key and for chosen values of Q, the most efficient way to gain knowledge on the key or on the complementary part of the keystream output, is exhaustive key search.

4 Discussion

4.1 The state updating transformation

γ is the simplest *nonlinear shift-invariant transformation*. Its propagation properties are described in detail in [4]. In short:

- The maximum correlation between its input and output diminishes exponentially with the Hamming Weight of the output selection vectors.
- The difference propagation probability diminishes exponentially with the Hamming Weight of the input difference vectors.

The transformation θ corresponds with the multiplication by a binary polynomial modulo $1 \oplus x^{17}$. It was selected from the invertible polynomials with Hamming weight 3 on the basis of its good diffusion properties. A single input difference gives rise to three output differences. For the vast majority of input difference vectors with a small (below 32) Hamming Weight, the Hamming Weight of the corresponding output vector is about three times higher.

The cyclic shift coefficients of π, described by the simple expression in (7), form an array of 17 different constants. The word permutation factor 7 is chosen to let every component of ρ depend on 9 state bits. For the chosen π parameters it has been verified that $\rho \circ \rho$ has propagation and correlation properties that are close to optimal with respect to the space of possible π parameters. On the average, a difference in a single bit diffuses to 6 bits after one iteration, 36 bits after two, 216 after three and all over the state after 4 iterations. Since γ, θ, π and σ are all invertible, the state updating transformation ρ is invertible.

σ includes the addition of a constant to a_0 to prevent symmetric properties. For the value of the constant 00000001_{hex} was chosen for its simplicity.

4.2   The hash function

The design of the PANAMA hash function differs from the currently popular MD4-derived designs such as MD5 [10], SHA-1 [7] and RIPEMD-160 [5] in three important ways:

- **Parallel iteration transformation:** the MD4-derived designs have an iteration function that consists in itself of a sequence of a large number of (simple) rounds, in PANAMA the iteration function consists of a single (more complicated) round with a parallel structure.
- **Large Chaining State:** the MD4-derived hash functions have a chaining variable that has the same size (16 or 20 bytes) as the hash result by design, the chaining variable of the PANAMA hash function comprising the internal state and buffer is over 1 Kbyte.
- **Presence of final transformation:** In the MD4-derived designs, the hash result is the final value of the chaining variable. For PANAMA the 32 blank Pull iterations form a final transformation mapping the final value of the chaining variable (state and buffer) to the hash result.

These differences are the consequence of a difference in design strategy. In the MD4-derived hash functions the iteration transformation is designed to be collision-resistant in itself. The iteration mechanism and the fact that the hash result is the value of the chaining variable after the last message block has been hashed, assure that the resulting hash function is collision-resistant. In the PANAMA hash function, it is the diffusion and nonlinearity realised by the successive application of the iteration function that is expected to prevent cryptographic weaknesses.

4.3   Collision resistance

In this section we explain the difficulties faced when trying to generate collisions for PANAMA.

The hash result is completely determined by the final value of the chaining variable: the state a^{V+1} and buffer contents b^{V+1}. The converse is however not true, and pairs of messages may be found with different values for the final chaining variable both consistent with the same hash result. In this case, the collision actually results from the fact that the hash result is not equal to the final hashing state. We call this a *terminal* collision. Collisions in which two different messages give rise to equal chaining variable at a certain point are called *internal*. In hash functions without a final transformation, such as MD4 and its descendents, terminal collisions cannot occur.

Generating an internal collision implies two different messages that give rise to a bitwise difference pattern in the hashing state *and* buffer that *dissolves*, i.e., that ends up being all-zero for some iteration. Realising this, while theoretically possible, is assumed to be infeasible because of the large diffusion and distributed nonlinearity of ρ combined with the fact that every message block affects the hashing state for a large number of iterations. Difference patterns in the buffer induce a so-called *differential characteristic* or equivalently *differential trail* [4] in the hashing state.

A possible strategy for finding internal collisions would be to look for an input difference pattern that can give rise to a differential trail that ends in a zero difference somewhere in the computation. Other strategies are

– finding fixed points of the Push mode or more general, fixed sequences of the Push mode,
– meet-in-the-middle attacks exploiting the invertibility of the iteration function.

These attacks have been considered in the design strategy and in the choice of the state and buffer updating transformations.

Because of the linear feedback in the buffer, a difference pattern in a single message block gives rise to an infinite difference propagation in the buffer. This is illustrated in the right-hand side of Fig. 3. Only difference patterns in the input sequence that meet a particular condition give rise to a finite difference propagation in the buffer.

The simplest difference pattern that gives rise to a finite difference propagation in the buffer is illustrated in the left-hand side of Fig. 3. All other message differences that give rise to a finite difference propagation in the buffer are superpositions (linear combinations) of shifted (in time and space) instances of this difference pattern.

Even the simplest difference pattern affects the state a during 5 different iterations, spread over a range of 32 iterations. Fig. 4 illustrates the difference propagation in σ resulting from a similar difference pattern in p that gives rise to a finite difference propagation and that has $p^{1'} = d_0, d_1, \ldots, d_7$. Four of the five difference vectors are in general different. For all other message difference patterns, the number of iterations with a non-zero difference pattern and the number of distinct difference vectors in σ are both larger.

If the difference pattern in the hashing state and/or buffer is not equal to zero after the loading of the last input block, the diffusion and nonlinearity of the transformation formed by the 32 final Pull iterations are assumed to make the controlled generation of terminal collisions infeasible.

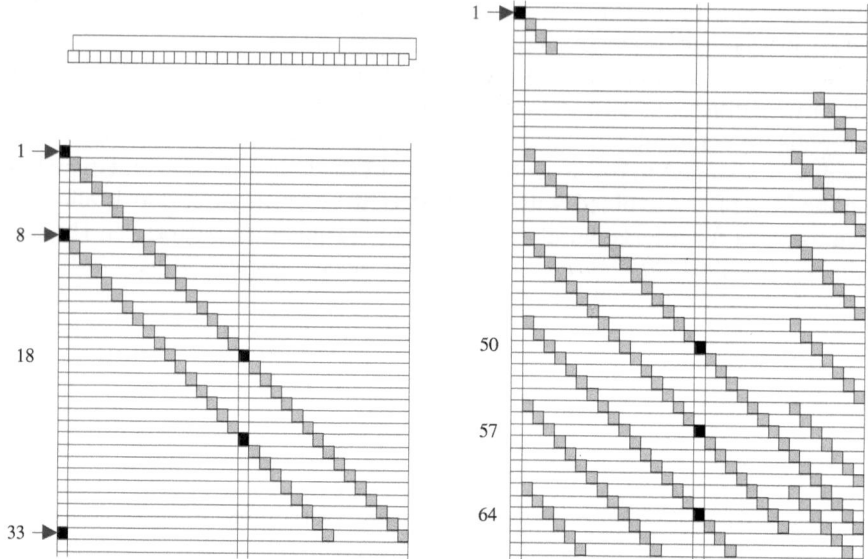

Fig. 3. Difference propagation in the buffer of PANAMA.

$t = 1$	-	d_0	d_1	d_2	d_3	d_4	d_5	d_6	d_7	-	-	-	-	-	-	-	-
$t = 8$	-	d_2	d_3	d_4	d_5	d_6	d_7	d_0	d_1	-	-	-	-	-	-	-	-
$t = 18$	-	-	-	-	-	-	-	-	-	d_0	d_1	d_2	d_3	d_4	d_5	d_6	d_7
$t = 25$	-	-	-	-	-	-	-	-	-	d_2	d_3	d_4	d_5	d_6	d_7	d_0	d_1
$t = 33$	-	d_0	d_1	d_2	d_3	d_4	d_5	d_6	d_7	-	-	-	-	-	-	-	-

Fig. 4. Difference patterns in σ.

In internal collisions the difference pattern in the state and buffer is zero before the final hashing state has been reached. Clearly, this restricts the difference pattern in the input sequence to superpositions of instances of the simple difference pattern given in the left-hand side of Fig. 3. These input difference patterns give rise to non-zero difference vectors in σ over a span of at least 32 iterations. We believe that controlling the differential trails in the hashing state over these large numbers of iterations to obtain a collision is too hard to be a threat to the security of the hash function.

One can avoid the need for this type of control over large numbers of iterations by eliminating differences in the state immediately after they have been created. As can be seen in Figure 4, this must however be repeated for 5 different instances, with the same set of input differences.

4.4   The stream encryption scheme

For every Pull iteration 16 words of the buffer are injected into the state and 8 state words are given at the output. In the short term, the number of buffer bits that are injected into the state is twice as large as the number of bits that are given at the output. It can be checked that this is the case for any number of iterations smaller than 12. The feedback from the state to the buffer causes the buffer contents to be renewed every 32 iterations. These factors cause the correlations between output bits and linear combinations of state and buffer bits to be too small to be of practical use for cryptanalysis.

Resynchronization attacks [4] should be made infeasible by the 32 blank Pull iterations after loading the initialization blocks. Because of the feedback from the state to the buffer in the Pull mode, the state and almost all buffer stages depend in a complicated way on these blocks at the end of the initialization phase. We expect that there are no exploitable 32-round differential trails.

5   Implementation aspects

PANAMA's heavy reliance on bitwise-logical operations on 32-bit words make it well suited to implementation on 8, 16, or 32-bit processors, except that its use of 32-bit rotations does somewhat favor 32-bit architectures.

Accesses to buffer b and operations between buffer b and state a can all be performed 64-bits at a time on processors that support this word size. Since a is not an integer multiple of 64-bits a dummy 32-bit word should be pre-pended to a_0 for properly aligning $\{a_1, a_2\}$ through $\{a_{15}, a_{16}\}$ to 64-bit boundaries. Buffer b should also be 64-bit aligned.

So, θ and γ can be performed efficiently with word sizes up to 32-bits, while π especially favors 32-bit architectures. σ and λ can both be performed efficiently with word sizes up to 64-bits. In the following analysis we concentrate on PANAMA's performance on 32-bit processors.

5.1   Theoretical performance limits

To determine the maximum theoretical performance of PANAMA in its hash and stream cipher modes on a suitably parallel processor architecture we seek to identify the software *critical path* through the algorithm. This is the closed path through the algorithm's computational flowgraph that has the largest *weighted* instruction count, the weighting being the number of cycles of latency associated with each type of instruction.

For instance, on most processors the result of a simple operation like an addition or XOR can be used in the subsequent cycle - these instructions are said to have a one cycle latency. On modern high performance processors it is also common for shifts and rotates to be single-cycle instructions. However reading from memory takes several cycles. Even when the data is in the CPU's local cache it commonly suffers a two or three cycle latency on modern deeply pipelined processors.

We start by examining the software critical path through ρ. For each of γ, π, θ, and σ, all seventeen 32-bit words a_0 to a_{16} can in principle be updated in parallel. The software critical path through these four routines is a total of 7 cycles (3 cycles for γ, 1 cycle for π, 2 cycles for θ, and 1 cycle for ρ), all corresponding to single-cycle instructions.

Many RISC processors have enough registers to hold state a in its entirety, so that when encrypting or hashing a long block of data we need not keep accessing any of a_0 through a_{16} from memory. However, buffer b is too large to be register based and is most efficiently implemented as a fixed circular-buffer in memory, with moving pointers used to create the appearance of a shift-register.

Notably, since the accesses to b are not data-dependent (i.e. not table-look-ups) all address calculations can be done well in advance and do not contribute to the software critical path. Also, since the stages which are read are several stages delayed from those that are written, the read-data can in principle be fetched one or more update cycle ahead of time, from which it becomes clear that updating the buffer is not on the software critical path.

We now explore the number of 32-bit instructions needed for each iteration of Push and Pull. If state a is fully held in registers and the rotation amounts in π are all hard-coded, then ρ entails a total of 16 reads (from buffer stages 4 and 16 for Pull, or input p and buffer stage 16 for Push) and 17×7 logical operations (actually one less than this because the zero-rotation need not be performed).

Updating buffer b involves 16 reads (buffer stages 24 and 31), 16 XOR operations, and 16 writes (buffer stages 0 and 25), plus a number of operations to update the pointers to the accessed stages in order to simulate a shift register. As a minimum these involve an increment and mask operation per pointer. Reading from stage 31 and writing to stage 0 takes only one pointer, likewise for stages 24 and 25 because of the way the circular-buffer is used to emulate a shift-register. Consequently there are four pointers to be updated for a Pull iteration, and three for a Push. The masking operation implements circular arithmetic on the pointers - a convenience arising from the buffer size being a power of two.

For each Pull iteration, applying the cipher output to the data stream involves 8 reads from the plaintext buffer, 8 XOR operations, and 8 writes to the ciphertext buffer, plus at least one additional instruction to update a pointer to these buffers. Each iteration ciphers 32 bytes of data.

In the case of Push we have already accounted for reading the input p under our discussion of ρ, so all that is left is updating the pointer to the input data.

Thus, ignoring for the moment the few extra instructions necessary for maintaining the loop, we have a workload of 189 instructions for each iteration of Push and 215 instructions for each Pull. This is equivalent to about 5.9 instructions per byte hashed or 6.7 instructions per byte enciphered.

An estimate of how many fully pipelined execution units the algorithm might usefully exploit can be obtained by dividing the total number of operations per iteration by the number of cycles in the critical path. This number ideally should be no less than the number of parallel execution paths in the target processors so that no CPU resources are left idle.

For PANAMA this works out to 189 or 215 instructions per iteration divided by 7 cycles per iteration, from which we estimate that the hashing and encryption modes might reasonably exploit processors with up to 27 and 31 parallel 32-bit datapaths respectively.

5.2   Benchmarked performance

The generous amount of parallelism in PANAMA lends itself naturally to efficient implementation on a processor capable of a high degree of instruction-level parallelism. To demonstrate the impressive throughputs achievable in such cases a highly optimized implementation was developed for the Philips TriMedia TM-1000 processor. The TriMedia processor is a Very Long Instruction Word (VLIW) CPU containing five 32-bit pipelined execution units sharing a common set of 128 32-bit registers. All five execution units can perform arithmetic and logical operations, but loads, stores, and shifts are each supported by only two of them. The two execution units that support shifts are distinct from the two that support loads and stores. Given an appropriate instruction mix the processor can issue up to five instructions per clock cycle.

The operations in PANAMA can be efficiently expressed in C-code except for the bitwise rotations. The optimized implementation was written completely in C-code except for resorting to a library call to access the processor's native 32-bit rotate instruction.

Since the parallelism present in the PANAMA algorithm is vastly more than is available in the TriMedia CPU we would hope to be able to completely saturate the processor, i.e. have very few vacant instruction slots. However, there may be other resource constraints that prevent this. For instance, π calls for the intensive use of rotate instructions, of which the TriMedia CPU can only issue two per cycle. Filling the other three instruction slots on each cycle requires overlapping the execution of π with that of γ and/or θ. For the benchmarked implementation each of these routines was expressed as fully unrolled in-line code, leaving the

TriMedia C-compiler to recognize and exploit the allowable overlap between γ, π, and θ.

The loop for iterating Pull mode compiled into 234 TriMedia assembly instructions. The difference between this and the 215 instructions previously counted comes from the instructions for maintaining the loop and from some additional overhead involved in the pointer updates associated with making the circular buffer appear as an LFSR. The scheduled code was tightly packed by the compiler into 47 instruction-issue cycles, i.e. a sustained 4.98 instructions per cycle were scheduled for issue, out of a theoretical maximum of 5. This is an unusually high utilization of the TriMedia CPU even compared to its efficiency on media-processing tasks for which it was designed.

Compiled code for Push showed comparable overhead and code density.

The optimized C-code was benchmarked on a 100 MHz TriMedia processor by encrypting or hashing a 128 Kbyte data buffer. We choose this buffer size as several times larger than the on-chip data cache so as to make the reported performance be representative of the sustainable encryption or hashing performance to external memory, in this case comprised of synchronous DRAM. At the level of performance achieved by PANAMA external memory bandwidth can become a significant factor in the overall performance. For the encryption benchmark the data buffer was encrypted in-place so as to minimize the performance loss arising from memory accesses. No off-chip cache was present (the Trimedia chip does not actually support off-chip cache).

An encryption throughput of 4.7 bits per cycle was achieved, equivalent to 1.7 cycles per byte, or 470 Mbps on a 100 MHz processor. This includes all loop-overhead and cycles lost to cache misses, memory accesses, etc. This is uncommonly fast among stream ciphers. For comparison, two other acknowledged fast software ciphers – RC4 [12], and SEAL [11], are reported as capable of 10.6 cycles per byte and 3.5 cycles per byte, or 75 Mbps and 230 Mbps respectively when benchmarked under these same conditions [3]. PANAMA is also slightly faster on this processor than the variants of WAKE described in [3].

Notably, PANAMA achieves its speed advantage not by having the lowest work-factor among these ciphers. For instance SEAL has a work factor of 4.25 instructions per byte on the TriMedia processor compared to PANAMA's 6.7. Rather, PANAMA's speed advantage comes from the substantial degree of parallelism present in the algorithm, an attribute that can be well exploited by a VLIW processor such as the TriMedia. Accordingly it should be noted that PANAMA's advantage may be diminished when running on processors having less instruction-level parallelism than the CPU reported here.

On the TriMedia processor PANAMA achieves a hashing throughput of 5.1 bits per cycle, equivalent to 1.6 cycles per byte, or 510 Mbps on a 100 MHz device. We are not aware of published performance figures for implementations of other currently popular hash functions on the TriMedia processor against which to directly compare PANAMA's speed. Still, a simple comparison shows that the per-byte workload of PANAMA is similar to that of MD4 [9], the fastest member of the family of hash functions to which MD5, SHA-1 and RIPEMD-160 all belong. Benchmarks for these popular hashes have been published for the Intel Pentium processor in [2] from which we can make comparisons to PANAMA.

Performance of an optimized C-code implementation of PANAMA on a 200 MHz Pentium Pro (using a library function for rotate) was measured at 198 Mbps for ciphering and 214 Mbps for hashing, i.e. a throughput of 0.99 bits per cycle for ciphering and 1.07 bits per cycle for hashing. This compares to hashing speeds reported in [2] for SHA-1 and RIPEMD-160 of 0.24 bits per cycle and 0.21 bits per cycle respectively for optimized C-code[1]. [2] also reports optimized assembly-code versions of SHA-1 and RIPEMD-160 as achieving 0.54 bits per cycle and 0.44 bits per cycle respectively. It is currently unknown what further speed improvement could be achieved for PANAMA by assembly coding it, but even without such improvement it shows about a 2× speed advantage over assembly coded versions of these other hashes.

Since the Pentium Pro can in principle issue two arithmetic or logical instructions per cycle compared to five for the TriMedia chip one may wonder why the throughput per cycle of PANAMA on the Pentium Pro is barely one fifth that achieved on the TM-1000. In part the reason is that PANAMA's large state cannot be maintained in the small register set of the 'x86 architecture, with the result that code for the Pentium Pro requires massively more load and store instructions than are required for the TriMedia, or for that matter other RISC processors with a generous complement of registers. Since both SHA-1 and RIPEMD-160 are substantially unhampered by the limited register set of the 'x86 architecture, we would expect PANAMA's advantage over them to be all the greater on processors not having this limitation.

In considering PANAMA's suitability to an application it should be borne in mind that the performance figures reported are for large block sizes. When hashing small blocks, or encrypting with frequent key changes or resynchronization, the overhead of the accompanying 32 blank Pull iterations may significantly impact the performance. A key change or resynchronization takes about as long as encrypting 1000 bytes. Similarly, each hashed block has a fixed overhead equivalent to hashing about 1000 bytes.

6   Conclusions

We have presented a new cryptographic module capable of cryptographic hashing and stream encryption suited for applications where large amounts of data must be protected. It has been shown that the inherent parallelism allows extremely fast software implementations on VLIW processors.

---

[1] [2] reports performance on a 90 MHz Pentium, while here we report performance on a 200 MHz Pentium Pro. By converting all results into the normalized measure of bits per cycle we attempt to provide a uniform basis for comparison, however the reader is cautioned that no allowance has been made for the architectural differences between the Pentium and Pentium Pro.

References

1. E. Biham and A. Shamir, "Differential cryptanalysis of DES-like cryptosystems," *Journal of Cryptology*, Vol. 4, No. 1, 1991, pp. 3–72.
2. A. Bosselaers, R. Govaerts, J. Vandewalle, "Fast Hashing on the Pentium", *Advances in Cryptology – Proceedings Crypto'96 LNCS 1109*, N. Koblitz, Ed., Springer-Verlag, 1996, pp. 298–312.
3. C.S.K. Clapp, "Optimizing a fast stream cipher for VLIW, SIMD, and superscalar processors," *Fast Software Encryption, LNCS 1267*, E. Biham, Ed., Springer-Verlag, 1997, pp. 273–287.
4. J. Daemen, "Cipher and hash function design strategies based on linear and differential cryptanalysis," *Doctoral Dissertation*, March 1995, K.U.Leuven.
5. H. Dobbertin, A. Bosselaers, B. Preneel, "RIPEMD-160: A Strengthened version of RIPEMD," *Fast Software Encryption, LNCS 1039*, D. Gollmann, Ed., Springer-Verlag, 1996, pp. 71–82.
6. FIPS 180, *Secure Hash Standard*, Federal Information Processing Standard (FIPS), Publication 180, National Institute of Standards and Technology, US Department of Commerce, Washington D.C., May 1993.
7. FIPS 180-1, *Secure Hash Standard*, Federal Information Processing Standard (FIPS), Publication 180-1, National Institute of Standards and Technology, US Department of Commerce, Washington D.C., April 1995.
8. B. Preneel and P.C. van Oorschot, "On the Security of Two MAC Algorithms", *Advances in Cryptology – Proceedings Eurocrypt'96 LNCS 1070*, U.M. Maurer, Ed., Springer-Verlag, 1996, pp. 19–32.
9. R.L. Rivest, *The MD4 message-digest algorithm*, Request for comments (RFC) 1320, Internet Activities Board, Internet Privacy Task Force, April 1992.
10. R.L. Rivest, *The MD5 message-digest algorithm*, Request for comments (RFC) 1321, Internet Activities Board, Internet Privacy Task Force, April 1992.
11. P. Rogaway and D. Coppersmith, "A Software-Optimized Encryption Algorithm," *Fast Software Encryption, LNCS 809*, R. Anderson, Ed., Springer-Verlag, 1994, pp. 56–63.
12. B. Schneier, *Applied Cryptography, Second Edition*, John Wiley & Sons, 1996, pp. 397–398.

Joint Hardware / Software Design of a Fast Stream Cipher

Craig S.K. Clapp

PictureTel Corporation, 100 Minuteman Rd., Andover, MA 01810, USA
craigc@pictel.com

Abstract. We explore the problem of designing a stream cipher that is fast in software yet may be efficiently implemented in hardware. We show that a keystream generator built as a word-wide non-linear-feedback shift register can offer both a high degree of parallelism and the hardware simplicity and flexible security of an iterated design. WAKE-ROFB is shown to be an example of this topology. A modified non-linear mixing function is proposed for WAKE-ROFB which makes it better suited to hardware implementation. The high degree of parallelism allows efficient implementation on processors having instruction-level parallelism, and leads naturally to high-speed pipelined hardware implementations. The recommended variant runs at 340 Mbps on a 266 MHz Pentium II and 270 Mbps on a 100 MHz TriMedia VLIW CPU, while a 2000 gate hardware implementation of the same cipher achieves 200 Mbps from a 50 MHz clock. A higher speed variant achieves 600 Mbps, 340 Mbps and 400 Mbps respectively with some loss of security, while needing slightly less hardware.

1 Introduction

The first Fast Software Encryption Workshop in December 1993 brought, perhaps for the first time, a major focus to the topic of cipher design for software based systems. The new degrees freedom afforded by this domain have led to many avenues of research that would previously not have been considered. Since then numerous ciphers have been proposed for software implementation, but of those that have the highest throughput in software none is especially well suited to hardware implementation. Typically they use large tables, e.g. SEAL [7], Blowfish [9], WAKE [10] and its derivatives [3], or have a large internal state, e.g. RC4 [8], STEPRIGHTUP [4].

In some applications there is a need for a high speed cipher that can be efficiently implemented in both hardware and software. One example comes from the trend to equip consumer electronics entertainment devices with high-speed digital interconnections.

High-speed digital interfaces, notably IEEE-1394, are expected to appear soon on digital video disc (DVD) players and digital television sets, and they have already appeared in personal computers. In response to the entertainment industry's concern to protect their content from unauthorized copying while

S. Vaudenay (Ed.): Fast Software Encryption – FSE'98, LNCS 1372, pp. 75–92, 1998.

being transmitted digitally between consumer electronics devices, it has been proposed to encrypt such connections whenever they carry copyrighted data.

In consumer electronics devices whose workload does not justify a high performance processor, such as a DVD player that simply passes the compressed data from the disc to the external digital interface encrypting it along the way, it is preferred to implement the encryption in hardware.

When compressed digital video content is re-played on a personal computer the computationally intensive task of video decompression already dictates the presence of a high-performance CPU, typically one with architectural optimizations for media-processing. Given its presence it is natural to ask it to also assume the task of decryption. Digital televisions will increasingly use embedded media-processors for these same tasks, so efficiency on these is also a consideration.

The performance requirements of the aforementioned application motivate our interpretation of 'high speed' and 'efficient hardware'. The consumer electronics industry's goals for this application equate to a throughput in software in excess of two bits per cycle on a general purpose computer, and a hardware implementation in 1000 to 2000 gates.

We know of no reported ciphers that simultaneously achieve these goals.

2   Joint optimization for hardware and software

We anticipate that the twin goals of fast software implementation and economical hardware are not mutually exclusive as might be concluded from the lack of ciphers sharing both characteristics. Rather we suspect the reason for the lack of good examples is that there has been a polarization of design philosophies to favor one or other of the environments in the extreme. Our strategy is to design a cipher while jointly optimizing for the respective constraints arising from both hardware and software implementations.

By presenting this work we hope to stimulate further work in jointly optimizing ciphers for both hardware and software.

2.1   Trade-off between efficiency in hardware versus software

Our goal is to find a topology that runs at high speed in software while being very efficient in hardware. Moreover, we would ideally like it to run efficiently not only on legacy architectures such as the 'x86, but also to take full advantage of the substantial instruction-level parallelism present in more advanced processors such as Intel's Pentium II and media-processors such as Philips' TriMedia chip.

The desire for a low gate-count tends to dictate a cipher with an efficient iterative implementation and relatively little state. Block ciphers such as DES typify such a strategy, but their iterated design tends to offer little parallelism that might be exploited in a software implementation.

Since even today's multimedia-optimized processors are capable of performing four or five 32-bit operations per cycle (or 128 to 160 bit-operations per cycle usable in a variety of word-lengths), it is highly desirable for a multimedia-optimized cipher to be capable of exploiting at least this much instruction-level parallelism. This tends to dictate that such a cipher have at least 128 to 160 bits of state, which of itself would consume a substantial proportion of our 1000 to 2000 gate hardware budget.

2.2   Fast software ciphers from hardware oriented ciphers

Many hardware oriented stream ciphers in the literature are shift-register based. However, characteristically these run very inefficiently in software since processors operate on words while these hardware oriented ciphers rely heavily on individual bit operations.

One way of utilizing the power of a word-oriented processor is to treat its words as n-bit vectors, and implement n independent copies of the cipher, one in each bit-lane, SIMD fashion. A variation on this theme is to allow some mild interaction between the bit lanes such as by replacing bit-wise addition of words (i.e. exclusive-OR) by regular addition of words, thereby converting n independent shift registers into an additive, or lagged Fibonacci, generator [5]. Such word-wide generators can then be used in combinations modeled after the favored topologies for combining bit-serial shift registers. Examples of this strategy can be seen in Fish [2], which is a cross between an additive generator and n concurrent shrinking generators [6], and Pike [1] which is a trio of additive generators whose bit-wise combination topology is patterned after A5 [8].

Common to these examples is that the increase in software speed comes at the expense of a corresponding increase in cipher state. Consequently these derivatives are no longer efficient in hardware. For instance, a 32-bit implementation of Fish has 3424 bits of state while Pike has 5440, a far cry from the few tens of bits of their hardware oriented prototypes.

An interesting question is whether by providing much stronger and non-linear interaction between the bits of the word (as opposed to the mild interaction in an additive generator) it is possible to achieve useful security with dramatically fewer register stages. In such a case, we can no longer expect the security to be established from some underlying single-bit shift-register cipher model.

2.3   Economical hardware ciphers from fast software ciphers

Taking fast software ciphers as our starting point, a natural inclination is to find a way to reduce their hardware complexity. Since for many such designs the hardware is dominated by look-up-tables an obvious question is how such look-up-tables might be reduced in size without crippling the cipher. An approach applicable to iterated block ciphers might be to increase the number of rounds to compensate for any loss of strength arising from reduced tables. However, there are no block ciphers at the speed we seek to achieve, leaving us to consider which of the fastest available stream ciphers might be modified to fit our needs.

Fast stream ciphers do not typically have a structure that supports incremental trade-off of speed versus security. A counter example comes from the derivatives of WAKE described in [3]. These ciphers have as one parameter the total number of mixing/register stages. Increasing the number of stages increases the security. Unfortunately it also increases the total amount of state associated with the cipher, resulting in an additional hardware burden that we would rather not incur.

3   WAKE, WAKE-OFB, and WAKE-ROFB

In [10] Wheeler introduced the cipher WAKE. It uses a mixing function **M** that combines two 32-bit inputs into one 32-bit output with the aid of a key-dependent 256×32-bit look-up-table, T. The particular construction of T makes the mixing function invertible in the sense that knowledge of the output word and one of the two input words is sufficient to uniquely specify the other input word.

WAKE consist of cascading four of these mixing functions having registered feedback around each one and overall feedback around the group. Four stages are chosen as the minimum number needed for complete diffusion but more stages can be added if desired for additional security.

As originally proposed, WAKE operated with cipher-feedback into the state machine, leading to the potential for chosen plaintext attacks as noted by Wheeler in [10]. In addition to its use in cipher-feedback mode, Wheeler also suggested WAKE as suitable for the production of a pseudo-random sequence for use as the keystream for a stream cipher. By changing the state machine from cipher-feedback to output-feedback mode the potential susceptibility to a chosen-plaintext attack is eliminated. This mode, referred to in [3] as WAKE-OFB, is shown in Fig. 1a. In the diagram all registers and signal paths are w-bits wide, with $w = 32$ for efficiency on 32-bit processors. This topology in conjunction with the invertibility of the mixing function results in a reversible generator.

In [3] it was shown that by running the keystream generator in reverse a threefold increase in computational parallelism could be achieved, resulting in very high throughput on processor architectures having instruction-level parallelism while claiming identical security to the non-reversed version. We call this form WAKE-ROFB.

4   WAKE-OFB and WAKE-ROFB as shift registers

While not originally portrayed as such, we can re-draw WAKE-OFB as an equivalent word-wide shift register. This is shown in Fig. 1b, again with all registers and signal paths being w-bits wide. In this form the shift register must be stepped four times to produce the new output that the parallel state machine of Fig. 1a produces in just one step. This is demonstrated by the state sequence listed in table 1.

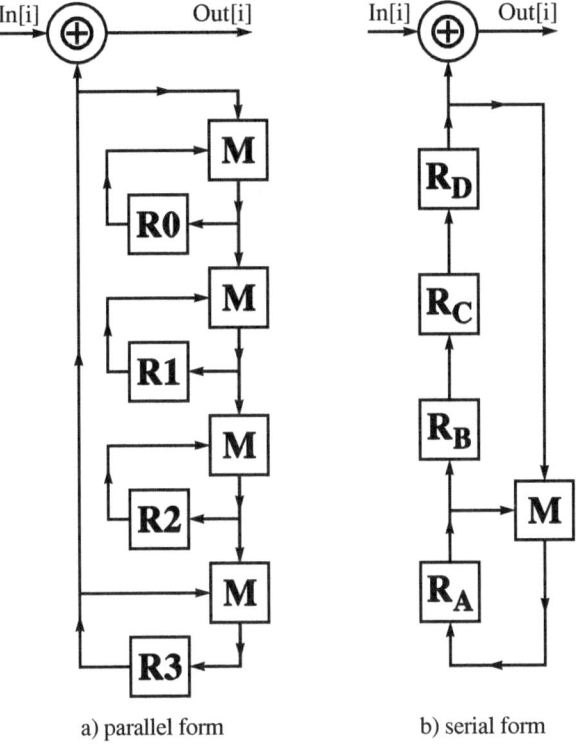

a) parallel form   b) serial form

Fig. 1. Alternate implementations of 4-stage WAKE-OFB

	T0	T1	T2	T3	T4=T0′
R_A	R3	R0′	R1′	R2′	R3′
R_B	R2	R3	R0′	R1′	R2′
R_C	R1	R2	R3	R0′	R1′
R_D	R0	R1	R2	R3	R0′
M_{Out}	R0′	R1′	R2′	R3′	R0″
Output	No	No	No	Yes	No

Table 1. Contents of registers R_A through R_D for time intervals T0 to T4

If we sub-sample the output of the shift-register version by four we produce the exact same output sequence as the original parallel version produces on consecutive steps. The serial form is particularly attractive for hardware implementation because, apart from there being just one mixing function to implement, the control logic for this topology is quite trivial. In hardware the serial form produces an average throughput of 8 bits per clock cycle which is still quite fast enough for most applications.

Note that in software *there is no distinction between the parallel and sub-sampled serial forms*. They are simply different but equally valid representations of a single computational flowgraph.

4.1   More registers, more iterations

For WAKE and its derivatives, adding more stages has been suggested as a convenient means of gaining extra security. Interpreting this strategy in the context of the serial form of WAKE-OFB we can see it as equivalent to increasing both the length of the shift register and the sub-sampling factor. i.e. 5-stage WAKE-OFB is equivalent to a five-stage word-wide shift register whose output is taken on every fifth step.

Now it becomes apparent that there are actually *two* degrees of freedom to be exploited. On the one hand we can increase the number of registers in the chain and independently we can select how many times to step the generator between outputs. Clearly, if we exercise these freedoms independently then the equivalent parallel hardware implementation, if we trouble ourselves to discover it, will be something other than the simple form in Fig. 1a. However this does not concern us since for hardware efficiency we will implement the serial form and this form also fully specifies the task in software.

Every time the shift register of Fig. 1b is stepped, the data in $\mathbf{R_A}$ participates in another iteration of the mixing function. If the mixing function is cryptographically useful then we can expect each additional iteration to contribute cryptographic strength to the contents of $\mathbf{R_A}$ in the same way that strength accumulates with the number of rounds in an iterated block cipher. Since the other registers are simply older copies of what was once one in $\mathbf{R_A}$, it does not matter which register is tapped for output from the viewpoint of cryptographic strength. Wherever the output is taken from it can reasonably be argued that the cryptographic strength generally increases with the sub-sampling factor.

Increasing the number of registers increases the amount of state that an attacker needs to deduce. It also increases the number of iterations of the mixing function by which the two inputs to the mixing function differ. In general, the greater this cryptographic separation the less chance there is of an unfortunate interaction between the two inputs, such as constructive reinforcement of some statistical bias. However, by analogy to differential characteristics in iterated block ciphers, if the mixing function is known to have an exploitable m-round characteristic, then it may be appropriate to avoid there being a small multiple of this number of register stages between the inputs of the mixing function.

We suggest that for a shift register with n-stages, as shown in Fig. 2a, it is ill advised to sub-sample by a factor of $n - 1$ since in this case the output stream reveals *both* inputs to the mixing function for some time steps, which may ease cryptanalysis. Also, any sub-sampling factor less than n will result in the contents of more than one register being known for at least some time steps, which diminishes one of the purposes for having more registers – i.e. increasing the amount of state that an attacker must deduce.

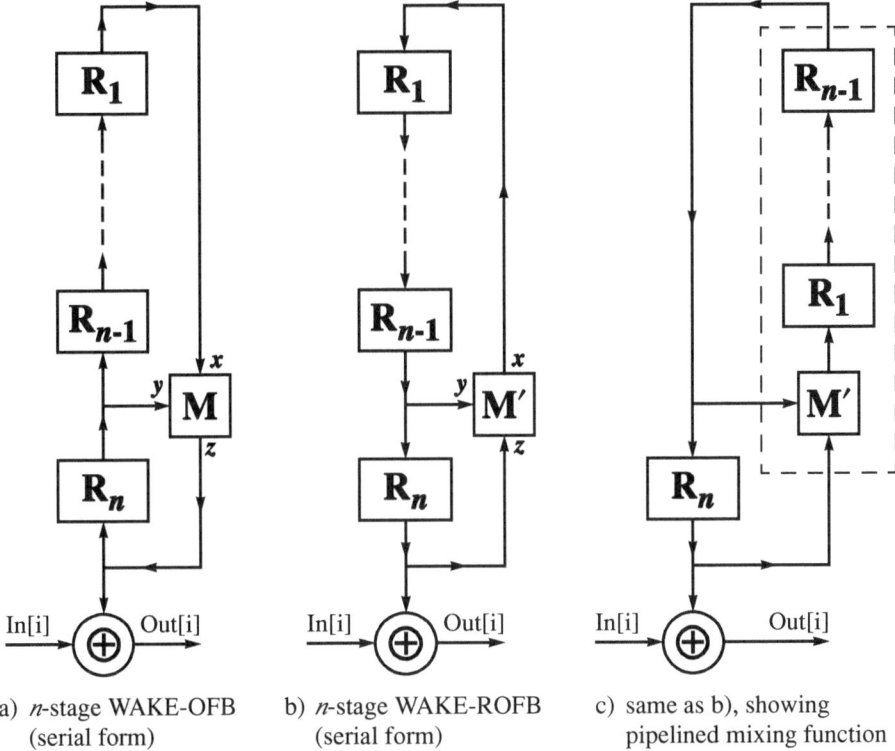

a) n-stage WAKE-OFB   b) n-stage WAKE-ROFB   c) same as b), showing
   (serial form)   (serial form)   pipelined mixing function

Fig. 2. Forward and reversed forms of generators with an arbitrary number of stages

4.2 Improving performance by running backwards

Cryptographic security of a pseudo-random sequence, as used for a stream cipher, demands that no one part of the sequence can be predicted from any other part of the sequence. This security property makes no distinction between the forward and time reversed versions of the pseudo-random sequence. Thus, if a *reversible* pseudo-random number generator, such as WAKE-OFB, produces a cryptographically secure sequence then that same generator running in reverse must also produce a cryptographically secure sequence.

In [3] it was observed that running the WAKE-OFB generator backwards exposed three times as much computational parallelism as was present in the forward version. We call the reversed version WAKE-ROFB. Since the forward and reversed versions produce the same keystream – simply in the opposite order from one another, we claim that their security is identical.

Just as we were able to draw an equivalent serial form for WAKE-OFB, we are able to draw an equivalent serial form for WAKE-ROFB. In Fig. 2a we show the serial form of WAKE-OFB with the number of registers generalized[1]. Fig. 2b shows the serial equivalent of WAKE-ROFB which is, not surprisingly, the time-reversed version of Fig. 2a. To get the time-reversed version we simply reverse the direction of data through each register and patch up the flowgraph accordingly. As part of this exercise the mixing function gets replaced by its inverse, $\mathbf{M}'$.

Now the reason for the increase in computational parallelism in WAKE-ROFB becomes obvious. If $\mathbf{R}_1$ through $\mathbf{R}_n$ in Fig. 2b are known then clearly the next $n-1$ pairs of inputs to the mixing function are already determined, so in principle the results of these $n-1$ mixing function evaluations can all be performed in parallel.

This provides us with another criterion to guide our choice of the number of register stages – it should be sufficient to let us effectively exploit the available instruction-level parallelism in the processors on which the cipher will run.

Previously, the available parallelism has been shown to be advantageous for efficiently exploiting instruction-level parallelism in software implementations, as reported in [3]. A notable advantage that this parallelism affords *hardware* implementations is the ability to pipeline the mixing function and thereby increase the maximum clock rate at which such implementations can operate.

Pipelining is a technique commonly used in hardware design as a means for increasing the maximum attainable clock rate of a system. The maximum clock rate of a system is bounded by the longest propagation path through combinational logic between any two registers. If a circuit topology allows additional registers to be placed in these long paths then the attainable clock rate can be increased.

In Fig. 2b we observe that all but one of the w-bit registers, i.e. $\mathbf{R}_1$ through $\mathbf{R}_{n-1}$, appear in series with the output of the mixing function. Thus, the mixing function can incorporate $n-1$ pipeline stages while maintaining the exact same pseudo-random sequence as the original circuit. This is re-drawn for clarity in Fig. 2c from which it is clear that evaluation of the pipelined mixing function can take as long as $n-1$ cycles so long as new inputs are accepted and a new output is produced, on every cycle. Of course in practice the registers would not simply be placed in series with the ouput of the mixing function, but instead would be buried inside. The optimum positioning of these registers when moved inside the mixing function is dependent on the specific propagation delays through its various paths and can be determined by standard hardware design techniques. We comment that speed-optimized positioning of the pipeline registers may place them at computational wavefronts crossed by more than w signals, each of which must be registered. So, the increase in speed due to pipelining may come at the cost of some small increase in register bits.

---

[1] The output tap has also been moved to the bottom to aid comparison with the other figures. This does not affect the function of the WAKE-OFB model, it simply advances the availability of the output.

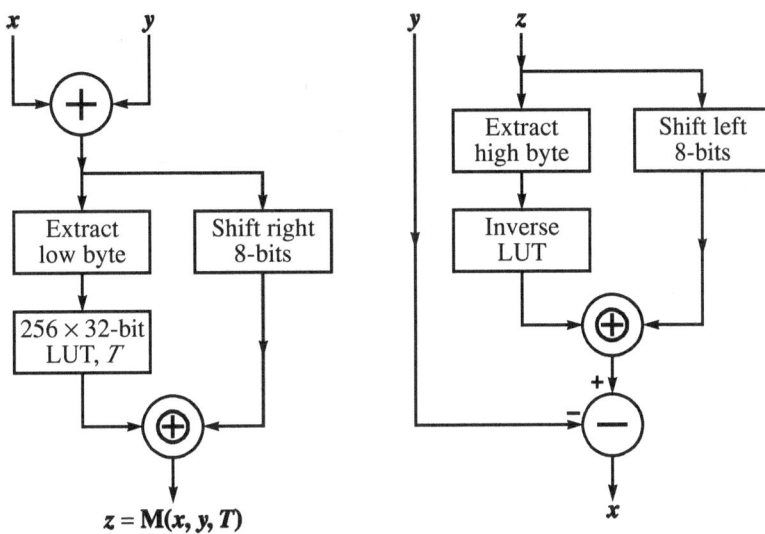

Fig. 3 a) WAKE's Mixing function   b) Inverse mixing function

4.3 Trade-off between speed, security, and hardware efficiency

From the preceding discussions we see that there is a three-way trade-off between hardware efficiency, security, and speed (both software speed and hardware
speed, with software speed being more of a concern).

The arrangement of Fig. 2a (WAKE-OFB) offers no cryptographic or performance advantages over that of Fig. 2b (WAKE-ROFB), and consequently
has no reason to be used. The fundamental distinction between the two is that
WAKE-ROFB allows multiple mixing functions to be evaluated in parallel while
WAKE-OFB does not. This distinction is analogous to the difference between
the Fibonacci and Galois forms of a linear feedback shift register.

In selecting the number of w-bit registers forming the shift register we trade
hardware efficiency and software state against the amount of available parallelism
(which up to a point translates into speed). For efficient operation on some
legacy processors, notably the 'x86 family, it is desirable to keep the amount
of state small so that it can stay resident in the very limited register set of
these processors. This can be less of a constraint if efficient operation is only
required on more recent family members like the Pentium II, since in them the
MMX registers can relieve pressure on the legacy registers, even if the algorithm
doesn't need the specialized MMX instructions.

In selecting the number of times to step the generator between outputs we
directly trade speed against security. Indirectly we are also trading them both
against hardware complexity, since in general we might expect to need fewer
iterations for the same level of security if the mixing function is made more
complex.

5   Design of the mixing function

WAKE's mixing function, $M(x, y, T)$, combines two 32-bit inputs, x and y, into one 32-bit output with the aid of a key-dependent 256×32-bit look-up-table, T. After summing the two input words the low byte becomes the index into the table. The other three bytes are shifted down by one byte and then XORed with the table's output. By constraining the values in the upper byte lane of the otherwise 'random' entries of T to form a permutation of the numbers 0 to 255, the mixing function is made invertible in the sense that knowledge of the output word and one of the two input words is sufficient to uniquely specify the other input word. The action of the table-look-up is to permute the bottom byte and place it in the top byte while also providing confusion for the other three bytes. The right-shift serves to propagate changes to the right while the addition operation provides propagation to the left. The mixing function, and its inverse, are shown in Fig. 3.

An attraction of this mixing function is that it is very fast in software. Unfortunately, for our purposes the 256×32-bit look-up-table presents an unacceptable hardware burden. Even if we count each bit as equivalent to just a single gate the table will take over 8000 gates. Our budget demands that we be an order of magnitude smaller than this.

5.1   Splitting the table

To address the hardware issue we modify the mixing function so that the look-up-table part may be compatibly implemented by either of two constructions. The first construction is as two 16×16-bit look-up-tables, while the second is as a single 256×32-bit table. The first construction is preferred for economic hardware implementation, while the second provides the fastest software implementation.

In the hardware case with two 16×16-bit tables, half of the original address bits go to one table while the second table receives the others. The pair of 16-bit outputs are combined to produce a 32-bit word by interleaving in a fashion that is discussed later.

This strategy results in a construction for which *there is always an equivalent 256×32-bit table* with a constrained form. Consequently *in software* we can continue to implement the mixing function by using a single 256×32-bit table, thereby avoiding degrading the speed of mixing function evaluation.

The resultant [inverse] mixing function, for use with the register topology of Fig. 2b, is shown in Fig. 4. Compared to the inverse mixing function of Fig. 3b we have also changed the shift direction to be that of Fig. 3a and adopted the corresponding table address field and arithmetic operation. This has no cryptographic significance nor impact on the hardware efficiency but there is a benefit in software to extracting the low byte as the table address rather than the high byte. Low byte extraction is simply a masking operation, which employs the processor's ALU, while high byte extraction needs a shift. In processors with multiple execution units it is common to have more ALUs than shifters. Since we

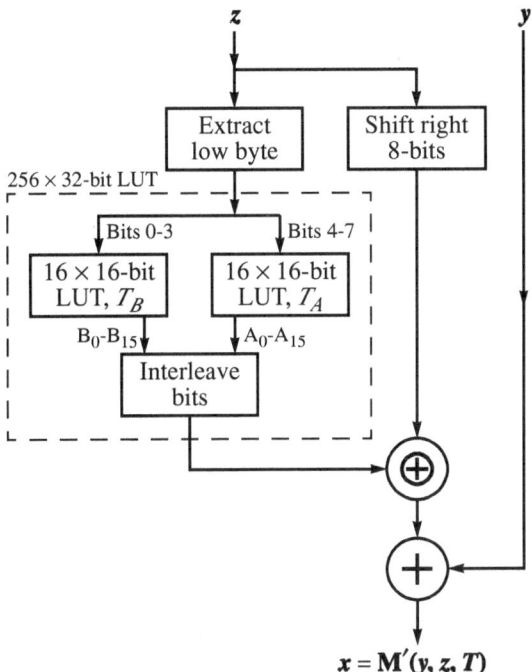

Fig. 4. Modified mixing function with split table

also need to shift the other three bytes we want to avoid creating a bottleneck here.

In order to maintain the reversibility of the generator we keep the mixing function invertible by applying Wheeler's trick to the upper nibble of each table, i.e. the upper nibble of each table contains a key-dependent pseudo-random permutation of that table's four-bit address input.

While the remaining contents of the tables could simply be pseudo-random entries, as in WAKE's original table, the table sizes in the new version are especially small so that statistical bias becomes more of a concern. To mitigate this concern we choose to fill each of the other three nibble lanes in both tables with their own key-dependent pseudo-random permutations.

5.2   Partitioning the address and re-combining the table outputs

A randomly filled 256-entry look-up-table will, with high probability, have outputs whose non-linear order is 8. However, outputs from a 16-entry table cannot have non-linear order greater than 4. In order to as rapidly as possible regain the non-linear complexity of the mixing function based on a single randomly filled 256-entry look-up-table, while incurring no additional hardware burden, we interleave the outputs of the two tables on a bit-by-bit basis, except that at byte boundaries the interleaving order is reversed.

This is illustrated below where bits $A_0 - A_{15}$ come from one table and $B_0 - B_{15}$ come from the other table:

$A_{15}B_{15}A_{14}B_{14}A_{13}B_{13}A_{12}B_{12}$ $B_{11}A_{11}B_{10}A_{10}B_9A_9B_8A_8$ $A_7B_7A_6B_6A_5B_5A_4B_4$ $B_3A_3B_2A_2B_1A_1B_0A_0$

Interleaving the table outputs on a bit-by-bit basis creates the maximum possible interaction between the two groups of bits in their next encounter with an adder, since any carries generated inside the adder then act to mix bits that came from independent tables. The deviation from strict bit-by-bit interleaving is so that the XOR stage always combines output bits from one table with those bits in the shifted path that have highest correlation with the other table (from an earlier mixing function). This again is to make all resulting bits as strongly dependent on both tables as possible.

We partition the original byte address into high and low nibbles for addressing the two smaller tables. This gives each table address a contribution from both tables in earlier operations, due to the interleaving of table outputs in a previous mixing function. The low nibble, which includes bit 0 that can have had no influence from table B through the action of a carry arising from the addition operation in the preceding mixing function, is used as the address for table B. Meanwhile the high nibble become the address for table A. In this way each table's address is composed of a roughly even contribution from both tables in earlier operations. Of the two groups of address bits, the one containing bit 0 has a greater bias towards table A, and by using bit 0 in table B's address we help to even out this bias.

Our partitioning of the address bits has no influence on the performance of a software implementation since however we choose to partition them we can always construct an equivalent 256×32-bit table.

5.3   Using a Feistel round as the mixing function

So far, reversibility of the word-wide shift register topology has been achieved by using a mixing function having as its non-linear element a pseudo-random permutation, thereby making the mixing function invertible. However, such a construction is not strictly necessary to ensure that the state machine is reversible.

In Fig. 5a we illustrate a more general form of the mixing function which guarantees the reversibility of the state machine. $\mathbf{P}_x$ and $\mathbf{P}_z$ are pseudo-random *permutations* while $\mathbf{F}_y$ is a pseudo-random *function*. While Fig. 5a shows $\mathbf{P}_z$ and $\mathbf{F}_y$ being combined by exclusive-OR, the only requirement on the combining operation is that it be invertible, so for instance addition modulo 2^w is equally suitable. Cryptographically useful mixing functions require at least one out of $\mathbf{P}_x$, $\mathbf{P}_z$, and $\mathbf{F}_y$ to be non-trivial.

We can now see that the earlier proposed mixing function is of the form having non-trivial $\mathbf{P}_z$, while $\mathbf{P}_x$ and $\mathbf{F}_y$ are identity operators (i.e. absent), with addition used as the combining operation.

If we consider the case where only $\mathbf{F}_y$ is present, as shown in Fig. 5b, we note that in essence the mixing function is then equivalent to the round function of

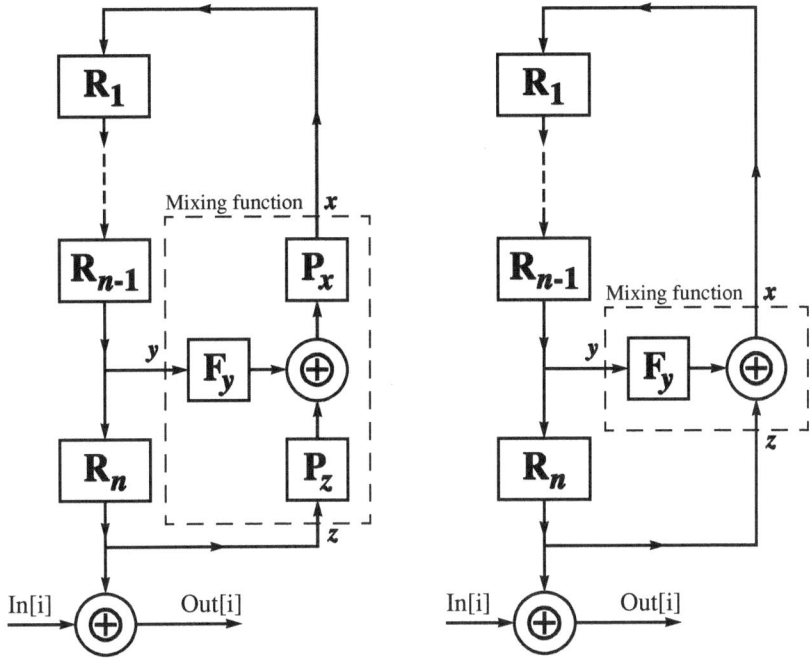

a) with generalized mixing function   b) with 'Feistel round' as mixing function

Fig. 5. Reversible generators with alternative types of mixing function

a Feistel network. This raises the interesting possibility to apply the techniques from block cipher design and analysis to this topology, knowing that the resulting stream cipher is guaranteed to have a high degree of parallelism. Of course, $\mathbf{F}_y$ is not required to be a *fixed* function, but could instead vary by shift-register step in the same way that common practice in Feistel-based block cipher design is for the key schedule to introduce some variation in the round function by round number.

6   Key scheduling and stream re-synchronization

For a complete stream cipher definition two further components – initialization of the key-dependent look-up-tables (key scheduling), and stream re-synchronization, need to be specified. We discuss these two operations in light of our goal of hardware efficiency.

We anticipate that even in applications where encryption is performed in hardware there will commonly exist a processor of some kind, such as a micro-controller used to orchestrate the operation of a consumer electronics device. While such a CPU cannot be relied upon to perform ciphering of high speed data streams, it is not unreasonable to call upon it to assist with initializing a hardware-based mixing function's key-dependent look-up-tables, so long as key changes are relatively infrequent. For this reason, we allow the key scheduling operation to be rather more complicated than would be desirable if it had to be performed strictly in hardware, instead concerning ourselves only to make it relatively easy to implement even on a low-end microcontroller. Specifically, only the construction of the pair of 16×16-bit tables need be straightforward for a microcontroller, not construction of the equivalent 256×32-bit table, since the large table is only required for efficient *software-only* cipher implementations.

On the other hand, we want to allow for the possibility of frequent stream re-synchronization without concern for unduly burdening the microcontroller in the case that encryption is being performed in hardware. For this reason we keep the re-synchronization process very simple so that its implementation in hardware alone is trivial. We perform re-synchronization by setting the shift register to a new state that is solely dependent on a supplied initialization vector and then step the generator until its state is thoroughly dependent on the contents of the look-up-tables, discarding the generator's output along the way.

7   Suggested configurations

To provide the amount of parallelism necessary to fully exploit a processor capable of performing four or five 32-bit operations per cycle we need five or six 32-bit register stages in our shift register. This is also an acceptable number for an efficient register-based implementation on a Pentium II.

A hardware implementation of this stream cipher, using the proposed split-table mixing function with a 32-bit $\times$ 5-stage shift register and the necessary control logic, takes about 2000 gates (with key scheduling assumed to be done in software).

We suggest two models:

- **5-stage shift register with the output sampled on every 8th step.**

  This model's 5-stage shift register allows four mixing functions to be evaluated concurrently, which lets it effectively exploit the instruction-level parallelism in media-processors such as the TriMedia TM-1000. By increasing the number of mixing function iterations between outputs to eight from WAKE-ROFB's original four we hope to amply compensate for the split look-up-table.

  Optimized implementations of this model achieve a throughput of 340 Mbps on a 266 MHz Pentium II, and 270 Mbps on a 100 MHz TriMedia chip, or 1.3 and 2.7 bits per cycle respectively. While the TriMedia performance can be achieved with good C-code, the reported Pentium II performance requires assembly coding. In particular, with the Pentium II's MMX registers being used to hold the 5-stage shift register and the standard registers being used for memory pointers and the loop counter, it is possible to avoid any redundant register-to-register moves or any memory accesses other than the data buffer being encrypted.

  In hardware, shift-register cycle times of under 20 ns are easily achievable by pipelining the mixing function. This gives a 200 Mbps throughput at a 50 MHz clock rate.

In the Appendix we provide specimen C-code for this model, which we refer to as WWNFSR-5/8. The Appendix includes illustrative routines for key scheduling and re-synchronization operations using a 128-bit key and 64-bit initialization vector. Re-synchronization of the form specified takes about as long as ciphering 32 bytes. Key scheduling software for hardware-assisted implementations need only perform the first half of the key scheduling routine in which the pair of 16×16-bit tables are constructed.

- **4-stage shift register with the output sampled on every 4th step.**

This model is equivalent to 4-stage WAKE-ROFB except that with our modified mixing function we expect it to be weaker. Four is the minimum number of iterations necessary for complete diffusion of the mixing function.

While this model only has three-fold parallelism this has not been found to be the limiting factor on the Pentium II. To get greater than two-fold parallelism out of the Pentium II requires taking advantage of the SIMD parallelism available through the MMX instructions. However, in practice we have not been able to take effective advantage of these because of our use of table-look-ups. The MMX instructions have no support for table-look-up, with the result that the benefit afforded by doing logical or arithmetic operations on two 32-bit words at a time (by using MMX) is negated by the inefficiency of re-arranging the data into the non-MMX registers from which table-look-ups can be performed. Similarly, the reported performance for WWNFSR-5/8 takes no advantage of the MMX instructions' SIMD parallelism, but simply uses the 64-bit MMX registers as 32-bit registers with the upper 32-bits going to waste.

Optimized implementations of this model achieve a throughput in excess of 600 Mbps on a 266 MHz Pentium II, and 340 Mbps on a 100 MHz TriMedia chip, or 2.3 and 3.4 bits per cycle respectively. In hardware it has a throughput of 400 Mbps at a 50 MHz clock rate.

The design intent of these non-linear finite state machines is that they behave as pseudo-random permutation generators. If such is the case then we expect the cycle length of a w-bit $\times$ n-stage version to be, on average, 2^{wn-1}. Cycle lengths have been determined experimentally for some artificially small models and found to be as expected. This is encouraging for the full-sized models because small models have less ability to hide statistical biases arising from 'boundary effects'. Statistical tests on full sized models have not revealed any weaknesses in these generators, however this alone should not be taken as an indication of their strength.

8   Conclusion

We have presented the word-wide non-linear-feedback shift register as a topology that offers a high degree of parallelism together with the ability to trade speed versus security by considering it as an iterated structure.

The topology can be used with any invertible non-linear mixing function, including the round function of a Feistel network. This represents fertile ground for future research – to identify suitable functions that maximize the security for a given amount of resources, and suggests the possibility that techniques developed for analyzing block ciphers may also be applicable to this topology.

The inherent parallelism present in the shift-register topology indicates that a stream cipher built this way is likely to be substantially faster than a block cipher used in OFB mode, even when both ciphers use the same round function and number of rounds/iterations. It remains to be demonstrated in this case that the speed increase does not in some way compromise security.

We suggest a non-linear mixing function derived from that of WAKE, but with the look-up-table split in two, thereby substantially reducing the hardware requirements. The potential loss of strength can be made up by increasing the number of iterations between outputs, albeit at a corresponding modest reduction in speed. We consider this to be an acceptable trade-off for the hardware efficiency gained.

References

1. R. J. Anderson, "On Fibonacci Keystream Generators", *Fast Software Encryption (Ed. B. Preneel), LNCS 1008,* Springer-Verlag, 1995, pp. 346-352
2. U. Blöcher, M. Dichtl, "Fish: A Fast Software Stream Cipher", *Fast Software Encryption (Ed. R. Anderson), LNCS 809,* Springer-Verlag, 1994, pp. 41-44
3. C. S. K. Clapp, "Optimizing a Fast Stream Cipher for VLIW, SIMD, and Superscalar Processors", *Fast Software Encryption (Ed. E. Biham), LNCS 1267,* Springer-Verlag, 1997, pp. 273-287
4. J. Daemen, *Cipher and Hash Function Design - strategies based on linear and differential cryptanalysis, Doctoral Dissertation,* K.U. Leuven, March 1995, pp. 172-180
5. D. Knuth, *The Art of Computer Programming, Volume 2 - Seminumerical Algorithms, Second edition,* Addison Wesley, 1981
6. D. Coppersmith, H. Krawczyk, Y. Mansour, "The Shrinking Generator", *Advances in Cryptology - Crypto '93, LNCS 773,* Springer-Verlag, 1994, pp. 22-39
7. P. Rogaway and D. Coppersmith, "A Software-Optimized Encryption Algorithm", *Fast Software Encryption (Ed. R. Anderson), LNCS 809,* Springer-Verlag, 1994, pp. 56-63
8. B. Schneier, *Applied Cryptography, Second Edition,* John Wiley & Sons, 1996
9. B. Schneier, "Description of a New Variable-Length Key, 64-bit Block Cipher (Blowfish)", *Fast Software Encryption (Ed. R. Anderson), LNCS 809,* Springer-Verlag, 1994, pp. 191-204
10. D. J. Wheeler, "A Bulk Data Encryption Algorithm", *Fast Software Encryption (Ed. R. Anderson), LNCS 809,* Springer-Verlag, 1994, pp. 127-134

Appendix A. WWNFSR–5/8 Reference Implementation

The following C-code implements the recommended 5-stage/8-step word-wide non-linear-feedback shift register model. Specimen routines for key scheduling and stream re-synchronization are included for the purposes of providing a complete cipher proposal for cryptanalytic investigation. This code is a functional reference only and does not represent all the performance optimizations embodied in the benchmarked code.

```c
/* WWNFSR-5/8, C-code reference, version 1.0                            */
/* 5-stage word-wide shift register with output sampled on every 8th step. */
/* Key-schedule illustrates mixing-function with split look-up-table.   */

typedef  unsigned long  UINT32;  /* 32-bit unsigned integer */

/* Array sizes - UINT32 T[256], key[4], IVec[2], state[5];  */


/* addition is modulo 2^32,       >> is right shift with zero fill */
#define M(y,z,T)  ((y) + (((z) >> 8) ^ T[(z) & 0xff]))

void ofb_crypt(UINT32 *In, UINT32 *Out, int length, UINT32 *T, UINT32 *state)
{
    UINT32 R1, R2, R3, R4, R5, Rt;
    int i;

    R1=state[0];  R2=state[1];  R3=state[2];  R4=state[3];  R5=state[4];

    for (i = 0; i < length; i++)
    {
        /* Naive implementation of shift-register obscures available parallelism */
        /*    for (shiftstep = 0; shiftstep < 8; shiftstep++)                     */
        /*       { Rt = M(R4,R5,T);  R5=R4; R4=R3; R3=R2; R2=R1; R1=Rt; }          */

        /* Logically equivalent flattened form makes four-fold parallelism clear */
        Rt = M(R2,R3,T);       /*                                    */
        R3 = M(R3,R4,T);       /*      All four mixing functions     */
        R4 = M(R4,R5,T);       /*      can be evaluated in parallel   */
        R5 = M(R1,R2,T);       /*                                    */

        R2 = M(Rt,R3,T);       /*                                    */
        R3 = M(R3,R4,T);       /*      All four mixing functions     */
        R4 = M(R4,R1,T);       /*      can be evaluated in parallel   */
        R1 = M(R5,Rt,T);       /*                                    */

        Out[i] = In[i] ^ R5;   /* Execution can overlap with second  */
                               /*      group of mixing functions     */
    }
    state[0]=R1;  state[1]=R2;  state[2]=R3;  state[3]=R4;  state[4]=R5;
}


void resync(UINT32 *IVec, UINT32 *T, UINT32 *state)
    /* set state from initialization vector and discard first eight words */
{
    UINT32 temp[8];  /* bit-bucket */

    state[0] = state[3] = state[4] = IVec[0];
    state[1] = state[2] = IVec[1];

    ofb_crypt(temp, temp, 8, T, state);
}
```

```
void key_sched(UINT32 *key, UINT32 *T)
  /* build 16 x 16 look-up-tables Tb, Ta, and 256 x 32 table T from 128-bit key */
{
  UINT32 i, j, k, n, mask, xordif;
  UINT32 TbTa[16];  /* pair of 16 x 16 look-up-tables stacked side-by-side */

  /* initialize permutations in each nibble-lane of TbTa */
  for (i = j = 0; i < 16; i++) { TbTa[i] = key[3] ^ j;  j = j + 0x11111111; }

  k = 0;  mask = 0x00f0f0ff;    /* mask selects four out of eight nibbles */

  for (j = 0; j < 8; j++)       /* do for each of 8 mask values */
  {
    for (i = 0; i < 16; i++)    /* do for each of 16 table entries */
    {
      /* scan across key and table nibbles to define table entry for swap */
      n = (TbTa[i] >> (j*4)) + (key[(j*2 + i/8) & 0x3] >> ((i*4) & 0x1c));
      k = (k + n) & 0x0f;

      /* swap masked nibbles between TbTa[i] and TbTa[k] */
      xordif = (TbTa[i] ^ TbTa[k]) & mask;
      TbTa[i] = TbTa[i] ^ xordif;
      TbTa[k] = TbTa[k] ^ xordif;
    }

    mask = (mask << 4) | (mask >> 28);  /* rotate mask left by 4 bits */
  }

  { /* build 256 x 32 table T by interleaving left and right halves of TbTa */

    UINT32 a, b;
    UINT32 expand[16] = { 0x00,0x01,0x04,0x05, 0x10,0x11,0x14,0x15,
                          0x40,0x41,0x44,0x45, 0x50,0x51,0x54,0x55 };

    for (i = 0; i < 256; i++) { T[i] = 0; } /* clear look-up-table T */
    for (j = 0; j < 16; j++)
    {
      k = TbTa[j];

      a =  (expand[(k >> 0 ) & 0xf] << 0 ) ^ (expand[(k >> 4 ) & 0xf] << 9 )
         ^ (expand[(k >> 8 ) & 0xf] << 16) ^ (expand[(k >> 12) & 0xf] << 25);

      b =  (expand[(k >> 16) & 0xf] << 1 ) ^ (expand[(k >> 20) & 0xf] << 8 )
         ^ (expand[(k >> 24) & 0xf] << 17) ^ (expand[(k >> 28) & 0xf] << 24);

      for (i = 0; i < 16; i++)   /* fill look-up-table T */
        { T[i+16*j] = T[i+16*j] ^ a;   T[16*i+j] = T[16*i+j] ^ b; }
    }
  }
}
```

Appendix B.   Test Case

```
void test(void)
{
  UINT32 key[4] = { 0x12345678, 0x98765432, 0xabcdef01, 0x10fedcba };
  UINT32 IVec[2] = { 0xbabeface, 0xf0e1d2c3 };  /* Initialization Vector */

  UINT32 text[4] = { 0x1234abcd, 0xa0b1c2d3, 0x1a2b3c4d, 0x55667788 };

  UINT32 T[256], state[5];
  int i;

  key_sched(key, T);              /* Schedule key */
  resync(IVec, T, state);         /* Initialize generator state using IV */

  for (i = 0; i < 256; i++)       /* Encrypt text buffer 256 times */
    { ofb_crypt(text, text, 4, T, state); }

  for (i = 0; i < 4; i++) { printf("0x%08lx ", text[i]); }  printf("\n");
}

/* final text[] == { 0xe5650b3d, 0xfdb4dca1, 0xc904b128, 0xd25f1934 } */
```

On the Security of the Hashing Scheme Based on SL_2

Kanat S. Abdukhalikov[1] and Chul Kim[2] **

[1] Institute for Pure and Applied Mathematics
Pushkin Str 125, Almaty 480021, Kazakhstan
`abdukh@alg.itpm.alma-ata.su, kanat@euler.kwangwoon.ac.kr`
[2] Department of Mathematics, Kwangwoon University
447-1 Wolgye-Dong, Nowoon-Gu, Seoul 139-701, Korea
`ckim@garam.kreonet.re.kr`

Abstract. Tillich and Zémor proposed a hashing scheme based on the group of unimodular matrices $SL_2(\mathbf{F}_q)$ over a finite field $\mathbf{F}_q$ of $q = 2^n$ elements. Charnes and Pieprzyk studied the security of this scheme. They showed that for $n = 131$ and for some irreducible polynomial $P_{131}(x)$ this scheme is weak. We show that with sufficiently high probability the polynomials $P_n(x)$ can be chosen in such a way that this type of attack can be avoided. Futhermore, we generalize the Tillich-Zémor hashing scheme for any finite field $\mathbf{F}_q$ and show that the new generalized scheme has similar properties.

1   Introduction

Tillich and Zémor [7] proposed a hashing scheme based on the group of unimodular matrices $SL_2(\mathbf{F}_q)$ over a finite field $\mathbf{F}_q$ of $q = 2^n$ elements. This scheme has several attractive properties: the algorithm can be easily implemented in software by using operations in $\mathbf{F}_q$, which allows fast computations; parallelization and precomputations are possible ; small modifications to the input text can be detected; the security of the scheme is equivalent to a precise mathematical problem, for which there exist several results in favor of its difficulty.

Tillich and Zémor recommended to use the range $130 \leq n \leq 170$ and the field $\mathbf{F}_q = \mathbf{F}_2[x]/\langle P_n(x)\rangle$, where $P_n(x)$ is an irreducible polynomial of degree n. Let α be a zero of the polynomial $P_n(x)$ (for example, class of the element x in $\mathbf{F}_2[x]/\langle P_n(x)\rangle$) and

$$A(\alpha) = \begin{pmatrix} \alpha & 1 \\ 1 & 0 \end{pmatrix}, \qquad B(\alpha) = \begin{pmatrix} \alpha & \alpha+1 \\ 1 & 1 \end{pmatrix}$$

be matrices from $G = SL_2(\mathbf{F}_q)$. When the polynomial $P_n(x)$ is fixed we denote $A = A(\alpha)$, $B = B(\alpha)$. Define the mapping $\pi = \pi_\alpha$:

$$\pi : \{0, 1\} \to \{A, B\},$$

** This author's research was partially funded by the Korea Science and Engineering Foundation, grant 961-0106-038-2

S. Vaudenay (Ed.): Fast Software Encryption – FSE'98, LNCS 1372, pp. 93–102, 1998.

$$\pi(0) = A, \qquad \pi(1) = B.$$

The hashcode of a binary message $x_1 x_2 \ldots x_k$ is just the matrix product

$$\pi(x_1)\pi(x_2)\ldots\pi(x_k).$$

Charnes and Pieprzyk studied the security of this scheme. They showed that for $n = 131$ and for some irreducible polynomial $P_{131}(x)$ this scheme is weak. By using properties of a dihedral subgroup in G, it is proven in [1, Theorem 6] that the following relation holds in G:

$$A^{-1}B(A^{-1}B^2 A^{-1})A^{-1}B = BA^{-2}B. \qquad (1)$$

This identity can be produced easier. It is sufficient to note that $A^{-1}B = \left(\begin{smallmatrix} 1 & 1 \\ 0 & 1 \end{smallmatrix}\right)$, which follows identities $(A^{-1}B)^2 = I$ and $BA^{-1}B = A$, where I is the identity matrix (the last identity is equivalent to (1)). Similarly, $B^{-1}A = \left(\begin{smallmatrix} 1 & 1 \\ 0 & 1 \end{smallmatrix}\right)$ follows $AB^{-1}A = B$.

Suppose that orders of A and B are s and t respectively. Then matrix identity $BA^{s-1}B = A$ (resp. $AB^{t-1}A = B$) means that binary strings $(1, 0^{s-1}, 1)$ and (0) (resp. $(0, 1^{t-1}, 0)$ and (1)) hash to the same value in the group G. On the other hand, the trivial factorization $A^s = I$ gives the similar result: the binary string (0^s) can be inserted into any message. But if the orders of elements A and B is approximately q then these identities are useless as actual forgeries (there is no such a case to be used as 2^{130} consecutive 0's (or 1's) in a hash input).

So, it turns out that to avoid this type of attack one has to choose an irreducible polynomial $P_n(x)$ in such a way that the orders of A and B would not be small. We will show that with sufficiently high probability the polynomials $P_n(x)$ can be chosen in such a way that the orders of elements A and B are equal to either $q-1$ or $q+1$ (maximal possible values), see Theorems 6, 13 and Table 1. We also propose an efficient algorithm for the determination of the orders of the elements $A = A(\alpha)$ and $B = B(\alpha)$ for any $P_n(x)$. We show the probability that the scheme is vulnerable against the Charnes and Pieprzyk attack is negligible (approximately 10^{-27}, see remark after theorem 7). Futhermore, we generalize the Tillich-Zémor $SL_2(\mathbf{F}_{2^n})$ hashing scheme for any finite field $\mathbf{F}_q$ and show that the new generalized scheme has similar properties.

2   Analysis of SL_2 hashing

We define recursively a sequence $f_i(x) \in \mathbf{F}_2[x]$ of functions

$$f_0(x) = 0, \quad f_1(x) = 1, \quad f_{i+2}(x) = xf_{i+1}(x) + f_i(x) \text{ for } i \geq 0. \qquad (2)$$

Let

$$A(x) = \begin{pmatrix} x & 1 \\ 1 & 0 \end{pmatrix}, \qquad B(x) = \begin{pmatrix} x & x+1 \\ 1 & 1 \end{pmatrix},$$

and α is a zero of the irreducible polynomial $P_n(x) \in \mathbf{F}_2[x]$ (e.g., class of the element x in $\mathbf{F}_2[x]/\langle P_n(x)\rangle$).

Lemma 1. $A(x)^m = \begin{pmatrix} f_{m+1} & f_m \\ f_m & f_{m-1} \end{pmatrix}$ for $m > 0$.

Proof. We have

$$A(x) = \begin{pmatrix} x & 1 \\ 1 & 0 \end{pmatrix} = \begin{pmatrix} f_2 & f_1 \\ f_1 & f_0 \end{pmatrix}.$$

Suppose

$$A(x)^m = \begin{pmatrix} f_{m+1} & f_m \\ f_m & f_{m-1} \end{pmatrix}.$$

Then

$$A(x)^{m+1} = A(x)^m \begin{pmatrix} x & 1 \\ 1 & 0 \end{pmatrix} = \begin{pmatrix} xf_{m+1} + f_m & f_{m+1} \\ xf_m + f_{m-1} & f_m \end{pmatrix} = \begin{pmatrix} f_{m+2} & f_{m+1} \\ f_{m+1} & f_m \end{pmatrix}.$$

Thus by induction we have the result.

Lemma 2. *The order of the element A in the group G is equal to the minimum positive number k, such that $f_k(\alpha) = 0$. Equivalently, the order of element A is equal to the minimum positive number k, such that $P_n(x)$ divides $f_k(x)$.*

Proof. If $A^m = I$ then by lemma 1 we have $f_m(\alpha) = 0$. Conversely, if $f_m(\alpha) = 0$ then by (2) one has $f_{m+1}(\alpha) = \alpha f_m(\alpha) + f_{m-1}(\alpha) = f_{m-1}(\alpha)$. Futhermore, $\det A^m = 1$, thus $f_{m+1}(\alpha) \cdot f_{m-1}(\alpha) - f_m(\alpha)^2 = f_{m+1}(\alpha) \cdot f_{m-1}(\alpha) = 1$. Consequently, $f_{m+1}(\alpha) = f_{m-1}(\alpha) = 1$ and $A^m = I$.

Corollary 3. *The order of the element B in the group G is equal to the minimum positive number k, such that $f_k(\alpha + 1) = 0$. Equivalently, the order of element B is equal to the minimum positive number k, such that $P_n(x + 1)$ divides $f_k(x)$.*

Proof. The order of $B = B(\alpha)$ is equal to the order of the element

$$A^{-1}BA = \begin{pmatrix} \alpha + 1 & 1 \\ 1 & 0 \end{pmatrix}.$$

Now we can apply lemma 2 to the element $\alpha + 1$.

So, if one chooses an irreducible polynomial $P_n(x)$, for the determination of the order of the element $A = A(\alpha)$ one has to sequentially calculate

$$f_n(x) \equiv xf_{n-1}(x) + f_{n-2}(x) \pmod{P_n(x)}$$

until it gets $f_k(x) \equiv 0 \pmod{P_n(x)}$. This value of k gives the order of A. Similarly, for the determination of the order of the element $B = B(\alpha)$ one has to sequentially calculate

$$f_n(x) \equiv xf_{n-1}(x) + f_{n-2}(x) \pmod{P_n(x + 1)}$$

until it gets $f_k(x) \equiv 0 \pmod{P_n(x + 1)}$. Note that the polynomial $P_n(x + 1)$ is irreducible.

Lemma 4. i) $\deg f_n(x) = n - 1$ *for* $n > 0$.

ii) *If* $n > 0$ *is even then* $f_n(x) = xg_n(x)^2$ *for some polynomial* $g_n(x) \in \mathbf{F}_2[x]$.

iii) *If* $n > 0$ *is odd then* $f_n(x) = h_n(x)^2$ *for some polynomial* $h_n(x) \in \mathbf{F}_2[x]$.

Proof. The case i) immediately follows from (2). Further, from (2) it is easy to see that $f_n(x)$ is a sum of monomials of odd degree for even n and a sum of monomials of even degree for odd n. Then

$$f_{2k+1}(x) = x^{2m_1} + x^{2m_2} + \cdots + x^{2m_s} = (x^{m_1} + x^{m_2} + \cdots + x^{m_s})^2,$$

$$f_{2k}(x) = x^{2m_1+1} + x^{2m_2+1} + \cdots + x^{2m_t+1} = x(x^{m_1} + x^{m_2} + \cdots + x^{m_t})^2.$$

Lemma 5. i) *Suppose* $\lambda_1 + \lambda_2 = x$, $\lambda_1\lambda_2 = 1$. *Then* $f_m(x) = \frac{1}{x}(\lambda_1^m + \lambda_2^m)$ *for* $m \geq 0$.

ii) $f_{2^m}(x) = x^{2^m - 1}$.

Proof. i) For $m = 0, 1$ we have $f_0(x) = \frac{1}{x}(\lambda_1^0 + \lambda_2^0) = 0$, $f_1(x) = \frac{1}{x}(\lambda_1 + \lambda_2) = 1$. Suppose our formulae is true for all $m \leq k$. Then

$$\begin{aligned}
f_{k+1}(x) &= x \cdot f_k(x) + f_{k-1}(x) \\
&= x \cdot \frac{1}{x}(\lambda_1^k + \lambda_2^k) + \frac{1}{x}(\lambda_1^{k-1} + \lambda_2^{k-1}) \\
&= \frac{1}{x}(\lambda_1 + \lambda_2)(\lambda_1^k + \lambda_2^k) + \frac{1}{x}(\lambda_1^{k-1} + \lambda_2^{k-1}) \\
&= \frac{1}{x}(\lambda_1^{k+1} + \lambda_2^{k+1} + \lambda_1\lambda_2(\lambda_1^{k-1} + \lambda_2^{k-1}) + \lambda_1^{k-1} + \lambda_2^{k-1}) \\
&= \frac{1}{x}(\lambda_1^{k+1} + \lambda_2^{k+1}).
\end{aligned}$$

ii) $f_{2^m}(x) = \frac{1}{x}(\lambda_1^{2^m} + \lambda_2^{2^m}) = \frac{1}{x}(\lambda_1 + \lambda_2)^{2^m} = \frac{1}{x}x^{2^m} = x^{2^m - 1}$. ∎

Remark. In fact, the elements λ_1 and λ_2 are eigenvalues of the matrix $A(x)$. They belong to an extension of the field of rational functions $\mathbf{F}_2(x)$. Thus eigenvalues of $A(x)^m$ are λ_1^m and λ_2^m. If Tr C denotes the trace of matrix C then

$$f_m(x) = \frac{1}{x}(f_{m+1}(x) + f_{m-1}(x)) = \frac{1}{x}\mathrm{Tr}\ A(x)^m = \frac{1}{x}(\lambda_1^m + \lambda_2^m).$$

The order of any nonidentity element from $SL_2(\mathbf{F}_q)$ either is equal to 2, or divides $q - 1$, or divides $q + 1$ (see [3]). So maximal possible values of the orders of A (and B) are $q - 1$ or $q + 1$. From lemmas 2 and 5 it is easy to see that for $n > 1$ the orders of A and B are not equal to 2.

Now let $P_n(x)$ be a random irreducible polynomial of degree $n > 1$ over $\mathbf{F}_2$. We are going to estimate the probability that the order of $A(\alpha)$ is equal to $q - 1$ or $q + 1$. Let $(2^n - 1)(2^n + 1) = p_1^{k_1} p_2^{k_2} \cdots p_r^{k_r}$ be the decomposition into a product of prime numbers, where $p_1, \ldots, p_r$ are different prime numbers. Set

$$d_2(n) = 1 - \frac{1}{2}(1 + 2^{-n/2+2}) \sum_{i=1}^{r} \frac{1}{p_i},$$

$$c_2(n) = 1 - (1 + 2^{-n/2+2}) \sum_{i=1}^{r} \frac{1}{p_i} = 2d_2(n) - 1.$$

Theorem 6. *Let $P_n(x)$ be a random irreducible polynomial of degree $n > 1$ with coefficients in $\mathbf{F}_2$. Then the probability, that the order of A is greater than or equal to $q - 1$, is greater than $d_2(n)$. Futhermore, the probability, that both the orders of A and B are greater than or equal to $q - 1$, is greater than $c_2(n)$.*

Proof. Since $2^n - 1$ and $2^n + 1$ are relatively prime numbers, they have different prime divisors. Suppose $2^n - 1 = p_1^{k_1} \cdots p_j^{k_j}$, $2^n + 1 = p_{j+1}^{k_{j+1}} \cdots p_r^{k_r}$. Then

$$\begin{aligned}
P &= \Pr\left(\operatorname{ord}(A) \geq q - 1\right) \\
&= 1 - \Pr\left(\operatorname{ord}(A) < q - 1\right) \\
&\geq 1 - \sum_{i=1}^{j} \Pr\left(A^{(q-1)/p_i} = I\right) - \sum_{i=j+1}^{r} \Pr\left(A^{(q+1)/p_i} = I\right) \\
&= 1 - \sum_{i=1}^{j} \Pr\left(P_n(x) \text{ divides } f_{(q-1)/p_i}(x)\right) \\
&\quad - \sum_{i=j+1}^{r} \Pr\left(P_n(x) \text{ divides } f_{(q+1)/p_i}(x)\right)
\end{aligned}$$

Since $\deg f_{(q-1)/p_i}(x) = (q-1)/p_i - 1$ and $f_{(q-1)/p_i}(x)$ is a square by lemma 4, we have

$$\Pr\left(P_n(x) \text{ divides } f_{(q-1)/p_i}(x)\right) \leq \frac{\frac{(q-1)/p_i-1}{2n}}{S_2(n)} < \frac{\frac{q}{2np_i}}{S_2(n)},$$

where $S_2(n)$ is the number of irreducible polynomials of degree n with coefficients in $\mathbf{F}_2$. But $S_2(n) > \frac{q}{n}(1 - \frac{1}{2^{n/2-1}})$ (see [6]), thus we have

$$\Pr\left(P_n(x) \text{ divides } f_{(q-1)/p_i}(x)\right) < \frac{\frac{q}{2np_i}}{\frac{q}{n}\left(1 - \frac{1}{2^{n/2-1}}\right)} \leq \frac{1}{2p_i}\left(1 + 2^{-n/2+2}\right).$$

We have the same estimation for $\Pr\left(P_n(x) \text{ divides } f_{(q+1)/p_i}(x)\right)$. So

$$P > 1 - \frac{1}{2}\left(1 + 2^{-n/2+2}\right) \sum_{i=1}^{r} \frac{1}{p_i}.$$

Finally,

$$\begin{aligned}
\Pr\left(\operatorname{ord}(A) \geq q - 1 \text{ and } \operatorname{ord}(B) \geq q - 1\right) \\
\geq 1 - \Pr\left(\operatorname{ord}(A) < q - 1\right) - \Pr\left(\operatorname{ord}(B) < q - 1\right) \\
= \left(1 - \Pr\left(\operatorname{ord}(A) < q - 1\right)\right) + \left(1 - \Pr\left(\operatorname{ord}(B) < q - 1\right)\right) - 1 \\
\geq 2d_2(n) - 1 = c_2(n). \blacksquare
\end{aligned}$$

Let M be a positive integer number such that there does not exist practically a binary message containing (0^M) (that is, M consecutive 0's) or (1^M). In order to avoid low order attack it would be sufficient to have A and B with orders that are greater than M.

Table 1 shows that it is not hard to find a polynomial which results in a large order. The user can check it by lemma 2.

Theorem 7. *Let $P_n(x)$ be a random irreducible polynomial of degree $n > 3$ with coefficients in $\mathbf{F}_2$. Then the probability, that both the orders of A and B are greater than M, is greater than $1 - \frac{M^2}{2^{n-1}}$.*

Proof.

$$\Pr\left(\mathrm{ord}(A) \leq M\right) \leq \sum_{i=1}^{M} \Pr\left(P_n(x) \text{ divides } f_i(x)\right)$$

$$\leq \sum_{i=1}^{M} \frac{\frac{i-1}{n}}{S_2(n)} < \frac{M^2}{2nS_2(n)},$$

$\Pr\left(\mathrm{ord}(A) > M \text{ and } \mathrm{ord}(B) > M\right)$

$> 1 - \Pr\left(\mathrm{ord}(A) \leq M\right) - \Pr\left(\mathrm{ord}(B) \leq M\right) > 1 - \frac{M^2}{nS_2(n)} > 1 - \frac{M^2}{2^{n-1}}.\blacksquare$

For example, let $M = 10^6$, $n = 131$. Then the probability, that both the orders of A and B are greater then M, is greater then $1 - 10^{-27}$. So the probability, that for a random irreducible polynomial $P_n(x)$ the scheme is vulnerable against the Charnes and Pieprzyk attack, is less then 10^{-27}.

We do not discuss important properties of this scheme (concatenation property, connections with associated Cayley graph, protections against local modifications, expanding properties), stability under subgroup and density attacks, easy computability, because it was done in [7,8] in detail.

Remark. One of referees drew our attention to the article [4].

3   Generalization of the hashing scheme for $p > 2$

In this section we introduce the analog of the Tillich-Zémor hashing scheme for $p > 2$. The new hash algorithm can be described as follows.

Defining Parameter. An irreducible polynomial $P_n(x)$ of degree n over a field $\mathbf{F}_p$ of p elements ($p > 2$ is prime, $q = p^n$, n is sufficiently large).

Algorithm. Let α be a zero of the polynomial $P_n(x)$ (e.g., the class of the element x in the field $\mathbf{F}_q = \mathbf{F}_p[x]/\langle P_n(x) \rangle$) and

$$A = A(\alpha) = \begin{pmatrix} \alpha & -1 \\ 1 & 0 \end{pmatrix}, \qquad B = B(\alpha) = \begin{pmatrix} \alpha & \alpha - 1 \\ 1 & 1 \end{pmatrix}$$

be matrices from $G = SL_2(\mathbf{F}_q)$. Define the mapping

$$\pi : \{0, 1\} \rightarrow \{A, B\},$$

$$\pi(0) = A, \qquad \pi(1) = B.$$

The hashcode of a binary message $x_1 x_2 \ldots x_k$ is the matrix

$$\pi(x_1)\pi(x_2)\ldots\pi(x_k)$$

from G.

This hashing scheme has all the cryptographic properties, analogous to the Tillich-Zémor hashing scheme [7].

The following theorem shows that the set of hashcodes is the whole group $SL_2(\mathbf{F}_q)$.

Theorem 8. *For $n > 2$ the elements A, B generate the group $SL_2(\mathbf{F}_q)$.*

Proof. It can be easily checked that

$$A^{-1}B^{-1}A^2 = \begin{pmatrix} 1 & 0 \\ 1 & 1 \end{pmatrix}, \qquad BA^{-2}BA = \begin{pmatrix} 1 & \alpha - 1 \\ 0 & 1 \end{pmatrix}.$$

According to Dickson's theorem (see [5,2]) the matrices $\begin{pmatrix} 1 & 0 \\ 1 & 1 \end{pmatrix}$ and $\begin{pmatrix} 1 & \alpha - 1 \\ 0 & 1 \end{pmatrix}$ generate the group $SL_2(\mathbf{F}_q)$. ∎

Note that $SL_2(\mathbf{F}_q)$ has nontrivial center $Z = \{\pm I\}$, where I is the identity matrix. That is, for any $g \in SL_2(\mathbf{F}_q)$ and $z \in Z$ we have $gz = zg$. If $A^s \in Z$ then binary strings $(0^s, w, v)$, $(w, 0^s, v)$, $(w, v, 0^s)$ hash to the same value. Consequently, we have to choose irreducible polynomial $P_n(x)$ in such a way that the condition $A^s \in Z$ follows s would not be small.

Define a sequence $f_i(x) \in \mathbf{F}_p[x]$ of functions

$$f_0(x) = 0, \quad f_1(x) = 1, \quad f_{i+2}(x) = x f_{i+1}(x) - f_i(x) \text{ for } i \geq 0, \qquad (3)$$

and define matrices

$$A(x) = \begin{pmatrix} x & -1 \\ 1 & 0 \end{pmatrix}, \qquad B(x) = \begin{pmatrix} x & x - 1 \\ 1 & 1 \end{pmatrix}.$$

Lemma 9. $A(x)^m = \begin{pmatrix} f_{m+1} & -f_m \\ f_m & -f_{m-1} \end{pmatrix}$ *for $m > 0$.*

Proof. By induction.

Lemma 10. $A^m \in Z$ *if and only if $f_m(\alpha) = 0$. Equivalently, $A^m \in Z$ if and only if $P_n(x)$ divides $f_m(x)$.*

Proof. If $A^m \in Z$ then by lemma 9 we have $f_m(\alpha) = 0$. Conversely, if $f_m(\alpha) = 0$ then by (3) one has $f_{m+1}(\alpha) = \alpha f_m(\alpha) - f_{m-1}(\alpha) = -f_{m-1}(\alpha)$. Futhermore, $\det A^m = 1$, thus $f_{m+1}(\alpha) \cdot f_{m-1}(\alpha) = -1$. Consequently, we have either $f_{m+1}(\alpha) = -f_{m-1}(\alpha) = 1$ or $f_{m+1}(\alpha) = -f_{m-1}(\alpha) = -1$, thus either $A^m = I$ or $A^m = -I$.

Corollary 11. $B^m \in Z$ *if and only if* $f_m(\alpha + 1) = 0$. *Equivalently,* $B^m \in Z$ *if and only if* $P_n(x - 1)$ *divides* $f_m(x)$.

Proof. It follows from equality

$$A^{-1}BA = \begin{pmatrix} \alpha + 1 & -1 \\ 1 & 0 \end{pmatrix}$$

and lemma 9.

Lemma 12. i) *Suppose* $\lambda_1 + \lambda_2 = x$, $\lambda_1 \lambda_2 = 1$. *Then* $f_m(x) = \frac{\lambda_1^m - \lambda_2^m}{\lambda_1 - \lambda_2}$ *for* $m \geq 0$.
ii) $f_{2p}(x) = x^p(x^2 - 4)^{(p-1)/2}$.

Proof. i) If $f_k(x) = \frac{\lambda_1^k - \lambda_2^k}{\lambda_1 - \lambda_2}$ and $f_{k-1}(x) = \frac{\lambda_1^{k-1} - \lambda_2^{k-1}}{\lambda_1 - \lambda_2}$ then

$$f_{k+1}(x) = xf_k(x) - f_{k-1}(x)$$
$$= \frac{1}{\lambda_1 - \lambda_2}((\lambda_1 + \lambda_2)(\lambda_1^k - \lambda_2^k) - (\lambda_1^{k-1} - \lambda_2^{k-1}))$$
$$= \frac{1}{\lambda_1 - \lambda_2}(\lambda_1^{k+1} - \lambda_2^{k+1} + \lambda_1\lambda_2(\lambda_1^{k-1} - \lambda_2^{k-1}) - (\lambda_1^{k-1} - \lambda_2^{k-1}))$$
$$= \frac{\lambda_1^{k+1} - \lambda_2^{k+1}}{\lambda_1 - \lambda_2}.$$

ii) We have

$$f_{2p}(x) = \frac{\lambda_1^{2p} - \lambda_2^{2p}}{\lambda_1 - \lambda_2} = \left(\frac{\lambda_1^2 - \lambda_2^2}{\lambda_1 - \lambda_2}\right)^p (\lambda_1 - \lambda_2)^{p-1}$$
$$= (f_2)^p(\lambda_1 - \lambda_2)^{p-1} = x^p(\lambda_1^2 - 2\lambda_1\lambda_2 + \lambda_2^2)^{(p-1)/2}$$
$$= x^p((\lambda_1 + \lambda_2)^2 - 4\lambda_1\lambda_2)^{(p-1)/2} = x^p(x^2 - 4)^{(p-1)/2}. \blacksquare$$

The group G has the unique element $-I$ of order 2 (see [3]). The order of any noncentral element from G is equal to p, $2p$ or is a divisor of $q - 1$ or $q + 1$. Lemmas 9 and 12 follow that for $n > 1$ the order of element A (resp. B) is equal to neither p nor $2p$. Consequently, the orders of the elements A and B are divisors of $q - 1$ or $q + 1$.

Let $P_n(x)$ be a random irreducible polynomial of degree $n > 1$ with coefficients in $\mathbf{F}_p$. We estimate the probability that the order of A is equal to $q - 1$ or $q + 1$. Let $(p^n - 1)/2 = p_1^{k_1} p_2^{k_2} \cdots p_j^{k_j}$, $(p^n + 1)/2 = p_{j+1}^{k_{j+1}} \cdots p_r^{k_r}$ be the decomposition into a product of primes, where $p_1, \ldots, p_r$ are different prime numbers. Set

$$d_p(n) = 1 - \frac{1}{2}\left(1 + p^{-n/2+2}\right) \sum_{i=1}^{r} \frac{1}{p_i},$$

$$c_p(n) = 1 - \left(1 + p^{-n/2+2}\right) \sum_{i=1}^{r} \frac{1}{p_i} = 2d_2(n) - 1.$$

Theorem 13. *Let $P_n(x)$ be a random irreducible monic polynomial of degree $n > 2$ with coefficients in $\mathbf{F}_p$. Then the probability, that the order of A is greater than or equal to $q - 1$, is greater than $d_p(n)$. Futhermore, the probability, that both the orders of A and B are greater than or equal to $q - 1$, is greater than $c_p(n)$.*

Proof. Since $(q - 1)/2$ and $(q + 1)/2$ are relatively prime numbers, they have different prime divisors. We have

$$P = \mathrm{Pr}\left(\mathrm{ord}(A) \geq q - 1\right)$$
$$= 1 - \mathrm{Pr}\left(\mathrm{ord}(A) < q - 1\right)$$
$$\geq 1 - \sum_{i=1}^{j} \mathrm{Pr}\left(A^{(q-1)/2p_i} \in Z\right) - \sum_{i=j+1}^{r} \mathrm{Pr}\left(A^{(q+1)/2p_i} \in Z\right)$$
$$= 1 - \sum_{i=1}^{j} \mathrm{Pr}\left(P_n(x) \text{ divides } f_{(q-1)/2p_i}(x)\right)$$
$$- \sum_{i=j+1}^{r} \mathrm{Pr}\left(P_n(x) \text{ divides } f_{(q+1)/2p_i}(x)\right).$$

On the other hand,

$$\mathrm{Pr}\left(P_n(x) \text{ divides } f_{(q-1)/2p_i}(x)\right) \leq \frac{\frac{(q-1)/2p_i - 1}{n}}{S_p(n)} < \frac{\frac{q}{2np_i}}{S_p(n)},$$

where $S_p(n)$ is the number of monic irreducible polynomials of degree n with coefficients in $\mathbf{F}_p$. But $S_p(n) > \frac{q}{n}\left(1 - \frac{1}{p^{n/2-1}}\right)$ (see [6]), thus we have

$$\mathrm{Pr}\left(P_n(x) \text{ divides } f_{(q-1)/2p_i}(x)\right) < \frac{\frac{q}{2np_i}}{\frac{q}{n}\left(1 - \frac{1}{p^{n/2-1}}\right)} < \frac{1}{2p_i}\left(1 + p^{-n/2+2}\right).$$

So

$$P > 1 - \frac{1}{2}\left(1 + p^{-n/2+2}\right) \sum_{i=1}^{r} \frac{1}{p_i}.$$

Finally,

$$\mathrm{Pr}\left(\mathrm{ord}(A) \geq q - 1 \text{ and } \mathrm{ord}(B) \geq q - 1\right)$$
$$\geq 1 - \mathrm{Pr}\left(\mathrm{ord}(A) < q - 1\right) - \mathrm{Pr}\left(\mathrm{ord}(B) < q - 1\right)$$
$$= (1 - \mathrm{Pr}\left(\mathrm{ord}(A) < q - 1\right)) + (1 - \mathrm{Pr}\left(\mathrm{ord}(B) < q - 1\right)) - 1$$
$$\geq 2d_p(n) - 1 = c_p(n). \blacksquare$$

Similarly to theorem 7 one can easily prove

Theorem 14. *Let $P_n(x)$ be a random irreducible monic polynomial of degree $n > 3$ with coefficients in $\mathbf{F}_p$. Then the probability, that both the orders of A and B are greater than M, is greater than $1 - \frac{M^2}{(p-1)p^{n-1}}$.*

p = 2			p = 3		
n	$d_p(n)$	$c_p(n)$	n	$d_p(n)$	$c_p(n)$
130	0.64...	0.28...	80	0.55...	0.10...
131	0.83...	0.66...	81	0.59...	0.18...
132	0.55...	0.10...	82	0.64...	0.28...
133	0.81...	0.63...	83	0.74...	0.49...
134	0.73...	0.46...	84	0.48...	-0.02...
135	0.65...	0.31...	85	0.68...	0.37...
136	0.69...	0.39...	86	0.64...	0.29...
137	0.83...	0.66...	87	0.62...	0.25...
138	0.60...	0.21...	88	0.57...	0.14...
139	0.83...	0.66...	89	0.74...	0.49...

Table 1. Bounds for probability.

For example, let $p = 3$, $n = 81$, $M = 10^6$. Then the probability, that both the orders of A and B are greater then M, is greater then $1 - 10^{-26}$.

Table 1 shows some examples for $p = 2, 3$.

4   Conclusion

We have proposed fixing of the SL_2 hashing scheme against an attack by Charnes and Pieprzyk. In fact, we have proposed an algorithm to decide whether the given irreducible polynomial leads to vulnerable (against the Charnes and Pieprzyk attack) hashing. We have also shown that the negligible part of the set of all polynomials is vulnerable to this attack. Finally, we give a generalization of the Tillich-Zémor hashing scheme.

References

1. C.Charnes and J.Pieprzyk. Attacking the SL_2 hashing scheme. In Advanced in Cryptology – Proceedings of ASIACRYPT'94 (1994). LNCS 917. Springer-Verlag pp. 322–330.
2. L.E.Dickson. Linear groups with an exposition of the Galois field theory. Leibzig: Teubner 1901 (New York: Dover Publ. 1958).
3. L. Dornhoff. Group representation theory, volume I. Marcel Dekker, Inc., New York 1971.
4. W. Geiselman. A note on the hash function of Tillich and Zémor. In Fast Software Encryption Workshop. LNCS 1039. Springer-Verlag pp. 51–52.
5. M.Suzuki. Group theory, volume I. Springer-Verlag 1982.
6. H.C.A. van Tilborg. An introduction to cryrtology. Klumer, 1989.
7. J-P. Tillich and G.Zémor. Hashing with SL_2. In Advanced in Cryptology – Proceedings of CRYPTO'94 (1994). LNCS 917. Springer-Verlag pp. 40–49.
8. J-P. Tillich and G.Zémor. Group-theoretic hash functions. In First French-Israeli workshop on algebraic coding (1994) LNCS 781. Springer-Verlag pp. 90–110.

About Feistel Schemes
with Six (or More) Rounds

Jacques Patarin

Bull PTS
68, route de Versailles - BP45
78431 Louveciennes Cedex - France
J.Patarin@frlv.bull.fr

Abstract. This paper is a continuation of the work initiated in [2] by
M. Luby and C. Rackoff on Feistel schemes used as pseudorandom permutation generators. The aim of this paper is to study the qualitative
improvements of "strong pseudorandomness" of the Luby-Rackoff construction when the number of rounds increase. We prove that for 6 rounds
(or more), the success probability of the distinguisher is reduced from
$\mathcal{O}\left(\frac{m^2}{2^n}\right)$ (for 3 or 4 rounds) to at most $\mathcal{O}\left(\frac{m^4}{2^{3n}} + \frac{m^2}{2^{2n}}\right)$. (Here m denotes
the number of cleartext or ciphertext queries obtained by the enemy in
a dynamic way, and $2n$ denotes the number of bits of the cleartexts and
ciphertexts).
We then introduce two new concepts that are stronger than strong pseudorandomness: "very strong pseudorandomness" and "homogeneous permutations". We explain why we think that those concepts are natural,
and we study the values k for which the Luby-Rackoff construction with
k rounds satisfy these notions.

1 Introduction

In their famous paper [2], M. Luby and C. Rackoff provided a construction of
pseudorandom permutations and strong pseudorandom permutations. ("Strong
pseudorandom permutations" are also called "super pseudorandom permutations": here the distinguisher can access the permutation *and* the inverse permutation at points of its choice.) The basic building block of the Luby-Rackoff
construction (L-R construction) is the so called Feistel permutation based on a
pseudorandom function defined by the key. Their construction consists of four
rounds of Feistel permutations (for strong pseudorandom permutations) or three
rounds of Feistel permutations (for pseudorandom permutations). Each round
involves an application of a different pseudorandom function. This L-R construction is very attractive for various reasons: it is elegant, the proof does not involve
any unproven hypothesis, almost all (secret key) block ciphers in use today are
based on Feistel schemes, and the number of rounds is very small (so that their
result may suggest ways of designing faster block ciphers).

The L-R construction inspired a considerable amount of research. One direction of research was to improve the security bound obtained in the "main

S. Vaudenay (Ed.): Fast Software Encryption – FSE'98, LNCS 1372, pp. 103–121, 1998.
© Springer-Verlag Berlin Heidelberg 1998

lemma" of [2] p. 381, *i.e.* to decrease the success probability of the distinguisher. It was noticed (in [1] and [7]) that in a L-R construction with 3 or 4 rounds, the security bound given in [2] was almost optimal. It was conjectured that for more rounds, this security could be greatly improved ([7], [10]). However, the analysis of these schemes appears to be very technical and difficult, so that some transformations in the L-R construction were suggested, in order to simplify the proofs ([1], [3], [4], [10]). However, by doing this, we lose the simplicity of the original L-R construction.

In this paper, we study again this original L-R construction. In [9], it was shown that the success probability of the distinguisher is reduced from $\mathcal{O}\left(\frac{m^2}{2^n}\right)$ for 3 or 4 rounds of a L-R construction, to at most $\mathcal{O}\left(\frac{m^3}{2^{2n}}\right)$ for 5 rounds (pseudorandom permutations) or 6 rounds (strong pseudorandom permutations) of a L-R construction. (In these expressions, m denotes the number of cleartext or ciphertext queries obtained by the enemy, and $2n$ denotes the number of bits of the cleartexts and ciphertexts).

In part I of this paper, we further improve this result: we show that, for 6 rounds (or more), the success probability of the distinguisher is at most $\mathcal{O}\left(\frac{m^4}{2^{3n}} + \frac{m^2}{2^{2n}}\right)$. Moreover, we know that a powerful distinguisher is always able to distinguish a L-R construction from a random permutation when $m \geq 2^n$ (as noticed in [1], [3], [7]). Then, in part II of this paper, we introduce two new concepts about permutation generators: "very strong pseudorandomness" and "homogeneous permutations". These concepts both imply that the generator is a strong pseudorandom generator. We explain why we feel that it is natural to introduce these notions, and we characterize the values k such that the L-R constructions with k rounds satisfy (or not) these notions.

Finally we formulate a few open problems and we conclude.

Part I: Improved security bounds for Ψ^6

2   Notations

(These notations are similar to those of [3], [9] and [10].)

- I_n denotes the set of all n-bit strings, $I_n = \{0,1\}^n$.
- F_n denotes the set of all functions from I_n to I_n, and B_n denotes the set of all such permutations ($B_n \subset F_n$).
- Let x and y be two bit strings of equal length, then $x \oplus y$ denotes their bit-by-bit exclusive-or.
- For any $f, g \in F_n$, $f \circ g$ denotes their composition.
- For $a, b \in I_n$, $[a, b]$ is the string of length $2n$ of I_{2n} which is the concatenation of a and b.
- Let f_1 be a function of F_n. Let L, R, S and T be elements of I_n. Then by definition:

$$\Psi(f_1)[L, R] = [S, T] \Leftrightarrow [(S = R) \text{ and } (T = L \oplus f_1(R))].$$

- Let f_1, f_2, ..., f_k be k functions of F_n. Then by definition:

$$\Psi^k(f_1, ..., f_k) = \Psi(f_k) \circ ... \circ \Psi(f_2) \circ \Psi(f_1).$$

(When f_1, ..., f_k are randomly chosen in F_n, Ψ^k is the L-R construction with k rounds.)
- We assume that the definitions of permutation generators, distinguishing circuits, normal and inverse oracle gates are known. These definitions can be found in [2] or [3] for example.
- Let ϕ be a distinguishing circuit. We will denote by $\phi(F)$ its output (1 or 0) when its oracle gates are given the values of a function F.

3   Our new theorem for Ψ^6 and related work

In [2], M. Luby and C. Rackoff demonstrated how to construct a pseudorandom permutation generator from a pseudorandom function generator. Their generator was mainly based on the following theorem (called "main lemma" in [2] p. 381):

Theorem 1 (M. Luby and C. Rackoff). *Let ϕ be a distinguishing circuit with m oracle gates such that its oracle gates are given the values of a function F from I_{2n} to I_{2n}. Let P_1 be the probability that $\phi(F) = 1$ when f_1, f_2, f_3 are three independent functions randomly chosen in F_n and $F = \Psi^3(f_1, f_2, f_3)$. Let P_1^* be the probability that $\phi(F) = 1$ when F is a function randomly chosen in F_{2n}. Then for all distinguishing circuits ϕ:*

$$|P_1 - P_1^*| \leq \frac{m^2}{2^n},$$

i.e. the security is guaranteed until $m = \mathcal{O}(2^{\frac{n}{2}})$.

In [9], we proved the following theorem:

Theorem 2 (J. Patarin, [9]). *Let ϕ be a super distinguishing circuit with m oracle gates (a super distinguishing circuit can have normal or inverse oracle gates). Let P_1 be the probability that $\phi(F) = 1$ when $f_1, f_2, f_3, f_4, f_5, f_6$ are six independent functions randomly chosen in F_n and $F = \Psi^6(f_1, f_2, f_3, f_4, f_5, f_6)$. Let P_1^{**} be the probability that $\phi(F) = 1$ when F is a permutation randomly chosen in B_{2n}. Then:*

$$|P_1 - P_1^{**}| \leq \frac{5m^3}{2^{2n}},$$

i.e. the security is guaranteed until $m = \mathcal{O}(2^{\frac{2n}{3}})$.

Moreover, in [7] p. 310, we presented the following conjecture:

Conjecture: *For Ψ^5, or perhaps Ψ^6 or Ψ^7, and for any distinguishing circuit with m oracle gates, $|P_1 - P_1^*| \leq \frac{30m}{2^n}$ (the number 30 is just an example).*

As far as we know, nobody has yet proved this conjecture (if the conjecture is true, then the security is guaranteed until $m = \mathcal{O}(2^n)$). As mentioned in [1] and [3], the technical problems in analysing L-R construction with improved bounds seem to be very difficult (moreover, our conjecture may be wrong...). However, this part I makes a significant advance in the direction of this conjecture:

Theorem 3 (J. Patarin, this conference FSE'98). *Using the same notations as in theorem 2:*

$$|P_1 - P_1^{**}| \leq \frac{47m^4}{2^{3n}} + \frac{17m^2}{2^{2n}},$$

i.e. the security is guaranteed until $m = \mathcal{O}(2^{\frac{3n}{4}})$.

To prove this theorem 3, we first prove this "basic result":

"Basic result": *Let $[L_i, R_i]$, $1 \leq i \leq m$, be m distinct elements of I_{2n} ("distinct" means that if $i \neq j$, then $L_i \neq L_j$ or $R_i \neq R_j$). Let $[S_i, T_i]$, $1 \leq i \leq m$, be also m distinct elements of I_{2n}. Then the number H of 6-uples of functions $(f_1, ..., f_6)$ of F_n^6 such that:*

$$\forall i, \ 1 \leq i \leq m, \Psi^6(f_1, ..., f_6)[L_i, R_i] = [S_i, T_i]$$

satisfies:

$$H \geq \frac{|F_n|^6}{2^{2nm}} \left(1 - \frac{47m^4}{2^{3n}} - \frac{16m^2}{2^{2n}}\right).$$

Proof of the "basic result": The proof of the "basic result" is given in the next section.

Proof of theorem 3: The proof of theorem 3 is a direct consequence of the "basic result" and the general theorems of the proof techniques given in [6] or [8] or [9].

Remark: It can be noticed that – to prove theorem 3 – we just need a general minoration of H (such as in the "basic result") and we do <u>not</u> need both a general minoration and majoration of H. This is particularly important since, as we will see in section 6, no general majoration of H exists near the value $\frac{|F_n|^6}{2^{2nm}}$.

4   Proof of the "basic result": $H \geq \frac{|F_n|^6}{2^{2nm}} \left(1 - \frac{47m^4}{2^{3n}} - \frac{16m^2}{2^{2n}}\right)$

4.1   Definition of (C)

Let $[X_i, P_i]$ and $[Q_i, Y_i]$, $1 \leq i \leq m$, be the values such that:

$$\Psi^2(f_1, f_2)[L_i, R_i] = [X_i, P_i]$$

and

$$\Psi^4(f_1, f_2, f_3, f_4)[L_i, R_i] = [Q_i, Y_i]$$

(i.e. $[L_i, R_i]$ are the inputs, $[X_i, P_i]$ are the values after two rounds, $[Q_i, Y_i]$ are the values after four rounds, and $[S_i, T_i]$ are the output values after six rounds).

We denote by (C) the following set of equations:

$$(C) \quad \forall i, j, \ 1 \le i \le m, \ 1 \le j \le m, \ i \ne j, \quad \begin{cases} R_i = R_j \Rightarrow X_i \oplus L_i = X_j \oplus L_j \\ S_i = S_j \Rightarrow Y_i \oplus T_i = Y_j \oplus T_j \\ X_i = X_j \Rightarrow P_i \oplus R_i = P_j \oplus R_j \\ Y_i = Y_j \Rightarrow Q_i \oplus S_i = Q_j \oplus S_j \\ P_i = P_j \Rightarrow X_i \oplus Q_i = X_j \oplus Q_j \\ Q_i = Q_j \Rightarrow P_i \oplus Y_i = P_j \oplus Y_j \end{cases}$$

Then, from [9], p. 145 or [8], p. 134, we know that the exact value for H is:

$$H = \sum_{(X,Y,P,Q) \text{ satisfying } (C)} \frac{|F_n|^6}{2^{6mn}} \cdot 2^{n(r+s+x+y+p+q)},$$

where:

- r is the number of independent equations $R_i = R_j$, $i \ne j$,
- s is the number of independent equations $S_i = S_j$, $i \ne j$,
- x is the number of independent equations $X_i = X_j$, $i \ne j$,
- y is the number of independent equations $Y_i = Y_j$, $i \ne j$,
- p is the number of independent equations $P_i = P_j$, $i \ne j$,
- and q is the number of independent equations $Q_i = Q_j$, $i \ne j$.

Remark: When m is small compared to $2^{n/2}$, and when the equalities in the R_i and S_j variables do not have special "patterns", then it is possible to prove that the dominant terms in the value of H above correspond to $x = y = p = q = 0$. Then the number of (X, Y, P, Q) satisfying (C) is about $\frac{2^{4nm}}{2^{n(r+s)}}$, so that:

$$H \simeq \frac{2^{4nm}}{2^{n(r+s)}} \cdot \frac{|F_n|^6}{2^{6nm}} \cdot 2^{n(r+s)} \simeq \frac{|F_n|^6}{2^{2nm}},$$

as expected.

However, we will see in section 6 that, when the equalities in R_i and S_j have special "patterns" (even for small values of m), then the value of H can be much larger than that (but never much smaller, as shown by the basic result).

Moreover, when m is not small compared to $2^{n/2}$, then the dominant terms in the value of H no longer correspond to $x = y = p = q = 0$.

These two facts may explain why the proof of the "basic result" is so difficult.

4.2   Plan of the proof

To prove the "basic result", we proceed as follows: we define two sets E and D, $D \subset E \subset I_n^4$, and a function $\Lambda : D \to I_n^4$ such that the three lemmas below are satisfied.

Lemma 4. $\forall (X, Y, P, Q) \in E$, $\Lambda(X, Y, P, Q)$ satisfies all the equations (C).

($\Lambda(X, Y, P, Q)$ will be often denoted by (X', Y', P', Q').)

Lemma 5. $\forall (X', Y', P', Q') \in \Lambda(E)$, the number of $(X, Y, P, Q) \in E$ such that $\Lambda(X, Y, P, Q) = (X', Y', P', Q')$ is $\leq 2^{n(r+s+x'+y'+p'+q')}$, where:

- r is the number of independent equations $R_i = R_j$, $i \neq j$,
- s is the number of independent equations $S_i = S_j$, $i \neq j$,
- x' is the number of independent equations $X'_i = X'_j$, $i \neq j$,
- y' is the number of independent equations $Y'_i = Y'_j$, $i \neq j$,
- p' is the number of independent equations $P'_i = P'_j$, $i \neq j$,
- q' is the number of independent equations $Q'_i = Q'_j$, $i \neq j$

Lemma 6.
$$|E| \geq 2^{4nm} \left(1 - \frac{47m^4}{2^{3n}} - \frac{16m^2}{2^{2n}} \right).$$

Then the "basic result" is just a consequence of these three lemmas, as follows.

As we said in section 4.1,

$$H = \sum_{(X,Y,P,Q) \text{ satisfying } (C)} \frac{|F_n|^6}{2^{6mn}} \cdot 2^{n(r+s+x+y+p+q)}.$$

Thus, from lemma 4:

$$H \geq \sum_{(X',Y',P',Q') \in \Lambda(E)} \frac{|F_n|^6}{2^{6mn}} \cdot 2^{n(r+s+x'+y'+p'+q')}.$$

Therefore, from lemma 5, H is greater than

$$\sum_{(X',Y',P',Q') \in \Lambda(E)} \frac{|F_n|^6}{2^{6mn}} \cdot |\{(X, Y, P, Q) \in E, \Lambda(X, Y, P, Q) = (X', Y', P', Q')\}|$$

i.e.

$$H \geq \frac{|E| \cdot |F_n|^6}{2^{6mn}}.$$

Finally, from lemma 6:

$$H \geq \frac{|F_n|^6}{2^{2nm}} \left(1 - \frac{47m^4}{2^{3n}} - \frac{16m^2}{2^{2n}} \right),$$

as claimed.

We will now below define Λ and prove lemma 4, lemma 5 and lemma 6.

4.3   General remarks

Remark 1:   Since the proof below is rather long and technical, we suggest the interested reader to first read the proof of theorem 2, which is more simple (this proof can be found in the extended version of [9], available from the author), because our proof of lemma 4, 5 and 6 below is essentially an improvement of this previous result.

Remark 2:   Figure 1 below shows how we define Λ (i.e. X', Y', P', Q') below. In a way, our aim can be described as follows: we must transform "most" (X, Y, P, Q) into a (X', Y', P', Q') that satisfies (C) (and the three lemmas). Roughly speaking, things can be seen as follows: we must handle the fact that *two* exceptional equations in X, P, Q or Y can occur (in order to have a proof in $\mathcal{O}\left(\frac{m^4}{2^{3n}} + \frac{m^2}{2^{2n}}\right)$ as wanted). However, the probability that *three* exceptional equations occur between four given indices i, j, k, ℓ is assumed to be negligible. (In Luby-Rackoff proof of theorem 1, the probability that *one* exceptional equation occurs between the intermediate variables was negligible, but no more here. Similarly, in our previous proof of theorem 2, the probability that *two* exceptional equations occur between the intermediate variable was negligible, but no more here.)

Remark 3:   Only *two* exceptional equations in X, P, Q or Y can occur between three of four given indices, but the total number of exceptional equations in X, P, Q or Y can be huge. For example, if $m = 2^{0.7n}$, then the number of equations $X_i = X_j$, $i \neq j$, is expected to be about $\frac{m^2}{2^n} = \frac{2^{1.4n}}{2^n} = 2^{0.4n}$.

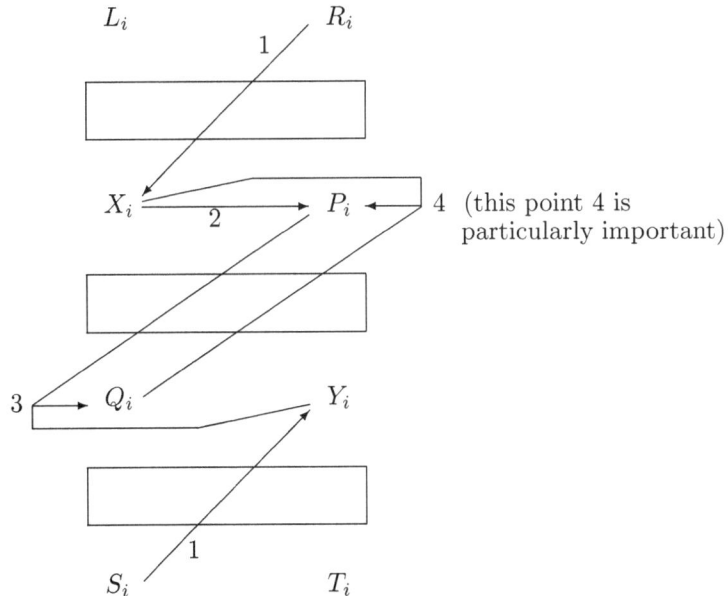

Figure 1: General view of the construction of Λ.

4.4   Definition of Λ

D is the domain of Λ (*i.e.* the set of all (X, Y, P, Q) for which Λ is defined). E will be a subset of D.

Definition of X'
Let $X = (X_1, ..., X_m)$ be an element of I_n^m. Similarly, let Y, P, Q be three elements of I_n^m. For all i, $1 \leq i \leq m$, let:

– i_R be the smallest integer, $1 \leq i_R \leq i$, such that $R_i = R_{i_R}$.
– i_S be the smallest integer, $1 \leq i_S \leq i$, such that $S_i = S_{i_S}$.

Then $X' = (X'_1, ..., X'_m)$ is (by definition) the element of I_n^m such that:

$$\forall i,\ 1 \leq i \leq m,\ X'_i = X_{i_R} \oplus L_i \oplus L_{i_R}.$$

Definition of Y'
Similarly, $Y' = (Y'_1, ..., Y'_m)$ is by definition the element of I_n^m such that:

$$\forall i,\ 1 \leq i \leq m,\ Y'_i = Y_{i_S} \oplus T_i \oplus T_{i_S}.$$

Note: These definitions of X' and Y' are shown with the two arrows numbered "1" in figure 1.

Definition of P^*
P^* is an intermediate variable that we use before defining P'. (In figure 1, the definition of P^* is shown with the arrow numbered "2", and the definition of P', that we do below, is shown with the arrow numbered "4"). For all i, $1 \leq i \leq m$, let i_X be the smallest integer, $1 \leq i_X \leq i$, such that $X'_i = X'_{i_X}$.
    Then $P^* = (P_1^*, ..., P_m^*)$ is (by definition) the element of I_n^m such that:

$$\forall i,\ 1 \leq i \leq m,\ P_i^* = P_{i_X} \oplus R_i \oplus R_{i_X}$$

Definition of Q'
Q' is now defined by a combined effect of P^* and Y'. (This is shown in figure 1 by the arrow numbered "3"). Before this, we need a definition of "Q^*-chain" and "Q^*-cycle".

Q^*-chain: Let i be an index, $1 \leq i \leq m$. Then, by definition, Q^*-$chain(i)$ is the set of all indices j, $1 \leq j \leq m$, such that it is possible to go from i to j by a chain of equalities of the type $(P_k^* = P_\ell^*)$ or $(Y'_\alpha = Y'_\beta)$.
    We also denote by $\min_{Q^*}(i)$ the smallest index in Q^*-$chain(i)$.

Remark: If we have $(P_j^* \neq P_i^*)$ and $(Y'_j \neq Y'_i)$ for all $j \neq i$, then $\min_{Q^*}(i) = i$.

Q^*-**cycles:**   Let ℓ be an even integer, $\ell \geq 2$. We call Q^*-ℓ-*cycle* a set of ℓ

equations of the form
$$
\begin{cases}
Y'_{i_1} = Y'_{i_2} \\
P^*_{i_2} = P^*_{i_3} \\
\quad\vdots \\
Y'_{i_{\ell-1}} = Y'_{i_\ell} \\
P^*_{i_\ell} = P^*_{i_1}
\end{cases}
, \text{ where } i_1, i_2, ..., i_\ell \text{ are } \ell \text{ pairwise distinct}
$$
indices.

We also call Q^*-*cycle* any Q^*-ℓ-*cycle*.

If (X, Y, P, Q) are such that a Q^*-*cycle* exists, then Q' and Λ are not defined (i.e. $(X, Y, P, Q) \notin E$). On the other hand, if no such Q^*-*cycle* exists, then from all the implications of the following type:

$$
\begin{cases}
P^*_\alpha = P^*_\beta \Rightarrow X'_\alpha \oplus Q'_\alpha = X'_\beta \oplus Q'_\beta & (*) \\
Y'_\gamma = Y'_\delta \Rightarrow Q'_\gamma \oplus S_\gamma = Q'_\delta \oplus S_\delta & (**)
\end{cases}
$$

it is possible to write all the Q'_i, $1 \leq i \leq m$, from the values $Q'_{\min_{Q^*}(i)}$, Y', P^*, S and X'. Q' is thus defined as follows:

1. $\forall i, 1 \leq i \leq m, Q'_{\min_{Q^*}(i)} = Q_{\min_{Q^*}(i)}$.
2. If $i \neq \min_{Q^*}(i)$, then Q'_i is uniquely defined from equations $(*)$ and $(**)$, and from the definition of $Q'_{\min_{Q^*}(i)}$ given in 1.

Definition of g: To simplify the notations, we write: $\forall i, 1 \leq i \leq m, Q'_i = Q_{\min_{Q^*}(i)} \oplus g(i, S, X')$. (Caution: g and $\min_{Q^*}(i)$ depend on Y' and P^*, and more precisely on the indices with equalities in Y' and P^*.)

Definition of P'

We now define P' (this definition of P' is particularly important, especially case 2 below) by a combined effect of X' and Q', and by keeping the equalities in P^* (i.e. if $P^*_i = P^*_j$, then $P'_i = P'_j$). Before this, we need a definition of "totalchain".

Totalchain: Let i be an index, $1 \leq i \leq m$. Then, by definition, $totalchain(i)$ is the set of all indices j, $1 \leq j \leq m$, such that it is possible to go from i to j by a chain of equalities of the type $(X'_\alpha = X'_\beta)$ or $(Q'_\gamma = Q'_\delta)$ or $(P^*_\varepsilon = P^*_\zeta)$.

For an integer i, $1 \leq i \leq m$, P'_i is now defined in 8 cases:

Case 1:   There is no equality of the type $Q'_\alpha = Q'_\beta$, with α and β in $totalchain(i)$ and $\alpha \neq \beta$. Then (by definition) $P'_i = P^*_i$.

Remark: If i is the only index of $totalchain(i)$, then we are in a particular case of this first case, and then $P'_i = P^*_i = P_i$.

Case 2:   There are exactly two elements i and j, $i < j$, in $totalchain(i)$, and they are linked only by the equality $Q'_i = Q'_j$. (This second case is particularly sensible: it is the most difficult case for the proof). Then there are two subcases:

Subcase 1: $\forall k,\ 1 \leq k \leq m,\ k \neq j,\ P_i^* \oplus Y_i' \oplus Y_j' \neq P_k^*$.

$$\text{Then (by definition):} \quad \begin{cases} P_i' = P_i^* \\ P_j' = P_i^* \oplus Y_i' \oplus Y_j' \end{cases}$$

Subcase 2: $\exists k,\ 1 \leq k \leq m,\ k \neq j,\ P_i^* \oplus Y_i' \oplus Y_j' = P_k^*$.

$$\text{Then (by definition):} \quad \begin{cases} P_i' = P_j^* \oplus Y_i' \oplus Y_j' \\ P_j' = P_j^* \end{cases}$$

Remark: This case 2 was the most difficult case to handle to improve theorem 2 in order to obtain theorem 3. The problem comes from the fact that $Q_i' = Q_j'$ might create an equality $P_a' = P_b'$, and $P_a' = P_b'$ might create $Q_i' = Q_j'$, and to prove lemma 5 we must know very precisely what equalities created what. In the definition given in this case 2, the problem is solved by introducing subcase 1 and 2, *i.e.* roughly speaking by selecting the subcase that creates the less trouble.

Case 3: There are exactly four distinct elements i, j, k, ℓ, in $totalchain(i)$, and they are linked only by the following three equalitites: $(Q_i' = Q_k')$ and $(X_i' = X_j')$ and $(X_k' = X_\ell')$.

$$\text{Then (by definition), if } i < k: \quad \begin{cases} P_i' = P_i^* \\ P_j' = P_j^* \\ P_k' = P_i^* \oplus Y_i' \oplus Y_k' \\ P_\ell' = P_i^* \oplus Y_i' \oplus Y_k' \oplus R_k \oplus R_\ell \end{cases}$$

$$\text{and if } k < i: \quad \begin{cases} P_k' = P_k^* \\ P_\ell' = P_\ell^* \\ P_i' = P_k^* \oplus Y_i' \oplus Y_k' \\ P_j' = P_k^* \oplus Y_i' \oplus Y_k' \oplus R_i \oplus R_j. \end{cases}$$

Case 4: There are exactly three distinct elements i, j, k in $totalchain(i)$, and they are linked only by the following two equalities: $(X_i' = X_j')$ and $(Q_i' = Q_k')$.

$$\text{Then (by definition):} \quad \begin{cases} P_i' = P_i^* \\ P_j' = P_i^* \\ P_k' = P_i^* \oplus Y_i' \oplus Y_k'. \end{cases}$$

Case 5: There are exactly three distinct elements i, j and k in $totalchain(i)$, and they are linked only by equalities in Q' (*i.e.* $Q_i' = Q_j' = Q_k'$).
Let $\alpha = \inf(i, j, k)$.

$$\text{Then (by definition):} \ \forall \beta \in \{i, j, k\},\ P_\beta' = P_\alpha^* \oplus Y_\alpha' \oplus Y_\beta'.$$

Case 6: There are exactly three distinct elements i, j, k in $totalchain(i)$, and they are linked only by the following two equations: $(P_i^* = P_j^*)$ and $(Q_i' = Q_k')$.

$$\text{Then (by definition):} \quad \begin{cases} P_i' = P_i^* \\ P_j' = P_j^*(= P_i') \\ P_k' = P_i^* \oplus Y_i' \oplus Y_k'. \end{cases}$$

Case 7: There are exactly three distinct elements i, j, k in $totalchain(i)$, and they are linked only by the two following equations: $(Q_i' = Q_j')$ and $(Y_i' = Y_k')$.

Then (by definition):
$$\begin{cases} P_i' = P_i^* \\ P_k' = P_k^* \\ P_j' = P_i^* \oplus Y_i' \oplus Y_j'. \end{cases}$$

Case 8: There are exactly four distinct elements i, j, k, ℓ in $totalchain(i)$, and they are linked only by the three following equations: $(Q_i' = Q_j')$ and $(Y_i' = Y_k')$ and $(Y_j' = Y_\ell')$.

Then (by definition):
$$\begin{cases} P_i' = P_i^* \\ P_j' = P_i^* \oplus Y_i' \oplus Y_j' \\ P_k' = P_k^* \\ P_\ell' = P_\ell^*. \end{cases}$$

If there exists an index i that lies in none of these eight cases, then Λ and P' are not defined (i.e. $(X, Y, P, Q) \notin E$).

4.5   Proof of the three lemmas

It is possible to prove that the function Λ defined above satisfies lemmas 4, 5 and 6 of section 4.2 (with a subset E of D). Due to the lack of space, we do not give details, but the complete proof is available from the author.

Part II: Homogeneous permutations, very strong pseudorandom permutations

5   Definitions

Let G be a permutation generator, such that G involves ℓ different pseudorandom functions of F_n to compute a permutation of B_{2n}. We denote by K the set of all ℓ-uples of functions $(f_1, ..., f_\ell)$ of F_n (i.e. $K = F_n^\ell$). Thus G associates to each $k \in K$ a permutation $G(k)$ of B_{2n}. K can be seen as the set of the keys of G, and $k \in K$ as a secret key.

Let $\alpha_1, ..., \alpha_m$ be m distinct elements of I_{2n}, and let $\beta_1, ..., \beta_m$ be also m distinct elements of I_{2n}. We denote by $H(\alpha_1, ..., \alpha_m, \beta_1, ..., \beta_m)$ the number of keys k of K such that:

$$\forall i, \ 1 \le i \le m, \ G(k)(\alpha_i) = \beta_i.$$

Definition 7. *We say that G is a "homogeneous" permutation generator if there exist a function $\varepsilon(m, n) : \mathbf{N}^2 \to \mathbf{R}$ such that, for any integer m:*

1. *For all α_1, ..., α_m being m distinct elements of I_{2n}, and for all β_1, ..., β_m being m distinct elements of I_{2n}, we have:*

$$\left| H(\alpha_1, ..., \alpha_m, \beta_1, ..., \beta_m) - \frac{|K|}{2^{2nm}} \right| \leq \varepsilon(m, n) \frac{|K|}{2^{2nm}}.$$

2. *For any polynomial $P(n)$ and any $\alpha > 0$, an integer n_0 exists such that:*

$$\forall n \geq n_0, \ \forall m \leq P(n), \ \varepsilon(m, n) \leq \alpha.$$

Remark: This notion of "homogeneous" permutations is a very natural notion: roughly speaking, a permutation generator is homogeneous when for all set of m cleartext/ciphertext pairs, there are always about the same number of possible keys that send all the cleartexts on the ciphertexts.

Definition 8. *We say that G is a "very strong" permutation generator if – with the same notations as above – the function $\varepsilon(m, n)$ satisfies condition 2, and the following condition 1' (instead of condition 1):*

1'. *For all α_1, ..., α_m being m distinct elements of I_{2n}, and for all β_1, ..., β_m being m distinct elements of I_{2n}, we have:*

$$H(\alpha_1, ..., \alpha_m, \beta_1, ..., \beta_m) \geq \frac{|K|}{2^{2nm}}(1 - \varepsilon(m, n)).$$

Theorem 9. *If G is a "homogeneous permutation generator", then G is a "very strong permutation generator".*

Theorem 10. *If G is a "very strong permutation generator", then G is a "strong permutation generator".*

Proof: Theorem 9 is an obvious consequence of the definitions. Theorem 10 corresponds to the technique of proof we used in part I. (This way of proving strong pseudorandomness was first explicitly used in [6].)

As a result, for permutation generators, we have:

$$Homogeneous \Rightarrow Very \ Strong \Rightarrow Strong \Rightarrow Pseudorandom.$$

Interpretations:
In order to distinguish (with a non-negligible probability) permutations generated by a homogeneous permutation generator, from truly random permutations of B_{2n}, an enemy must know a large number of cleartext/ciphertext pairs. (More precisely, this number must increase faster than any polynomial in n, **whatever** the cleartext/ciphertext pairs may be.)

Remark 1: Related (but not equivalent) notions can be found in [11] ("multipermutations") and in [5].

Remark 2: In some very special cases, this property of "homogeneity" may be useful and "strong peusorandomness" is not enough. For example, let us assume that the enemy has a spy inside the encryption team. Let us also assume that the aim of the enemy is to distinguish the encryption algorithm from a truly random permutation, and that his spy has access to the whole database of cleartext/ciphertext pairs, but can only send very few such pairs to help distinguishing. In such a case, "homogeneity" may be a more natural property than strong pseudorandomness. However, we introduced the concepts of "homogeneity" and "very strong pseudorandomness" because they are very natural in the proofs, and not with applications in mind.

6   Examples

6.1 Ψ^4 is not homogeneous

Example 1 (with $m = 2$):
   As shown in [7] p. 314 (or in [1] p. 309), if $\Psi^4[L_1, R_1] = [S_1, T_1]$ and $\Psi^4[L_2, R_2] = [S_2, T_2]$, and $R_1 = R_2$, $L_1 \neq L_2$, then the probability that $S_1 \oplus S_2 = L_1 \oplus L_2$ is about twice what it would be with a truly random permutation of B_{2n} (instead of Ψ^4). In [7] (and [1]), this result was used to show that the security bound given by Luby and Rackoff for Ψ^4 in a chosen-cleartext attack is tight (the attack requires $\simeq \sqrt{2^n}$ messages to ensure $S_i \oplus S_j = L_i \oplus L_j$).
   Here, we use this result to show that Ψ^4 is not homogeneous, and the non-homogeneity property appears with only two (very special) messages.

Remark: However, Ψ^4 is a very strong permutation generator (and for Ψ^4, we can take $\varepsilon(m, n) = \frac{m^2}{2^n}$). (As mentioned above, the proof of strong pseudorandomness of Ψ^4 given in [6] is also a proof of very strong pseudorandomness.)

Example 2 (with $m = 4$):
   Let $R_1 = R_3$, $R_2 = R_4 = R_1 \oplus \alpha$, $S_1 = S_2$, $S_3 = S_4 = S_1 \oplus \alpha$, $L_1 = L_2$, $L_3 = L_4 = L_1 \oplus \alpha$, $T_1 = T_3$, $T_2 = T_4 = T_1 \oplus \alpha$.
   Then the value H for Ψ^4 with these R, L, S, T is at least about $\frac{|F_n|^4}{2^{6n}}$ (instead of about $\frac{|F_n|^4}{2^{8n}}$ as expected if it was homogeneous). The proof of a similar property will be done in details for Ψ^6 below.

6.2 Ψ^5 is not homogeneous

If $\Psi^5[L_1, R_1] = [S_1, T_1]$ and $\Psi^5[L_2, R_2] = [S_2, T_2]$, and if $R_1 = R_2$ and $L_1 \neq L_2$, then the probability that $S_1 = S_2$ and $L_1 \oplus L_2 = T_1 \oplus T_2$ is about twice what it would be with a truly random permutation of B_{2n} (instead of Ψ^5). Therefore Ψ^5 is not homogeneous, and the non-homogeneity property appears with only two (very special) messages.

Remark: However, since here we have two equations and two indices ($S_i = S_j$ and $L_i \oplus L_j = T_i \oplus T_j$), this non-homogeneity property would require about $m = 2^n$ messages in a chosen-cleartext attack (instead of the $\sqrt{2^n}$ messages above for Ψ^4).

6.3 Ψ^6 is not homogeneous

Example 1 (with $m = 4$):
Let $\Psi^6[L_i, R_i] = [S_i, T_i]$ for $i = 1, 2, 3, 4$. Then if $L_1 = L_2$, $L_3 = L_4$, $R_1 = R_3$ and $R_2 = R_4$, it is possible to prove that the probability that $S_1 = S_2$, $S_3 = S_4$, $L_1 \oplus L_3 = S_1 \oplus S_3$ and $T_1 \oplus T_2 = T_3 \oplus T_4 = R_1 \oplus R_2$ is at least about twice what it would be with a truly random permutation of B_{2n} (instead of Ψ^6). The proof can be done as explained in example 2 below. Therefore, Ψ^6 is not homogeneous.

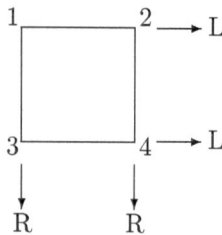

Figure 2: Modelisation of the equations $L_1 = L_2$, $L_3 = L_4$, $R_1 = R_3$ and $R_2 = R_4$.

Example 2 (with $m = 9$):

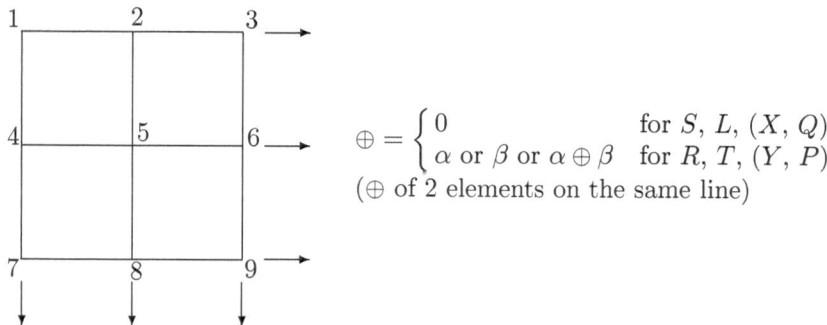

$$\oplus = \begin{cases} 0 & \text{for } S, L, (X, Q) \\ \alpha \text{ or } \beta \text{ or } \alpha \oplus \beta & \text{for } R, T, (Y, P) \end{cases}$$
($\oplus$ of 2 elements on the same line)

$$\oplus \text{ of 2 elements on the same column} = \begin{cases} 0 & \text{for } R, T, (Y, P) \\ \alpha' \text{ or } \beta' \text{ or } \alpha' \oplus \beta' & \text{for } S, L, (X, Q) \end{cases}$$

Figure 3: Modelisation of the equations in S, L, R, T (and in the X, Y, P, Q that we will consider).

Let $\Psi^6[L_i, R_i] = [S_i, T_i]$ for $1 \le i \le 9$. Let $\alpha \neq 0$ and $\beta \neq 0$ be two distinct values of I_n. Similarly, let $\alpha' \neq 0$ and $\beta' \neq 0$ be two distinct values of I_n. We study the values of H when

$$\begin{cases} L_1 = L_2 = L_3 \\ L_4 = L_5 = L_6 = L_1 \oplus \alpha' \\ L_7 = L_8 = L_9 = L_1 \oplus \beta' \\ R_1 = R_4 = R_7 \\ R_2 = R_5 = R_8 = R_1 \oplus \alpha \\ R_3 = R_6 = R_9 = R_1 \oplus \beta \end{cases} \text{ and } \begin{cases} S_1 = S_2 = S_3 \\ S_4 = S_5 = S_6 = S_1 \oplus \alpha' \\ S_7 = S_8 = S_9 = S_1 \oplus \beta' \\ T_1 = T_4 = T_7 \\ T_2 = T_5 = T_8 = T_1 \oplus \alpha \\ T_3 = T_6 = T_9 = T_1 \oplus \beta. \end{cases}$$

(All these relations are represented in figure 3).

Then – as we will see below – for such L, R, S, T values, the value of H is at least $\frac{|F_n|^6}{2^{14n}}$, instead of $\frac{|F_n|^6}{2^{18n}}$ as expected if it was homogeneous Therefore, Ψ^6 is not homogeneous.

Proof: We consider (X, Y, P, Q) values such that:

$$\begin{cases} X_1 = X_2 = X_3 \\ X_4 = X_5 = X_6 = X_1 \oplus \alpha' \\ X_7 = X_8 = X_9 = X_1 \oplus \beta' \\ Y_1 = Y_4 = Y_7 \\ Y_2 = Y_5 = Y_8 = Y_1 \oplus \alpha \\ Y_3 = Y_6 = Y_9 = Y_1 \oplus \beta \end{cases} \text{ and } \begin{cases} Q_1 = Q_2 = Q_3 \\ Q_4 = Q_5 = Q_6 = Q_1 \oplus \alpha' \\ Q_7 = Q_8 = Q_9 = Q_1 \oplus \beta' \\ P_1 = P_4 = P_7 \\ P_2 = P_5 = P_8 = P_1 \oplus \alpha \\ P_3 = P_6 = P_9 = P_1 \oplus \beta. \end{cases}$$

(All these relations are also represented in figure 3).

All the L_i, R_i, S_i, T_i, X_i, Y_i, Q_i, P_i, $1 \le i \le 9$, can be written from L_1, R_1, S_1, T_1, X_1, Y_1, Q_1, P_1.

Moreover, whatever the values are for L_1, R_1, S_1, T_1, X_1, Y_1, Q_1 and P_1, it is easy to verify that all the conditions (C) are satisfied (these conditions were explicitly written in section 4.1 for Ψ^6).

For example, $R_1 = R_4 \Rightarrow X_1 \oplus L_1 = X_4 \oplus L_4$, since $X_1 \oplus X_4 = \alpha' = L_1 \oplus L_4$. Similarly, $Q_7 = Q_9 \Rightarrow P_7 \oplus Y_7 = P_9 \oplus Y_9$, since $P_7 \oplus P_9 = \beta = Y_7 \oplus Y_9$.

Therefore, from the exact value of H (given in section 4.1), and by considering only such (X, Y, P, Q), we have:

$$H \ge 2^{4n} \cdot \frac{|F_n|^6}{2^{54n}} \cdot 2^{n(6+6+6+6+6+6)} = \frac{|F_n|^6}{2^{14n}},$$

as claimed (instead of $H \simeq \frac{|F_n|^6}{2^{18n}}$ if Ψ^6 was homogeneous).

6.4 $\forall k \in \mathbf{N}^*$, Ψ^k is not homogeneous

For simplicity, we assume that k is even (the proof is very similar when k is odd). Let $k = 2\lambda$. Let $\Psi^k[L_i, R_i] = [S_i, T_i]$ for $1 \le i \le m$. We essentially generalize to Ψ^k the construction given in example 2 for Ψ^6.

The exact value of H is:

$$H = \sum_{(X^{(1)}, \ldots, X^{(k-2)}) \text{ satisfying } (C)} \frac{|F_n|^k}{2^{knm}} \cdot 2^{n(r+s+x^{(1)}+\ldots+x^{(k-2)})},$$

where the $X^{(1)}$, ..., $X^{(k-2)}$ variables are the intermediate round variables, and where (C) denotes the conditions on the equalities (i.e. $R_i = R_j \Rightarrow X_i^{(1)} \oplus L_i = X_j^{(1)} \oplus L_j$, etc). The proof of this formula is not difficult and is given in [8], p. 134.

We take $m = \lambda^2 (= \frac{k^2}{4})$.

Let α_1, ..., $\alpha_{\lambda-1}$ be $\lambda - 1$ pairwise distinct and non-zero values of I_n. Let α_1', ..., $\alpha_{\lambda-1}'$ be also $\lambda - 1$ pairwise distinct and non-zero values of I_n.

We study the value H when L_i, R_i, S_i, T_i, $1 \leq i \leq m$, satisfy the equalities modelised in figure 4. (For simplicity, we do not write these equalities explicitly).

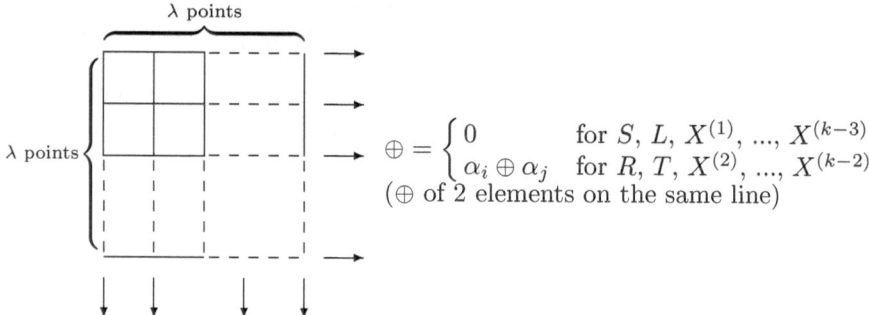

$$\oplus = \begin{cases} 0 & \text{for } S, L, X^{(1)}, ..., X^{(k-3)} \\ \alpha_i \oplus \alpha_j & \text{for } R, T, X^{(2)}, ..., X^{(k-2)} \end{cases}$$
($\oplus$ of 2 elements on the same line)

$$\oplus \text{ of 2 elements on the same column} = \begin{cases} 0 & \text{for } R, T, X^{(2)}, ..., X^{(k-2)} \\ \alpha_i' \oplus \alpha_j' & \text{for } S, L, X^{(1)}, ..., X^{(k-3)} \end{cases}$$

Figure 4: Modelisation of the equations in S, L, R, T (and in the $X^{(1)}$, ..., $X^{(k-2)}$ that we will consider).

In the exact formula given above for H, we study the corresponding terms for values of $X^{(1)}$, ..., $X^{(k-2)}$ that satisfy the equalities represented in figure 4. We find

$$H \geq 2^{(k-2)n} \cdot \frac{|F_n|^k}{2^{knm}} \cdot 2^{nk\lambda(\lambda-1)},$$

so that, with $m = \lambda^2 = \frac{k^2}{4}$,

$$H \geq 2^{(k-2)n} \cdot \frac{|F_n|^k}{2^{2mn}}$$

(instead of $\frac{|F_n|^k}{2^{2nm}}$ if Ψ^k was homogeneous). Therefore, Ψ^k is not homogeneous, as claimed.

In conclusion:

Ψ^k is very strong pseudorandom $\Leftrightarrow k \geq 4$.

Ψ^k is never homogeneous (this was a surprise for us).

Remark 1: The fact that Ψ^k is never homogeneous may explain why the proofs about the quality of pseudorandomness of the Ψ^k construction (such as theorem 3 of section 3) are so difficult.

Remark 2: In section 6.4, in order to give an explicit construction with a non homogeneous property, we have taken $m = \frac{k^2}{4} = \mathcal{O}(k^2)$, where k is the number of rounds of the L-R construction, so m increases when k increases. It is possible to prove that this increase was a necessity: when m is fixed, then all the values of H are converging to the same value when k tends to infinity. (This property can be proved with "Markov chain" theory for example).

7   Open problems

	Pseudorandom	Strong pseudo-random	Very strong pseudorandom	Homogeneous
Ψ	No	No	No	No
Ψ^2	No	No	No	No
Ψ^3	$= \mathcal{O}(\frac{m^2}{2^n})$	No	No	No
Ψ^4	$= \mathcal{O}(\frac{m^2}{2^n})$	$= \mathcal{O}(\frac{m^2}{2^n})$	$= \mathcal{O}(\frac{m^2}{2^n})$	No
Ψ^5	$\leq \mathcal{O}(\frac{m^3}{2^{2n}})$	$\leq \mathcal{O}(\frac{m^2}{2^n})$	$\leq \mathcal{O}(\frac{m^2}{2^n})$	No
$\Psi^k,\ k \geq 6$	$\leq \mathcal{O}(\frac{m^4}{2^{3n}} + \frac{m^2}{2^{2n}})$	$\leq \mathcal{O}(\frac{m^4}{2^{3n}} + \frac{m^2}{2^{2n}})$	$\leq \mathcal{O}(\frac{m^4}{2^{3n}} + \frac{m^2}{2^{2n}})$	No

Figure 5: Known results about the qualities of the Ψ^k pseudorandom permutations.

In figure 5, we represented the known results about the qualities of the L-R constructions with k rounds. For example, we see in this figure that Ψ^3 is not strong pseudorandom (this is written "No"), but that it is pseudorandom with an advantage of $\mathcal{O}(\frac{m^2}{2^n})$ for the best chosen-cleartext attack.

We also see that Ψ^5 is very strong pseudorandom, with an advantage of at most $\mathcal{O}(\frac{m^3}{2^{2n}})$ in a chosen-cleartext attack, and of at most $\mathcal{O}(\frac{m^2}{2^n})$ in a chosen-ciphertext and chosen-cleartext attack. "At most" means that we do not know if these $\mathcal{O}(\frac{m^3}{2^{2n}})$ and $\mathcal{O}(\frac{m^2}{2^n})$ bounds are reached or not: it is an open problem.

Similar open problems are shown in figure 5, when the "$\leq$" symbol appears.

It was conjectured in 1991 that, for Ψ^6 or Ψ^7, the advantage is negligible as long as m is negligible compared to 2^n. This is still unproven, as well as the following property:

When $k \to +\infty$, m must be $\Omega(2^n)$ to obtain a non-negligible advantage.

Another open problem that we mentioned is the following:

Is it possible to design homogeneous permutation generators ?

8   Conclusion

In order to improve the proved security bounds of pseudorandom permutations or pseudorandom functions, various authors have suggested new designs for the permutation generators ([1], [3], [4], [10]). This comes from the fact that proofs are much easier to obtain in these modified schemes than in the original L-R construction.

However, in [1] and [4], the functions with improved security bounds are no longer bijections, and in [3] and [10], the design of the permutations if sensibly less simple, compared to the L-R construction. Should we conclude that these new constructions really have better security properties than the L-R construction ? Should we therefore develop new, fast, and secure encryption schemes based on these new constructions ? Or is it only a "technical problem", and is the L-R construction in fact as secure as these constructions, but with more difficult proofs ? This question is not completely solved yet. However, we have seen in this paper that the security properties of the L-R construction with six (or more) rounds are in fact better than what was proved before about them.

Nevertheless, we have defined two new natural notions about the quality of strong pseudorandom permutations: the concept of "very strong pseudorandomness" and the concept of "homogeneous permutations". We have seen that no L-R construction gives homogeneous permutations. This result may be surprising, since it shows that – whatever the number of rounds of the L-R construction may be – there are still some "non-random places" in the resulting permutations (however, after a few rounds, the enemy is not able to choose the cleartexts or ciphertexts of his attack in order to be in one of these places: the scheme is pseudorandom).

We have finally given a few still open questions about Luby-Rackoff-like analysis of Feistel schemes.

References

1. W. Aiello, R. Venkatesan, *Foiling birthday attacks in length-doubling transformations*, EUROCRYPT'96, Springer-Verlag, pp. 307-320.
2. M. Luby, C. Rackoff, *How to construct pseudorandom permutations from pseudorandom functions*, SIAM Journal on Computing, vol. 17, n. 2, pp. 373-386, April 1988.
3. M. Naor, O. Reingold, *On the Construction of Pseudo-Random Permutations: Luby-Rackoff revisited*, Electronic Colloquium on Computational Complexity (ECCC), Report TR 97-005. Preliminary version in: Proc. 29th Ann. ACM Symp. on Theory of Computing, 1997, pp. 189-199. To appear in the Journal of Cryptology.
4. U. Maurer, *A simplified and generalized treatment of Luby-Rackoff pseudorandom permutation generators*, EUROCRYPT'92, Springer-Verlag, pp. 239-255.
5. U. Maurer, J.Massey, *Local randomness in pseudorandom sequences*, Journal of Cryptology, vol. 4, pp. 135-149, 1991.
6. J. Patarin, *Pseudorandom Permutations based on the DES Scheme*, EUROCODE'90, LNCS 514, Springer-Verlag, pp. 193-204.
7. J. Patarin, *New results on pseudorandom permutation generators based on the DES scheme*, CRYPTO'91, Springer-Verlag, pp. 301-312.
8. J. Patarin, *Étude des Générateurs de Permutations Pseudo-aléatoires basés sur le schéma du DES*, Ph.D. Thesis, Université Paris VI, November 1991.

9. J. Patarin, *Improved security bounds for pseudorandom permutations*, 4th ACM Conference on Computer and Communications Security, April 1-4, 1997, pp. 142-150.

10. J. Pieprzyk, *How to construct pseudorandom permutations from single pseudorandom functions*, EUROCRYPT'90, Springer-Verlag, pp. 140-150.

11. S. Vaudenay, *La Sécurité des Primitives Cryptographiques*, Ph.D. Thesis, École Normale Supérieure, April 1995, section II.8: "Les multipermutations".

Monkey: Black-Box Symmetric Ciphers Designed for MON*opolizing* KEY*s*

Adam Young[1] and Moti Yung[2]

[1] Dept. of Computer Science, Columbia University
ayoung@cs.columbia.edu.
[2] CertCo New York, NY, USA.
moti@certco.com, moti@cs.columbia.edu

Abstract. We consider the problem of designing a black-box symmetric cipher that leaks information *subliminally* and *exclusively* to the designer. We show how to construct a cipher which we call 'Monkey' that leaks one key bit per output block to the designer of the system (in any mode). This key bit is leaked only if a particular plaintext bit is known to the designer (known bit/message attack which is typically available in plain ASCII). The attack is of kleptographic nature as it gives a unique advantage to the designer while using strong (e.g., externally supplied) keys. The basic new difficulty with the design of spoofable block ciphers is that it is a *deterministic function* (previous attacks exploited randomness in key generation or message encryption/signature), and the fact that we do not want easy (statistical) observability of the spoofing (e.g., the variability of ciphertexts should be noticeable when keys change etc.).

We distinguish between three entities: the designer, the reverse-engineer and the user. We show a design methodology that assures that: (1) if the device is not reverse-engineered, the attack is secure (namely, the cipher is good) and undetectable, (2) if the device is reverse-engineered, then the reverse-engineer learns at most one plaintext bit from every ciphertext (but no past/future keys), and (3) the designer learns one plaintext bit and one key bit from each ciphertext block (say in ECB mode). The method is therefore highly robust against reverse-engineering.

Key words: design methodologies for symmetric ciphers, secret cryptographic algorithms, spoofing, kleptographic attacks, trust, software vs. tamper-proof hardware designs, tamper-proof reverse engineering, public scrutiny.

1 Introduction

The US government has proposed recently a classified secret block cipher called SkipJack as part of the Clipper Initiative. Furthermore, since the mid 80's the NSA's Commercial COMSEC Endorsement Program has been active trying to base cryptography for secure computer and communication based on tamper-proof devices (see [Sch], page 598). The motivation of this paper is to investigate

S. Vaudenay (Ed.): Fast Software Encryption – FSE'98, LNCS 1372, pp. 122–133, 1998.
© Springer-Verlag Berlin Heidelberg 1998

the possibilities of designing secret symmetric ciphers with sophisticated trapdoors that are hard to detect and are immune to reverse engineering, and at the same time maintain the basic properties of block ciphers. The issue is essentially methodological, as it points at potential non-trivial leakage attacks which are possible with black-box cipher designs as opposed to public designs. (Our goal is neither to undermine SkipJack, nor to claim any concrete attack – we merely point at what we believe is a threat of secret designs that is beyond giving trivial known advantages.)

We first note that it is easy to mount attacks on secret devices, e.g. by fixing their keys. Such trivial advantage of easily reverse-engineered secret cipher (reverse engineering has been shown to be a concrete possibility, recently and can be done by a company with a well-equipped micro-electronic laboratory). This risks the designer's unique advantage (of getting other past/ future keys). Thus, one may argue that such designs will not be put to use (e.g., by an agency which is concerned about losing out to the resourceful companies). On the other hand, with a unique advantage, even after reverse engineering, a designer of a secret algorithm will have less hesitation to put it in general use. The above simplistic attack is also easily detectable when encryptions under supposedly "different keys" turn out to be identical. Another attack is by adding an encryption of the key under a secret designer key to ciphertexts; this will be easily noticeable due to data expansion. It will not be possible to classify such design as a block cipher. Yet another attack is by designing secret devices using pseudorandomness known to the attacker. However, block ciphers are deterministic functions such that when given the same input with the same key, the same result is expected. Thus, one may employ pseudorandomness for encryption which is derived from the key and a message, but then the encryption depends strongly on the key which is unknown to the designer (attacker). Ignoring the key or using only a partial key is statistically detectable.

Moving ahead, we then noticed that when we resort to known message attacks, we may once in a while leak key bits, so such attacks can be more powerful in attacking. This leakage should not destroy the quality of the cipher (e.g., make it insecure w.r.t. differential or linear cryptanalytic attacks or other statistical attacks). We would like to go even further and have the cipher be immune to reverse engineering, which is a characteristic of "kleptographic attacks" on black box devices (e.g. tamper resistant hardware). This means, for example, that the pseudorandom function's permanent key is not exposed after reverse engineering. If it were exposed, then the secret keys used with other devices would be exposed. We hope that the above discussion reveals some of the reasons behind the methodology we propose herein.

There has been much recent work on designing cryptosystems to leak secret information securely and subliminally to the designers. The basic notions underlying these attacks as well as tools that accomplish them were developed in [YY96,YY97a,YY97b]. Specifically, they introduced the notion of a SETUP attack, where SETUP stands for a "Secretly Embedded Trapdoor with Universal Protection". In their attack it is a secretly embedded trapdoor (public key) that is used to securely leak the secret information out of the cryptosystem. Their attacks are geared specifically towards public key systems and exploit randomness and subliminal channels [S94] in key generation, message encryption, and signing. Here, we show how to perform these attacks on deterministic symmetric block ciphers that are secret. We will propose a design we call "Monkey" [1].

The 'Monkey' is a general design (a methodology). For concreteness and clarity of presentation we give an algorithm which uses an 80 bit key and has a block size of 64 bits.

The design has an aspect which is inherently more challenging than SETUP attacks. Namely, we want to allow a strong cipher and the use of the actual large key space (that is, to design a real block cipher)– and block ciphers in turn are deterministic algorithms (implementing a permutation which is length-preserving). The attacker cannot control the choice of (strong and random) keys used by the cipher (we want to allow external source keying). Further, the attacker cannot know when it has access to mount the attack (we cannot assume some partial control over the device operation). In all the early work on SETUPs, the fact that the public key algorithms (key generation or signing) were probabilistic in nature was exploited. Thus, the natural question is: how can we design kleptographic attacks in this fully deterministic setting? In the present paper, we show that if we are able to mount minimalistic known-plaintext attacks (known bit per message), then a close approximation to a SETUP attack (called Quasi-SETUP) can be performed on a symmetric cipher with secret specification (the notion of Quasi-SETUP may be of independent interest). The attack gives exclusive advantage to the designer as in a SETUP and has strong protection against reverse engineering (which is needed given the potential of reverse engineering of tamper resistant devices which has been demonstrated recently in some settings [AK96]).

2   Background and Definitions

Our attack bears a resemblance to setup attacks let us recall what is a setup and define our quasi-setup attack. The following is the definition of a regular setup [YY97a].

Definition 1. *Assume that C is a black-box cryptosystem with a publicly known specification. A (regular) SETUP mechanism is an algorithmic modification made to C to get C' such that:*

1. *The input of C' agrees with the public specifications of the input of C.*
2. *C' computes efficiently using the attacker's public encryption function E (and possibly other functions as well), contained within C'.*

---

[1] We call our method "Monkey", because: (1) it allows the attacker to [**Mon**]opolize [**key**]s, (2) it "monkeys around with the users secret key" in a way that a trusted cipher shouldn't, (3) jokingly, we can say that the design reassures that one has to be a monkey to still trust secret unscrutinized designs, and finally (4) the naming follows a "recent tradition" of calling cipher designs after beasts of various kinds!

3. The attacker's private decryption function D is not contained within C' and is known only by the attacker.
4. The output of C' agrees with the public specifications of the output of C. At the same time, it contains published bits (of the user's secret key) which are easily derivable by the attacker (the output can be generated during key-generation or during system operation like message sending).
5. Furthermore, the output of C and C' are polynomially indistinguishable to everyone except the attacker.
6. After the discovery of the specifics of the setup algorithm and after discovering its presence in the implementation (e.g., reverse-engineering of hardware tamper-proof device), users (except the attacker) cannot determine past (or future) keys.

The kleptographic attack presented in this paper is what we call a "quasi-setup". The following is a formal definition of a quasi-setup:

Definition 2. *A QUASI-SETUP is a black-box cryptosystem C (inaccessible to its regular user but may be attacked by reverse engineering). It has secret specifications and we require that:*

1. *The size of the input key, input plaintext, and output ciphertext is publicly known (and perhaps other extraneous and possibly erroneous information is given); key may be loaded by the user.*
2. *If C is a symmetric cipher, then, for efficiency, C does not use public-key operations in real-time (to avoid being recognized by time measurement). C pre-computes using the designer's secret public key E (and possibly other functions as well), contained within C.*
3. *The designer's private key D is not contained within C and is known only by the designer.*
4. *The output of C is secure based on security of an underlying block cipher and other tools. At the same time it contains published bits (of the user's secret key) which can be derived in poly-time by the attacker (designer), assuming the attacker has a sufficient amount of known-plaintext.*
5. *The fact that C leaks key bits is undetectable to everyone except the designer (attacker) and a successful reverse-engineer.*
6. *After the discovery of the specification of C (e.g., after reverse-engineering of the device), no one except the designer can determine past or future keys, though the successful reverse-engineer may be able to determine some fraction of plaintexts via known-plaintext attacks.*

2.1 Related Work

In [YY96] a setup attack was given on RSA key generation devices. It was shown how an RSA key generation device could be designed to output a public and private key pair, such that the upper order bits of the public key can be used by the designer to compute the corresponding private key. More generally, they proved that any cryptosystem that contains a subliminal channel contains a SETUP version. In [YY97a] it was shown how one of the exponents in a Diffie-Hellman key exchange can be securely and subliminally leaked to the designer over the course of two (wlog) consecutive key exchanges. This setup attack was the first setup attack that did not make use of explicit subliminal channels but rather generated channels for leakage due to repeated executions. In [YY97b], setup attacks were given for the ElGamal public key cryptosystem, the ElGamal Digital Signature algorithm, the Digital Signature Algorithm, Schnorr, and other systems. The attacks on the signature schemes leak the private signing key over the course of two (wlog) consecutive signatures. All the above attacks used randomness employed by the cryptosystem.

Rijmen and Preneel [RP97] gave a much more ambitious direction which is different from ours. They suggested the construction of the first example trapdoor ciphers, demonstrating that even an open design has to be justified or pseudo-randomly generated to avoid potential spoofings. (How secure is their design and if strong trapdoors exist at all, are still open). Indeed, the potential existence of trapdoor ciphers already points at a difficulty with secret design. We will show that with Monkey, the attack is assured, and also the attack can be based on minimal knowledge (known bit attack), and the length of a key recovery attack can be much shorter than the specific trapdoor ciphers based attack in [RP97]; (these are of course advantages of attacks on hidden designs which are easier in nature).

An approach to "generating trust" in a secret cipher design was attempted by producing a report of an inspection team in [BD93] in the context of SkipJack (more about that report, see [R94]). The subtleties presented here may potentially cast some extra doubts on a secret way to assure trust by a known team (since one does not know how much the team knows and what information was made available to it). However, we must mention that, by the same token, we cannot have any concrete complaint against the specific report above.

Finally, let us mention that in the context of the recent NIST initiative to design the next generation of block cipher standard (AES), the work here re-enforces the notion of public scrutiny of suggested standards (to avoid various trapdoors which are indeed possible).

3 Monkey Implementation

The Monkey secret symmetric cipher takes an 80 bit symmetric key as input, in addition to 64 bits of plaintext. It outputs a 64 bit ciphertext block. Monkey uses for pre-computations the SMALL1 public key cipher to be described. We assume that Monkey is a secret cipher. This is of course nonsensical, since we are publishing it now. What we mean is that our suggestion is an instance of a methodology of design and we can assume that a cipher like it (a variant) may be kept secret and implemented. We assume that it is tamper-resistant so that it is hard to get. Methodologically what is important is this last fact (of being hard to get, i.e. black-box to the user) and not the exact physical assumption about tamper resistance (we are fully aware of recently discovered weaknesses of certain such claimed designs while we are aware of other designs which were not reversed-engineered so far).

3.1   Some tools

In [SOMS] a fast key exchange algorithm was given that uses elliptic curves. In their scheme the curve E is a set of points (x, y) with x and y lying in the field $F_{2^{155}}$. They implement Diffie-Hellman over this curve using a publicly defined point P on E. This scheme appears to be secure as long as the Index Calculus method cannot be extended to elliptic curves.

It is a trivial matter to define a public key encryption algorithm based on Diffie-Hellman over E. Suppose Alice wishes to encrypt the message m where m is 80 bits in size. Alice wishes to send this message to Bob, whose private key is x, where x is in the range [2,order(E)-2]. Bob's corresponding public key is the point $y = xP$. To send the encryption of m, Alice chooses a random integer r in the range [2,order(E)-2] and computes kP by iterating the addition of P using the "double and add" scheme. Alice then computes $z = ry$. Alice can then use some or all of the shared secret string $H(z)$ to encipher the value m to get the value c. Here H is a suitable hash function. Note that c need only be as large as m. Alice then sends (rP, c) to Bob. Decryption is straightforward. Note that Alice sends Bob 310 bits corresponding to rP plus 80 bits that constitute c. Hence, Alice sends Bob 390 bits of information. Note that the ciphertext size is smaller than what is possible using RSA with a 512 bit modulus. In this paper, we define this to be the $SMALL1$ public key cryptosystem. Hence, $SMALL1$ uses a random parameter r that is as least 2 and at most order(E)-2, and takes a public key y as input. $SMALL1$ takes 80 bits of input data m_a and produces a 390 bit ciphertext c_a. The operation of $SMALL1$ is denoted by $c_a = SMALL1(r, y, m_a)$. Let $SMALL1'$ denote decryption. Hence, $m_a = SMALL1'(x, c_a)$.

3.2   System Setup

The designer chooses a keyed pseudorandom function F_{63} that takes as input a large seed (key) s and a 63 bit input x, and produces a 63 bit output y. For an explanation on how to construct pseudorandom functions, see [GGM86]. The operation of F_{63} is denoted by $y = F_{63}(s, x)$. The designer also chooses two seeds, s_1 and s_2 uniformly at random (from the seeds of the given length). The designer chooses a private key x randomly for use in SMALL1, and computes the corresponding public key y. The designer puts (F_{63}, s_1, s_2, y) in the black-box device and keeps x private.

3.3   System Operation: Encryption

Let K denote the 80 bit key of the user. The user wishes to use the black-box device to encrypt the 64 bit plaintext message m, to get the corresponding 64 bit ciphertext c. The operation of Monkey is denoted by $c = $ Monkey(K, m). The device contains a secret symmetric block cipher called $CIPHER1$. $CIPHER1$ takes a 63 bit symmetric key k_b and a 63 bit plaintext m_b. $CIPHER1$ produces a 63 bit ciphertext c_b. We denote the operation of $CIPHER1$ by $c_b = CIPHER1(k_b, m_b)$. The corresponding decryption operation is $CIPHER1'$. Hence, $m_b = CIPHER1'(k_b, c_b)$. Let H denote a cryptographically secure hash function (e.g., a collision intractable function, or a pseudo-random function) that maps $\{0,1\}^{80}$ to $\{0,1\}^{63}$. G is a random function that maps $\{0,1\}^{80}$ to a value in the interval $[2, \text{order}(E)\text{-}2]$. The following is the computation of Monkey(K,m) to get c:

1. m_b is set to the lower order 63 bits of m
2. $k_b = H(K)$
3. $c_b = CIPHER1(k_b, m_b)$
4. the lower order 63 bits of c is set to c_b.
5. $r = G(K)$
6. $c_a = SMALL1(r, y, K)$
7. $i = F_{63}(s_1, c_b) \bmod 390$
8. b is set to the ith bit of c_a.
9. $z = F_{63}(s_2, c_b) \bmod 2$
10. p is set to the most significant bit of m
11. the most significant bit of c is set to be b XOR z XOR p

3.4   System Operation: Decryption

The following is the operation of Monkey'(K,c) which returns m:

1. c_b is set to the lower order 63 bits of c
2. $k_b = H(K)$
3. $m_b = CIPHER1'(k_b, c_b)$
4. the lower order 63 bits of m is set to m_b
5. $r = G(K)$
6. $c_a = SMALL1(r, y, K)$
7. $i = F_{63}(s_1, c_b) \bmod 390$
8. b is set to the ith bit of c_a.
9. $z = F_{63}(s_2, c_b) \bmod 2$
10. p' is set to the most significant bit of c
11. the most significant bit of m is set to be b XOR z XOR p'

3.5   Secret Key Recovery

Suppose that we manage to obtain 390 ciphertexts $c_0, c_1, ..., c_{389}$ such that we know the least significant bits of the 390 corresponding plaintexts. Suppose further that $F_{63}(s_1, c_j) \bmod 390$ for $0 \leq j \leq 389$ is a permutation on $0,1,...,389$. The following algorithm computes the ciphertext bit in c_a corresponding to c_j.

1. c_b is set to the lower order 63 bits of c_j

2. $i = F_{63}(s_1, c_b) \bmod 390$
3. $z = F_{63}(s_2, c_b) \bmod 2$
4. p is set to the most significant bit of m_j
5. p' is set to the most significant bit of c_j
6. outputs p XOR z XOR p'

The above algorithm is applied to $c_0, c_1,...,c_{389}$, to recover the ciphertext c_a. We then decrypt c_a using x to recover K. We choose the most significant bit since if the plaintext is ASCII for instance, this bit is known to be zero. In fact, since the block size is 64 bits, if the plaintext is ASCII we can leak using a bandwidth of 8 bits. Another possibility is to compress the plaintext (when possible) and then add high order bits which are known, decryption will decompress.

3.6   Needed Strengthening

The above cipher is a concatenation of two secure ciphers. The problem is the separability of the ciphers, so that messages that differ only at the last bit has a ciphertext which also differs on that bit. (This separability in general may help in attacks like chosen message attacks. This is the case here as messages come in pairs and when they differ on the last bit the ciphertext of one implies that of the other in the pair). This can be overcome by a cascading of ciphers. Namely by a post-processing (pre-processing in decryption) by CIPHER2 which employs (say, at least four) Feistel transformations based on Luby-Rackoff's construction [LR88] with a fixed secret pseudorandom function. This spreads the local difference in the last bit only, uniformly over the resulting ciphertext.

Another aspect that we did cover carefully are the sizes of the various keys (of the pseudorandom functions). If we need larger keys in total we may derive these keys pseudorandomly (using a secret fixed internal seed) from the given key.

4   Security of the Black-Box Monkey Implementation

There are two perspectives with which to analyze the security. Those perspectives are the black-box perspective, and the perspective of an attacker who is able to reverse-engineer the device (hence, no black-box assumption). In this section we consider both of these in turn. We assume that the ciphers and pseudorandom functions used CIPHER1, SMALL1, G, H, and F_{63} are secure.

Now we assume that a given user is unable to reverse-engineer the black-box device. It follows that C is a black-box cryptosystem with private specification from the user's perspective.

We note that a specific "security of a block cipher" is not defined here (of course, such generic definition does not exist!). Our general methodology therefore attempts to preserve within the overall Monkey design, whatever "security notion" CIPHER1 has, based on the strong security properties of the other building blocks (pseudorandom functions and permutations). One may argue that what we assume are heavy cryptographic tools; but this should not be a problem at this stage of the development of the methodology. Indeed, we leave open the issue of minimal assumptions needed, as well as the efficiency of design of ciphers with Quasi-SETUP. What we claim is:

Claim 1 *Assuming that CIPHER1, CIPHER2, SMALL1, G, H, and F_{63} are secure, then if Monkey is a black-box cryptosystem with private specification, then C constitutes a secure symmetric cipher.*

Proof. c_b constitutes a secure encryption of m_b since CIPHER1 is a secure symmetric cipher. It remains to show that the least significant bit of c constitutes a secure encryption of the least significant bit of m. Since F_{63} is a secret pseudorandom function, and since s_2 is unknown, it follows that $F_{63}(s_2, c_b)$ is random and unknown to the user. Thus $z = F_{63}(s_2, c_b) \bmod 2$ is a random secret bit with respect to the user. Since this bit is exclusive-or'ed (one-time padded) with the least significant bit of m, the least significant bit of c constitutes a secure encryption.

Now, the two secure values give two separable encryptions and can be viewed as a secure block cipher on 63 bits, concatenated to a one-bit strong stream cipher encryption. The further Feistel like transformations in CIPHER2 based on a pseudorandom function which strengthen the design, assure a strong inseparable encryption (due to the "avalanche" properties of pseudorandom functions) and prevents easy chosen message attacks. The overall cipher can be viewed as two ciphers cascaded. The cascade is as secure as the first cipher (or as each of the ciphers for weaker attacks) [MM93,EG85]. QED.

Recall that if the pseudorandom function (its key) is not known then the value of the function at a point cannot be approximated even in a very liberal sense even if the values of the function at polynomially many other points is also given [GGM86]. It is this property of pseudorandom functions that makes this attack secure; practical designs of pseudorandom functions can be based on iterated strong block ciphers with large keys. Only a minimal amount of security at best is sacrificed by using the 63 bit block cipher CIPHER1, as opposed to a true 64 bit block cipher (e.g., it is very easy to modify DES to work on 63 bits). The additional bit encryption of the last bit is a strong cipher as well. Note also that cipher designs that are tunable to each size in small granularity (e.g. as in the NIST's AES specifications) have this disadvantage of being a potential CIPHER1 component in a Monkey design – where the difference (one bit in our case) is the amount of required known bits per message.

Now suppose that an attacker manages to reverse-engineer the device (thus, CIPHER2 can be ignored hereafter). The attacker therefore knows (F_{63}, s_1, s_2, y) and the complete specification of the cipher Monkey. In this case, the reverse-engineer is able to recover at most the least significant bit corresponding to the ciphertexts that are output by the device, as long as a sufficient number of known-plaintext bits are gathered. To see this, note that the reverse-engineer can recover the bit z from the secret key recovery algorithm in the same way as the designer. Since the reverse-engineer also knows p' and presumably the least significant bit of m, he or she can exclusive-or these three bits together to recover one of the bits of c_a. If the reverse-engineer gets a sufficient number of known plaintexts, the reverse-engineer can reconstruct c_a. When equipped with the value c_a for a given key K (not available in the device at time of reverse engineering i.e. a past or future key), the reverse-engineer can decrypt the least significant plaintext bits of all values encrypted with K. Can the reverse-engineer also recover the bits of K, like the designer? The answer to this is no, as we will claim next.

Claim 2 *Assuming that CIPHER1, SMALL1, G, H, and F_{63} are secure, and that the adversary has a polynomial number of plaintext/ciphertext pairs, the successful reverse-engineer is unable to learn anything about K.*

Proof. Consider the ciphertext values that result from a particular K. Since we assume that CIPHER1 is secure, the reverse-engineer is unable to learn anything about k_b and hence K, from the 63 upper order bits of the ciphertexts alone. Note that the application of the pseudorandom function to c_b to derive the least significant ciphertext bits has the effect of the application of a random oracle to c_b to get the least significant ciphertext bits. Thus, anything that can be deduced about K from all of the bits of the ciphertexts can be deduced from the least significant bits alone. So, it remains to consider what can be deduced from the least significant bits of the ciphertexts alone. Since m, c, and (F_{63}, s_1, s_2) are already known to the reverse-engineer, the reverse-engineer knows c_a. It remains to show that nothing about K can be learned from c_a. Since we assumed that G is a secure random function, $G(K)$ is a random string to the reverse-engineer. So, since SMALL1 is a secure public key cryptosystem, SMALL1($G(K)$,y,K) $=$ c_a is a secure public key encryption of K. QED.

Claim 2 hinges on the fact that F_{63} and G are random functions, and that SMALL1 is a secure public key encryption function. At the same time Claim 1 argues that in the event that the device is never reverse-engineered, and in the event that the designer never abuses his or her power, Monkey *is* a secure symmetric cipher. Thus, in summary, the capability of users with respect to Monkey can be broken down into three different categories:

1. Users who are unable to reverse-engineer the device are unable to learn any plaintext.
2. Users who are able to reverse-engineer the device, when given enough known-plaintext, are able to learn one plaintext bit of every ciphertext.
3. The designer, when given enough known-plaintext, is able to learn all plaintext bits of every ciphertext (since it monopolizes the keys in use).

Given the above two claims, recall also that Monkey can be loaded with external keys. Monkey therefore constitutes a quasi-setup.

Note that this paper provides motivation for having new public key cryptosystems that output very small ciphertexts (such schemes, with yet hard to analyze security, based on polynomial manipulation have been design, e.g. in [P96]). If a public key cryptosystem exists that outputs ciphertexts that are, say, 200 bits in size, then far fewer known-plaintexts need to be gathered to leak the secret key K securely. (The size of the public key block is related to the number of known messages required).

5   Conclusion

We introduced the notion of a quasi-setup and demonstrated a symmetric cipher (Monkey) which constitutes a quasi-setup. We showed how to design a secret cipher that gives an unfair advantage to the designer, and that is very robust against reverse-engineering. Our results imply that secret symmetric ciphers implemented in black-box settings should only be used if they come from trusted sources and cannot be simply trusted based on extensive statistical testing. It strengthens the need for open cipher design efforts. We did not attempt to hide which "known bit" is required for the attack. It may be the case that in more convoluted ciphers where this needed-bit is not specified, the combination of internal stream cipher operations, pseudorandom operations like S-boxes and Feistel transformations and exponentiation operations, can even hide which bits are needed to be known (and may evade an inspecting team lacking the original design documents).

Finally, efficient and minimal designs and designs that maintain specific properties of block ciphers while enabling a quasi-SETUP attacks are left as open questions.

References

AK96.   R. Anderson and M. Kuhn. Tamper Resistance – a Cautionary Note. In *The 2-d Usenix Workshop on Electronic Commerce*. 1996, pages 1–11.

BD93.   E. F. Brickell, D. E. Denning, S. T. Kent, D. P. Maher, and W. Tuchman SkipJack review: the SkipJack Algorithm, Interim Report, July 28, 1993.

EG85.   S. Even and O. Goldreich, On the Power of Cascade Ciphers. *ACM Tra. on Comp. Systems*, V. 3, 1085, 108 116.

GGM86. O. Goldreich, S. Goldwasser, and S. Micali, How to Construct Random Functions. *J. of the ACM*, 33(4), pp 210-217, 1986.

LR88.   M. Luby and C. Rackoff, How to Construct Pseudorandom Permutations from Pseudorandom Functions. In *SIAM J. on Computing*, V. 17, 1988, pages 373–386.

MM93.   U. Maurer and J. Massey, Cascade ciphers: the importance of being first. In *J. of Cryptology*, V. 6, 1993, pages 55–61.

P96.   J. Patarin, Hidden Field Equations (HFE) and Isomorphisms of Polynomials (IP): Two New Families of Asymmetric Algorithms, In *Advances in Cryptology—Eurocrypt '96*, pages 33–48, 1996. Springer-Verlag.

RP97.   V. Rijmen and B. Preneel, A Family of Trapdoor Ciphers, Fast Software Encryption 97, (Ed. E. Biham).

R94.   M. Roe, How to Reverse Engineer an EES Device, Fast Software Encryption 94, Dec. 94, (Ed. B. Preneel), LNCS 1008, Springer-Verlag.

Sch.   B. Schneier. Applied Cryptography, 1994. John Wiley and Sons, Inc.

SOMS.   R. Schroeppel, H. Orman, S. O'Malley, O. Spatscheck. Fast Key Exchange with Elliptic Curve Systems. In *Advances in Cryptology—CRYPTO '95*, pages 43–56, 1995. Springer-Verlag.

S94.   G. J. Simmons. Subliminal Channels: past and present. *European Tra. on Telecommunications* V. 5, 1994, pages 459–473.

YY96.   A. Young, M. Yung. The Dark Side of Black-Box Cryptography. In *Advances in Cryptology—CRYPTO '96*, pages 89–103, Springer-Verlag.

YY97a.   A. Young, M. Yung. Kleptography: Using Cryptography against Cryptography. In *Advances in Cryptology—Eurocrypt '97*, pages 62–74. Springer-Verlag.

YY97b.   A. Young, M. Yung. The Prevalence of Kleptographic Attacks on Discrete-Log Based Cryptosystems. In *Advances in Cryptology—CRYPTO '97*, Springer-Verlag.

MRD Hashing

Rei Safavi-Naini [*], Shahram Bakhtiari, and Chris Charnes

School of IT and CS, University of Wollongong, Northfields Ave
Wollongong 2522, Australia
[rei, shahram, charnes]@uow.edu.au

Abstract. We propose two new classes of hash functions inspired by Maximum Rank Distance (MRD) codes. We analyze the security and describe efficient hardware and software implementations of these schemes. In general, the system setup remains computationally expensive. However, for a special class of parameters we show that this computation can be completely avoided.

1 Introduction

A *Message Authentication Code (MAC)* is a symmetric key cryptographic primitive that ensures message integrity against active spoofing. A MAC consists of two algorithms. A *MAC generation* algorithm, G_k, that takes an arbitrary message s from a set $\mathcal{S}$ of messages and produces a *tag*, $t = G_k(s)$. The tag is appended to the message to produce an *authenticated* message, $m = s\|t$. A *MAC verification* algorithm takes an authenticated message of the form $s\|t$, and produces a *true/false* value depending on whether the message is authentic or fraudulent. The secret key k is only known to the legitimate communicants and hence a valid tag can only be computed by them. An outsider tries to generate a fraudulent message that is acceptable by the receiver.

Computationally secure MACs are usually constructed from hash functions by using a secret key during the hashing process. Secure MACs are constructed by examining the computational complexity of various attacks [10,11], and by choosing the system parameters so that the complexity of the best attack exceeds the computational resources of the enemy. Constructions of these type are always subject to revision as new attacks become available.

In provably secure MACs an intruder has a provably small chance to tamper with the message, and no limit on the computational resources of the enemy is assumed. Wegman and Carter [16] investigated unconditionally secure MACs and gave a construction for a key-efficient MAC with provable security. This construction uses an ϵ-almost strongly universal$_2$ (ϵ-ASU$_2$) class of hash functions. By Stinson's [14] composition method a new ϵ-ASU$_2$ can be created starting with a given ϵ-ASU$_2$ class and an ϵ-AU$_2$ class. Hence the question of constructing secure MACs reduces to the problem of constructing computationally efficient ϵ-AU$_2$ functions.

[*] Support for this project was partly provided by Australian Research Council.

S. Vaudenay (Ed.): Fast Software Encryption – FSE'98, LNCS 1372, pp. 134–149, 1998.
© Springer-Verlag Berlin Heidelberg 1998

Wegman and Carter's construction was refined by Krawczyk [6], who showed that a secure MAC only requires an ϵ-almost XOR universal$_2$ (ϵ-AXU$_2$) hash function.

In this paper we make two contributions. Firstly, we introduce two new classes (ϵ-AU$_2$ and ϵ-AXU$_2$) of hash functions, which are inspired by MRD codes and demonstrate their efficiency it terms of key size and the ease and speed of hashing. These schemes are examples of Shoup's [13] evaluation hashing, where polynomial evaluation over $GF(2^n)$ is replaced by matrix multiplication over $GF(2)$, resulting in a fast software implementation. These schemes have desirable properties such as small key size and flexibility in the block size of the hashed messages. However the system setup is in general a computationally expensive. Next, we completely describe the 2-polynomials in Galois fields of prime degree in which 2 is primitive. This result allows us to avoid the setup phase computation.

Section 2 has the preliminaries. In section 3 we describe the two classes of hash functions. Section 4 presents a MAC based on MRD hashing and compares it with one based on bucket hashing, and summarises the results on the 2-polynomials. Section 5 concludes the paper.

2 Preliminaries

Unconditionally secure MAC systems are especially important because they give efficient authentication systems that have provable security properties. With this approach, the security of a MAC system is assessed by the best chance that active spoofer has in constructing a fraudulent message-tag pair, $s\|t$, which is acceptable to the receiver, after observing a sequence of p message-tag pairs, $s_1\|t_1, \cdots s_p\|t_p$ sent over the communication channel.

Definition 1. *[16] A MAC for which the best chance of the enemy in the above described attack is at most ϵ, is called ϵ-secure.*

Wegman and Carter proposed the construction of a ϵ-secure MAC using ϵ-ASU$_2$ class of hash functions. Let Σ^m and Σ^n denote the set of binary strings of length m and n, respectively. A *hash function* h is a mapping from Σ^m to Σ^n. Let $\mathcal{H}$ denote a class of hash functions, $\mathcal{H} = \{h : \Sigma^m \to \Sigma^n\}$.

Definition 2. *[14] A class of hash functions, $\mathcal{H}$, is ϵ-AU$_2$ if for all $x \neq x' \in \Sigma^n$, we have $|\{h \in \mathcal{H} : h(x) = h(x')\}| \leq \epsilon|\mathcal{H}|$.*

Definition 3. *[14] A class of hash functions, $\mathcal{H}$, is ϵ-ASU$_2$ if the following two conditions are satisfied:*

- *for every $x \in \Sigma^m$ and $y \in \Sigma^n$, $|\{h \in \mathcal{H} : h(x) = y\}| = |\mathcal{H}| \times 2^{-n}$;*
- *for every pair $x, x' \in \Sigma^m$ and $x \neq x'$, and every pair $y, y' \in \Sigma^n$ we have $|\{h \in \mathcal{H} : h(x) = y, h(x') = y'\}| \leq \epsilon|\mathcal{H}| \times 2^{-n}$.*

These definitions can be expressed in terms of probabilities. For example, for an ϵ-AU$_2$ class hash function, $\Pr_h[h(x) = h(x')] \leq \epsilon$ for all $x, x' \in \Sigma^m$, $x \neq x'$.

Wegman and Carter's construction, which is the basis of our system is the following. Let $\mathcal{H}$ be an ϵ-ASU$_2$ class of hash function from Σ^m to Σ^n. The transmitter and the receiver share a secret key that consists of two parts. The first part identifies an element $h \in \mathcal{H}$, and the second part is a randomly generated one-time pad $r = r_1, r_2 \cdots$ of n-bit numbers. The transmitter and receiver maintain a counter, *count*, which is initialised *count* := 1 and is incremented by one after each message. The tag value of the ℓth message, x_ℓ, is $h(x_\ell) + r_\ell$. The receiver can reconstruct this tag and verify the authenticity of the received message. Wegman and Carter proved that this construction is ϵ-secure and the key size is asymptotically optimal. Replacing the one-time pad with a pseudorandom sequence generator in the MAC, reduces unconditional security to computational security.

Definition 4. *[6] A class of hash functions is called ϵ-otp-secure if it is ϵ-secure in the one-time pad construction of Wegman and Carter.*

Krawczyk showed in [6] that provable security in the above sense does not really require the ϵ-ASU$_2$ property:

Definition 5. *[6] A class of hash functions is called ϵ-almost XOR universal$_2$, or ϵ-AXU$_2$, if for all $x \neq x' \in \Sigma^n$ and any $c \in \Sigma^m$, we have $\Pr_h[h(x) \oplus h(x') = c] \leq \epsilon$.*

Theorem 6. *[6] A class of hash functions is ϵ-otp-secure if and only if it is ϵ-AXU$_2$.*

Stinson proved that composition of hash functions can be effectively used to replace the construction of ϵ-ASU$_2$ by ϵ-AU$_2$ hash functions.

Proposition 7. *[14] Let $\mathcal{H}_1 = \{h : \Sigma^m \to \Sigma^n\}$ be ϵ_1-AU$_2$ and $\mathcal{H}_2 = \{h : \Sigma^n \to \Sigma^k\}$ be ϵ_2-ASU$_2$. Then $\mathcal{H}_1 \circ \mathcal{H}_2 = \{h : \Sigma^m \to \Sigma^k\}$ is $(\epsilon_1 + \epsilon_2)$-ASU$_2$.*

A similar result holds for composition of ϵ_1-AU$_2$ and ϵ_2-AXU$_2$ classes of hash functions [12].

The advantage of the composition method is that to achieve computationally efficient hashing, it suffices to construct an efficient ϵ-AU$_2$. This result has shifted the emphasis of research in the recent years to the construction of computationally efficient ϵ-AU$_2$ functions. Johanson [4], Taylor [15], Krawczyk [6,7], Rogaway [12], and Shoup [13] have investigated computationally efficient MACs that have relatively small key size. The most efficient construction is bucket hashing [12].

3   MRD-MAC

In this section we firstly introduce a class of functions $\mathcal{P}_{MRD}(t)$ from F_n to F_n which is inspired by MRD codes, and describe its properties. We describe two classes of hash functions: $\mathcal{H}^1_{MRD}(t)$ and $\mathcal{H}^2_{MRD}(t)$ which are based on $\mathcal{P}_{MRD}(t)$. Finally, we make some remarks on the implementation of the two classes.

In the rest of this paper we use the correspondence between binary strings $s = s_0 s_1 \cdots s_{n-1}$ and elements of $GF(2^n)$, represented as binary n-tuples $(s_0 s_1 \cdots s_{n-1})$.

3.1 $\mathcal{H}^1_{MRD}(t)$ and $\mathcal{H}^2_{MRD}(t)$

Maximum Rank Distance (MRD) codes were studied in [3]. They were used by Chen [2] for the purpose of identification, and Johansson [4] for arbitrated authentication. Although our proposed MAC system is inspired by MRD codes, we will give an independent presentation of these because we do not use directly any results from the theory of these codes. Throughout the section we assume a basic knowledge of the theory of finite fields. Some of the important definitions are included in Appendix 5. We refer the reader to [8] for an excellent introduction to this topic.

Let $L(x)$ be a linearized polynomial (more precisely a 2-polynomial; see [8]) $\beta = (\beta_1, \beta_2, \ldots, \beta_n) = (\beta, \beta^{2^1} \cdots \beta^{2^{n-1}})$, denotes a normal basis for $F_n = GF(2^n)$. The n-tuple $(L(\beta_1), L(\beta_2), \ldots L(\beta_n))$, defines an $n \times n$ binary matrix C_L, where the i-th column is the binary representation of $\beta^{2^{i-1}}$. So a linearized polynomial $L(x)$ defines a mapping $f_L : F_n \to F_n$, given by $f_L(u) = C_L.u = v$, where '.' denotes matrix multiplication and $u, v \in F_n$ are $n \times 1$ binary vectors.

Proposition 8. *Let* $u = (u_0 u_1 \cdots u_{n-1}) \in F_n$. *Then* $f_L(u)$ *can be evaluated by finding the value of* $L(x)$ *at* $u.\beta$ *where* u *and* β *are* n-*dimensional row and column vectors, respectively.*

The set of all mappings f_L defined above when degree of $L(x)$ is at most 2^t is denoted by $\mathcal{P}_{MRD}(t)$. The important properties of $\mathcal{P}_{MRD}(t)$ are stated in Lemma 9.

Define an array C whose rows are labelled by f_L, columns are labelled by the elements of F_n, and the entry in row f_L and column α is $f_L(\alpha)$. I.e., $C(f_L, \alpha) = f_L(\alpha)$, $\alpha \in F_n$. Let V_n denote the n-dimensional vector space over $GF(2)$. For a mapping $f : V_n \to V_n$, the *null space* $\mathcal{N}_f$, is the collection of vectors $v \in V_n$ such that $f(v) = 0$. $\mathcal{N}_f$ is a subspace of V_n in the case of linear mappings. Let 2^{d_α} denote the degree of the minimum linearized polynomial of α.

Lemma 9. *Array* C *has the following properties.*

1. *The column labelled by* 0 *contains only* $0 \in F_n$. *The column labelled by the all one vector contains only* 0, *or* $1 \in F_n$.
2. *If* $\alpha \in F_n$ *occurs in a row of* C *labelled by* f_L, *then it occurs* $|\mathcal{N}_L|$ *times, where* $\mathcal{N}_L$ *is the null space of* f_L.
3. *If* $L(x)$ *has even number of terms then* $|\mathcal{N}_L| \geq 2$. *In this case* $f_L(x) = f_L(\overline{x})$ *where* $\overline{x}$ *is the binary complement of* x.
4. *The number of zeros in a column of* C *labelled by* $\alpha \in F_n$ *is* $M_\alpha = 2^{t-d_\alpha}$. *Moreover if an element* β *occurs in this column, then it occurs* M_α *times.*
5. *For any two elements* $x, y \in F_n$, $x \neq y$, *the number of rows with* $f_L(x) = f_L(y)$ *is equal to* M_{x+y}.

The proof of this lemma is sketched in Appendix 5.

In Lemma 9, property 3 shows that two values that are complements of each other are mapped to the same value. This is an undesirable property for hashing. It can be removed by restricting $L(x)$ to polynomials with an odd number of terms, or alternatively, by restricting input to a subset F_n' of elements of F_n consisting of elements of the form $(\alpha_1, \alpha_2, \cdots \alpha_{n-1}, 0)$.

Let $d_{min} = \min_{\alpha \in F_n} d_\alpha$. We can assume that hash values of complements are distinct in view of the two proposed methods. In $\mathcal{P}_{MRD}(t)$, if we only consider linearized polynomials of degree less than $2^{d_{min}}$, then the following properties hold; they are direct consequences of Lemma 9.

Corollary 10. *Let $t < d_{min}$. Then array C satisfies the following properties.*

1. *In every row of C an element of F_n occurs at most once.*
2. *$\forall x \in F_n$ and $x \neq 0, 1$, $|\{f_L : f_L(x) = 0\}| = |\{f_L : f_L(x) = 1\}| = 0$. That is, a column labelled by $x \in F_n$ and $x \neq 0, 1$ does not contain 0 or $1 \in F_n$.*
3. *$\forall x, y \in F_n$ and $y \neq 0, 1 \in F_n$, $|\{f_L : f_L(x) = y\}| \leq 1$. That is, in a column of C, an element $x \in F_n$ and $x \neq 0, 1$ occurs at most once.*
4. *$\forall x, y, z \in F_n$ and $z \neq 0, 1 \in F_n$, $|\{f_L : f_L(x) \oplus f_L(y) = z\}| \leq 1$.*

Property 3 shows that every element of F_n occurs at most once in a column. Equivalently, two mappings give the same value when evaluated on the same element of the field. Property 4 suggests that $\mathcal{P}_{MRD}(t)$ is an ϵ-AXU$_2$ class of function.

However elements of $\mathcal{P}_{MRD}(t)$ are mappings from F_n to F_n and cannot compress the input, which is a basic requirement in hashing.

In the following we define two classes of hash functions based on the class $\mathcal{P}_{MRD}(t)$. We assume that the complementary property is removed by restricting the domain of elements of $\mathcal{P}_{MRD}(t)$ to F_n'.

Definition 11. *[6] $\mathcal{H} = \{h : \Sigma^m \to \Sigma^n\}$ is $\oplus$-linear if for all $h \in \mathcal{H}$ and all $x, x' \in \Sigma^m$, we have $h(x \oplus x') = h(x) \oplus h(x')$.*

$\mathcal{H}^1_{MRD}(t)$:

For a mapping $f_L \in \mathcal{P}_{MRD}(t)$, define a mapping $h_L^{(1)} : F_n' \times F_n' \to F_n$ as $h_L^{(1)}(x) = f_L(x_1) \oplus x_2$, where $x_1, x_2 \in F_n'$ and $x \in F_n' \times F_n'$. That is, the value of $h_L^{(1)}$ for a $2n$-tuple $x = x_1 \| x_2$ is obtained by applying f_L to x_1 and XOR-ing the result with x_2. We note that effectively x contains $2n - 2$ information bits

Theorem 12. *$\mathcal{H}^1_{MRD}(t)$ is a $\oplus$-linear ϵ-AU$_2$ class of hash function with $\epsilon = \frac{1}{2^{t+1}}$. (Proof in Appendix 5.)*

$\mathcal{H}^2_{MRD}(t)$:

For a mapping $f_L \in \mathcal{P}_{MRD}(t)$ and a binary string $s = s_0 s_1 \cdots$ of length $2^n - 2$ define the hash value $h_L^{(2)}(s) = \sum_{j=0}^{2^n-1} s_j f_L(\alpha^j)$ where α is a primitive element of F_n.

In other words $h_L^{(2)}(s)$ is a weighted sum of the values of $f_L(x)$ at all nonzero $x \in F_n$.

Theorem 13. $\mathcal{H}^2_{MRD}(t)$ *is a $\oplus$-linear ϵ-AXU$_2$ class of hash function with $\epsilon = \frac{1}{2^{t+1}}$.* (Proof in Appendix 5.)

3.2   Practical Considerations

$\mathcal{H}^1_{\mathbf{MRD}}(\mathbf{t})$

The compression ratio of this ϵ-AU$_2$ hash function is 2:1. To obtain higher compression ratios, the composition method of Proposition 7 can be used. We give the details in section 4.

Evaluating the hash values: is achieved by finding the product of a binary matrix C_L and a binary input vector. This can be efficiently implementated in hardware and software.

The key: is a randomly chosen binary vector of length d_{min}. Once a key $k_0 k_1 \cdots$ is chosen, a linearized polynomial $L(x) = \sum_{i=0}^{d_{min}-1} k_i x^{2^i}$ is determined and the matrix C_L is calculated. This is a fast computation.

Set-up phase: The main cost of the system is during the set-up phase in order to calculate d_{min} for a given n. The results of our experiment for $n < 19$ are given in Table 1. We have only considered prime values for n although theoretically this restriction is not necessary.

n	d_{min}	ϵ
5	4	2^{-4}
7	3	2^{-3}
11	10	2^{-10}
13	12	2^{-12}
17	8	2^{-8}

Table 1. *Parameters for $\mathcal{H}^1_{MRD}(t)$ for $n < 19$. d_{min} is the smallest degree of minimal linearized polynomials of all field elements, which is the same as the the number of key bits, and ϵ is the security parameter of the ϵ-AXU$_2$ class.*

It can be seen that some values like $n = 11$ and 13 give the maximum value for d_{min} (that is, $d_{min} = n - 1$), but others like $n = 17$ give small t. This table was produced by calculating the degree of the minimum linearized polynomial of the representative field elements of the conjugate groups. For higher values of n this approach becomes increasingly inefficient. But since it needs to be done only once during the life time of the system, the overhead of the required computation is acceptable for $n < 40$. For higher values of n, we require more efficient algorithms for finding d_{min}. In section 4.3 we present results which allows us to avoid this computation in extensions $GF(2^p)$, where p is a prime and 2 is a primitive mod p.

For practical applications a range of suitable values for n must be calculated and published. The user can then choose the n that gives the required level of security and is most suitable for the message sizes considered.

$\mathcal{H}^2_{\mathbf{MRD}}(\mathbf{t})$

This is an ϵ-AXU_2 with a compression ratio of $2^n - 1$ to n. To avoid the generation of fraudulent messages by appending extra zeros at the end of a message, we assume that s_{2^n-2} is always 1.

Implementation: can be efficiently done in hardware and software. Evaluating $h_L^{(2)}(s)$ could be similar to LFSR hashing used in [6]. Figure 1 gives a schematic diagram for this evaluation. The LFSR has fixed and public taps and produces a maximum length sequence (feedback polynomial is primitive). As the LFSR goes through consecutive states $\sigma_1, \sigma_2, \cdots$, the circuit c_L which implements the linear transformation defined by C_L produces a sequence of elements of F_n corresponding to the values of $f_L(\sigma_1), f_L(\sigma_2), \cdots$. The last part is an accumulator that calculates the weighted sum of $f_L(\sigma_i)$.

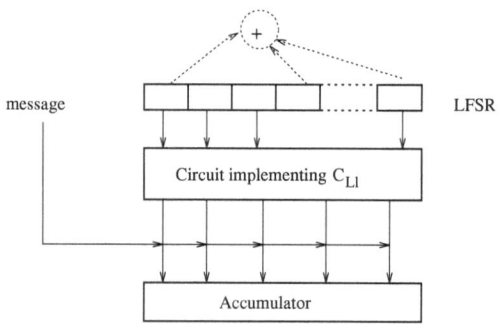

Fig. 1. *Implementation of* $\mathcal{H}^2_{MRD}(t)$

The circuit c_L can be implemented as an accumulator as shown in Figure 2.

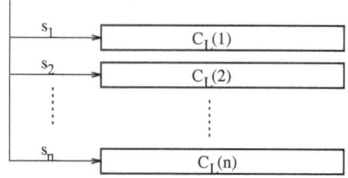

Fig. 2. *Implementation of* c_L

Each column $C_L(i), i = 1, \cdots, n$ of C_L is stored in a register and the binary n-tuple, which is the state of the LFSR, determines the subset of the registers that are XOR-ed to produce the output.

$C_L(i), i = 1, \cdots n$ form the key information (they are computable from the d_{min} key bits) and so must be kept secret.

The hashing method can also be efficiently implemented in software. Storing the consecutive outputs of the LFSR allows a parallel software evaluation of the hash function.

4   MAC from $\mathcal{H}^1_{MRD}(t)$

We use Proposition 7 to construct an ϵ-AXU$_2$ class of hash functions.

Two such classes are $\mathcal{H}_M[m, n]$ defined in [1], and $\mathcal{H}_K[m, n]$ defined in [6]. The first class maps Σ^m to Σ^n. Its elements can be described by binary $m \times n$ matrices. Let $h \in \mathcal{H}_M[m, n]$. Then $h(x) = h.x$ where h is an $m \times n$ matrix and $x \in V_n$.

Theorem 14. *[1] $\mathcal{H}_M[m, n]$ is an ϵ-AXU$_2$ class with $\epsilon = 2^{-n}$.*

The second class maps Σ^m to Σ^n. Its elements are described by irreducible polynomials of degree n over $GF(2)$. To find the hash value of $b \in \Sigma^m$ using a hash function described by $p(X)$, which is assumed to be an irreducible polynomial of degree n, a polynomial $b(X) = \sum_{i=0}^{n-1} b_i X^i$ in a formal variable X is formed. The remainder of dividing $X^n b(X)$ by $p(X)$ yields an n-tuple which is the required hash value.

Theorem 15. *[6] $\mathcal{H}_K[m, n]$ is an ϵ-AXU$_2$ class with $\epsilon = \frac{m+n}{2^n-1}$.*

Using proposition 7 we have the following compositions.

1. $\mathcal{H}_K[n, k] \circ \mathcal{H}^1_{MRD}(t) : F'_n \times F'_n \to \Sigma^k$ is a $(2^{1+t} + \frac{n+k}{2^k-1})$-AXU$_2$ class of hash function.
2. $\mathcal{H}_M[n, k] \circ \mathcal{H}^1_{MRD}(t) : F'_n \times F'_n \to \Sigma^k$ is a $(2^{1+t} + 2^{-k})$-AXU$_2$ class of hash function.

In the rest of this section we will use the second composition; in which the digest is evaluated by matrix multiplication over $GF(2)$.

4.1   A Comparison

The following comparison between the MAC described above, and the MAC obtained from Bucket Hashing (BH) gives some insight into the actual values of the parameters and application of the proposed hash functions.

In this comparison we assume the word size of 1 bit although in an efficient software implementation word size corresponds to the word size supported by the hardware. The following proposition shows that the results of our analysis are directly applicable to larger word sizes.

Proposition 16. *If $\mathcal{H} = \{h : A \to B\}$ is ϵ_1-AXU$_2$, then $\mathcal{H}' = \{h' : A^m \to B^m\}$, with $h'(x) = h(x_1)|\cdots|h(x_m)$, where $x = (x_1|\cdots|x_m)$, is ϵ-AXU$_2$. (The proof is a straightforward extension of Proposition 2 in [12].)*

We consider a typical value for BH parameters and consider an example of our scheme with a similar value for ϵ. In particular we consider a BH that maps 1024-bit messages to 140-bit digests and has a collision probability of around 2^{-31}. The family of hash function is (2^{-31})-AU_2 with $(420 \times \log_2 140)$ bits of key. Since BH is not ϵ-AXU_2, it needs to be composed with an ϵ-AXU_2 to obtain the required security (cf. Proposition 7). In our scheme we should choose $n > 32$. In Section 4.3 we show that to find a suitable value for n, one needs only to verify that 2 is a primitive element modulo the prime n. That is, 2^i, for $i = 1, \ldots, n - 1$, should generate all the elements of the set $\{1, 2, \ldots, n - 1\}$. Table 2 shows that the first suitable value for n is 37. This is the least prime greater than 32 for which 2 is a primitive element. Therefore in our comparison we assume $n = 37$.

n	g	n	g	n	g	n	g	n	g
5	2	7	3	11	2	13	2	17	3
19	2	23	5	29	2	31	3	37	2
41	6	43	3	47	5	53	2	59	2
61	2	67	2	71	7	73	5	79	3
83	2	89	3	97	5	101	2	103	5
107	2	109	6	113	3	127	3	131	2
137	3	139	2	149	2	151	6	157	5
163	2	167	5	173	2	179	2	181	2
191	19	193	5	197	2	199	3	211	2
223	3	227	2	229	6	233	3	239	7
241	7	251	6	257	3				

Table 2. *List of the smallest primitive elements (g) of prime numbers (n) in the range* $[5 \ldots 257]$.

Message block length: In bucket hashing, for a given level of security, the size of the digest is lower bounded and is usually larger than what is used in MAC systems. Although other hash functions can be applied to reduce the size of the digest, the size of the original message block will remain large. For example for a 140 word digest a 1024 word message size must be used. For 32 bit machines this results in a large value for the minimum size of the message. If the message length is 20 words then the digest is 7 times longer than the message, and for 32 bit machines maps 640 bits to 4480 bits. So in BH, there is a large overhead in the computation when the message length is small [13]. Several layers of hashing in Rogaway's TOY-MAC makes no sense when the message length is small and the operations on the padding of the message (to increase its length to the acceptable minimum) is wasteful.

In our method for the same value of ϵ the message block length can be as small as 74 words (when $n = 37$). In this case $d_{min} = 2^{36}$ and so the key is

$k = (\log_2 d_{min} - 1) = 35$ bits. By choosing a larger suitable n we can increase the message block length and also the required level of security.

Key length: Similarly to message lengths, our system can be used for various key lengths which can be as small as 35 bits. BH uses 420 bits which is too long when used in a computational secure model where the one-time pad is replaced with a pseudorandom generator, which typically has a key length of around 128 bits. Rogaway [12] suggested the use of a pseudorandom number generator to obtain the required key for hashing.

Digest length: In BH, the digest length is 140 bits which is much longer than the required length (32-64 bits) in a MAC. Rogaway presented a TOY-MAC in which there exist four extra hashing levels to reduce the digest length to 64 bits. These extra levels result in an overall loss of efficiency of the system (particularly when the message is not very long). In our scheme, the digest length can be as low as 37 bits.

Memory: Our scheme does not require a large memory. With $n = 37$, it only needs 172 bytes for C_L. We note that for key distribution only the n-tuple determining the polynomial $L(x)$ can be used, but for hashing C_L must be calculated and saved. The message block requires 5 bytes, and 5 bytes are needed for the digest.

4.2 Implementation

Our scheme was implemented and tested on different data. The implementation can be divided into two parts. In the fist part, the key which is the n binary coefficients of the a linearized polynomial of degree less than 2^{n-1}, is extended to an $n \times n$ binary matrix. This matrix is stored in memory and does not need to be recalculated for each MAC calculation. In the second part, this matrix is used for hashing a given message (cf. the example given in Appendix 5). Once calculated, it is stored in the system for multiple use. The complexity of this part is equivalent to the complexity of binary matrix multiplication, which has been extensively studied in the literature [5].

4.3 Setup Phase

We can calculate the minimal linearized polynomials of elements in $GF(2^n)^*$ using the σ-orbits (conjugacy groups). Since each element of an orbit has the same minimal linearized polynomial. We only state the results here, the proofs will be given elsewhere.

The Galois automorphisms of $GF(2^n)$ are generated by: $\sigma : x \to x^2$. This group acts on $GF(2^n)^*$ and partitions this set into orbits. Recall that a *modulus* M of $GF(q^n)$ is a $GF(q)$-subspace which is invariant under σ, i.e. $M^\sigma = M$. A modulus is a union of σ-orbits: $M = \Omega_1 \cup \cdots \cup \Omega_r$. A q-polynomial is a linearized polynomial $L(z)$ whose coefficients lie in the ground field $GF(q)$.

Lemma 17. *Suppose $L(x)$ is a linearized polynomial over $GF(q^n)$. The zeroes of $L(x)$ form a modulus iff $L(x)$ is a q-polynomial.*

Thus each modulus M determines a q-polynomial, which is defined as follows

$$L(x) = \prod_{\beta \in M} (x - \beta).$$

For these facts consult [8].

Let $\Omega = \{\alpha^{2^i}\}$ be a σ-orbit. We define the following $GF(2)$-span of the vectors in this orbit

$$L(\Omega) =< \alpha, \alpha^2, \dots > .$$

Proposition 18. *Let* $L(\Omega) =< \alpha, \alpha^2, \dots >$ *where* $\Omega = \{\alpha^{2^i}\}$ *is an* σ-*orbit, then* $L(\Omega)^\sigma = L(\Omega)$, *i.e.* $L(\Omega)$ *is a modulus.*

$L(\Omega)$ not always irreducible. It can have a 1-dimensional σ-invariant subspace namely: $V =< \alpha + \alpha^2 + \cdots >$.

Proposition 19. $L(\Omega) \simeq V \oplus W$, *where* V *and* W *are* σ-*invariant* $GF(2)$-*subspaces.*

Let $\mathrm{Tr}(\beta)$, be the absolute trace of $\beta \in GF(2^n)$ over $GF(2)$, then $\mathrm{Tr}(\beta) = 0$, or 1. Moreover, $V =< \alpha + \alpha^2 + \cdots >$ is the zero subspace if $\mathrm{Tr}(\alpha) = 0$; it is a 1-dimensional σ-invariant subspace if $\mathrm{Tr}(\alpha) = 1$.

Suppose now that p is a prime such that 2 is a primitive mod p, i.e. $2^{p-1} \equiv 1 \bmod p$, and $p - 1$ is the least power of 2 for which this is true.

Theorem 20. *Let* $L(\Omega) \simeq V \oplus W$, *be decomposed into* σ-*invariant subspaces as in Proposition 19. If 2 is a primitive* mod p, *then* W *is an irreducible modulus.*

An analysis of the linearized polynomials corresponding to the irreducible moduli in Theorem 20 leads to the following result.

Theorem 21. *For any prime degree extension* $GF(2^p)$ *in which 2 is primitive* mod p, *the minimal 2-polynomial of any* $\beta \in GF(2^p)^*$ *is* $x^{2^p} + x$ *if* $\mathrm{Tr}(\beta) = 1$. *It is* $x^{2^{(p-1)}} + x^{2^{(p-2)}} + \cdots + x^2 + x$, *if* $\mathrm{Tr}(\beta) = 0$.

5   Conclusions

We have introduced two new evaluation hashing schemes that have a number of important and useful properties. In particular, the evaluation of digests reduces to matrix multiplication over $GF(2)$ which can be efficiently implemented. Developing an efficient algorithm for calculating d_{min} in general requires further research. However we have characterized a class of $GF(2^n)$ which require no computation. We have also compared a MAC based on $\mathcal{H}^1_{MRD}(t)$ with one based on BH.

A Proofs

Proof of Lemma 9: *(sketch)*

1. For all $L(x)$, x is a factor and hence $L(0) = 0$. For $x = (11, \cdots, 1)$ we have $x.\beta \in GF(2)$ (page 52, [8]).

2. In a linearized polynomial the set of zeros form a subspace $\mathcal{N}_L$ of $GF(2^n)$ (Theorem 3.50, [8]).
   Now if $L(x) = y$ for some y, then for all $z \in \mathcal{N}_L$ we have $L(x + z) = L(x) + L(z) = y$. Conversely, if $L(x) = L(u) = y$ then $L(x + u) = 0 = L(z)$ where $z \in \mathcal{N}_L$. Hence if an element occurs once, it occurs exactly $|\mathcal{N}_L|$ times.

3. If $L(x)$ has even number of terms then $L(x) = x(x + 1)f(x)$ and we have $L(0) = L(1) = 0$ and hence $L(11, \cdots, 1) = 0$. This results in the following complementary property

$$L(\overline{x}) = L(x + 11 \cdots 1) = L(x) + L(11 \cdots 1) = L(x)$$

4. Let $u \in F_n$. There is a unique minimum linearized polynomial $L(x)$ such that $L(u) = 0$. An example computation of this polynomial is given in Appendix 5. Now if $L'(x)$ is a linearized polynomial such that $L'(u) = 0$, there exists a linearized polynomial $L''(x)$ such that $L''(x) = L'(x) \otimes L(x)$ (Theorem 3.68, page 113 [8]). Let $\mathcal{M}_u$ denote the collection of linearized polynomials for which u is a root and $|\mathcal{M}_u| = M_u$. Then we have $M_u = 2^{t-d_u}$, where d_u is the degree of $L(x)$. Now if L_1 is a linearized polynomial satisfying $L_1(u) = v$ and $L_2(x) \in \mathcal{M}_u$, then $(L_1 + L_2)(u) = L_1(u) + L_2(u) = v$ and hence v occurs 2^{t-d_u} in the column labelled by u.

5. For a pair $x, y \in GF(2^n)$, the number of rows, $L(x)$, of C, such that $L(x) = L(y)$ is the same as the number of rows of C with $L(x + y) = 0$ and hence the result.

$\square$

This completes the first part.

Proof of Theorem 12. We only need to show that,

$$\frac{|\{h^{(1)} \in \mathcal{H}^1_{MRD}(t) : h^{(1)}(M) = 0\}|}{|\mathcal{H}^1_{MRD}(t)|} \leq \frac{1}{2^{t+1}}, \quad \forall M \neq 0 \in \Sigma^{2n}.$$

Let $M = (M_1 | M_2)$. Since $h_L^{(1)}(M) = f_L(M_1) \oplus M_2$ and $|\mathcal{H}^1_{MRD}(t)| = 2^{t+1}$, we have to prove that,

$$|\{f_L : f_L(M_1) = M_2\}| \leq 1.$$

This is true because of Corollary 10. Note that since $M \neq 0$, M_1 and M_2 cannot be both zero. We consider two cases.

- If $M_1 = 0$, then $f_L(M_1) = 0 \neq M_2$ and so $|\{f_L : f_L(M_1) = M_2\}| = 0 \leq 1$.
- If $M_1 \neq 0$, then based on Corollary 10 $|\{f_L : f_L(M_1) = M_2\}| \leq 1$.

This proves the theorem. □

Proof of Theorem 13. Because of the linearity of f_L we have

$$h^{(2)}(s) = \Sigma_{j=0}^{2^n-2} s_j f_L(\alpha^j) = f_L(\Sigma_{j=0}^{2^n-2}\alpha^j) = f_L(\gamma_s)$$

for $\gamma = \Sigma_{j=0}^{2^n-2} s_j \alpha^j \in F_n$. To show that the class is ϵ-AXU$_2$ we note that for arbitrary $x, y \in F_{2^n-1}$ and $z \in F_n$, we have

$$\frac{|\{h^{(2)} \in \mathcal{H}^2_{MRD}(t) : h^{(2)}(x) \oplus h^{(2)}(y) = z\}|}{|\mathcal{H}^1_{MRD}(t)|} =$$

$$\frac{|\{h^{(2)} \in \mathcal{H}^2_{MRD}(t) : h^{(2)}(x \oplus y) = z\}|}{|\mathcal{H}^1_{MRD}(t)|}$$

But $h_L^{(2)}(x \oplus y) = f_L(\gamma_x \oplus \gamma_y)$ and so

$$\frac{|\{h^{(2)} \in \mathcal{H}^2_{MRD}(t) : h^{(2)}(x \oplus y) = z\}|}{|\mathcal{H}^1_{MRD}(t)|} \leq \frac{1}{2^{t+1}} .$$

□

B Hashing Example for $\mathcal{H}^1_{MRD}$

Let $n = 5$, $F_5 = GF(2^5)$, $f(x) = x^5 + x^2 + 1$ be a primitive polynomial of degree n, and $f(\alpha) = 0$ for some $\alpha \in F$. We group the elements of F_5 into conjugate groups as shown in Table 3 and find the minimum linearized polynomial of each group. Since the smallest degree of the minimum linearized polynomials is 16, therefore we can use all linearized polynomial of degree less than 16 with odd number of terms, and construct a matrix which represents our hash function.

Conjugate groups	Minimal polynomials	Linearized minimal polynomials
$\alpha, \alpha^2, \alpha^4, \alpha^8, \alpha^{16}$	$x^5 + x^2 + 1$	$x + x^2 + x^4 + x^8 + x^{16}$
$\alpha^3, \alpha^6, \alpha^{12}, \alpha^{24}, \alpha^{17}$	$x^5 + x^4 + x^3 + x^2 + 1$	$x + x^{32}$
$\alpha^5, \alpha^{10}, \alpha^{20}, \alpha^9, \alpha^{18}$	$x^5 + x^4 + x^2 + x + 1$	$x + x^{32}$
$\alpha^7, \alpha^{14}, \alpha^{28}, \alpha^{25}, \alpha^{19}$	$x^5 + x^3 + x^2 + x + 1$	$x + x^2 + x^4 + x^8 + x^{16}$
$\alpha^{11}, \alpha^{22}, \alpha^{13}, \alpha^{26}, \alpha^{21}$	$x^5 + x^4 + x^3 + x + 1$	$x + x^{32}$
$\alpha^{15}, \alpha^{30}, \alpha^{29}, \alpha^{27}, \alpha^{23}$	$x^5 + x^3 + 1$	$x + x^2 + x^4 + x^8 + x^{16}$

Table 3. *Conjugate groups with the corresponding minimal and linearized minimal polynomials. (Cf. Theorem 21.)*

One can verify that $\{\alpha^3, \alpha^6, \alpha^{12}, \alpha^{24}, \alpha^{17}\}$ are linearly independent and thus a normal basis for $GF(2^5)$. Let $L(x) = x^4$. The vector corresponding to $L(x)$ is:

$$\begin{aligned}
c_L &= (L(\alpha^3), L(\alpha^6), L(\alpha^{12}), L(\alpha^{24}), L(\alpha^{17})) \\
&= ((\alpha^3)^4, (\alpha^6)^4, (\alpha^{12})^4, (\alpha^{24})^4, (\alpha^{17})^4) \\
&= (\alpha^{12}, \alpha^{24}, \alpha^{17}, \alpha^3, \alpha^6) \\
&= (\alpha^3 + \alpha^2 + \alpha, \alpha^4 + \alpha^3 + \alpha^2 + \alpha, \alpha^4 + \alpha + 1, \alpha^3, \alpha^3 + \alpha)
\end{aligned}$$

The matrix representation of the function f_L is:

$$C_L = \begin{pmatrix} 0 & 0 & 1 & 0 & 0 \\ 1 & 1 & 1 & 0 & 1 \\ 1 & 1 & 0 & 0 & 0 \\ 1 & 1 & 0 & 1 & 1 \\ 0 & 1 & 1 & 0 & 0 \end{pmatrix}$$

and $h_L^{(1)} = \begin{pmatrix} C_L \\ I_5 \end{pmatrix}$ where I_5 is the identity matrix. Now let $M = [11010\,01100]$ which results in $M_1 = [11010]^t$ and $M_2 = [01100]$. We have, $f_L(M_1) = C_L \cdot M_1 = [00011]$ and so,

$$h_L^{(1)}(M) = f_L(M_1) \oplus M_2 = [00011] \oplus [01100] = [01111].$$

C   Finite Fields

Consider F_n, the finite field with 2^n elements. We consider only the binary fields, although most of the results hold for general q-ary fields.

F_n is an n-dimensional vector space over $GF(2)$. It can be constructed using an irreducible polynomial, $f(x)$, of degree n. Let α denote a root of $f(x)$. Then $1, \alpha, \alpha^2 \cdots \alpha^{n-1}$ forms a basis of F_n. An element of F_n is a *primitive element* if its powers generate all non-zero elements of F_n.

Elements of F_n are partitioned into *conjugate groups* (the σ-orbits). The conjugate group of β consists of $\{\beta, \beta^2, \cdots\}$. If the elements of $\{\beta, \beta^2, \cdots \beta^{2^{n-1}}\}$ are linearly independent then they form a basis for F_n. A *normal basis* is a basis of the form $\{\beta, \beta^2 \cdots \beta^{2^{n-1}}\}$. Every field has at least one normal basis.

Let ψ denote an element of a finite field F_n. The *minimum polynomial* of ψ is an irreducible polynomial $f(x)$ over F_n such that $f(\psi) = 0$ and for any other polynomial $g(x)$ with $g(\psi) = 0$ has $f(x)$ as a factor. It can be shown that every element of the field has a unique minimum polynomial and all the conjugate elements have the same minimum polynomial.

A polynomial of the form $L(x) = \sum_i \alpha_i x^{2^i}$ with $\alpha_i \in GF(2)$ is called a 2-polynomial; we refer to these as *linearized polynomials*.

Linearized polynomials satisfy the following two properties:

$$L(\alpha + \beta) = L(\alpha) + L(\beta), \quad \alpha, \beta \in F;$$
$$L(c\alpha) = cL(\alpha), \quad \alpha, \in F, c \in GF(2).$$

The ordinary product of linearized polynomials is not a linearized polynomial. The *Symbolic product* of two polynomials $L_1(x)$ and $L_2(x)$, defined

$$L_1(x) \otimes L_2(x) = L_1(L_2(x))$$

gives a linearized polynomial.

Let $\psi \in F$ and $f(x)$ denote its minimum polynomial. The *minimum linearized polynomial* of ψ is a linearized polynomial $L(x)$ such that $f(x)$ is a factor of $L(x)$ and any other linearized polynomial for which $L_1(\psi) = 0$ can be written as $L_1(x) = L_2(x) \otimes L(x)$. It can be shown that $L(x)$ is unique.

D How to Calculate Minimal Linearized Polynomials

Suppose that $\alpha \in F_n$ and $f(\alpha) = 0$. To find the minimum linearized polynomial $p(x)$ for α we proceed as follows. Let $q(x)$ denote the minimum polynomial of α. We have

$$p(x) = \sum_{i=0}^{n} t_i x^{2^i} = m(x)q(x) \tag{1}$$

where $m(x)$ is a polynomial over $GF(2)$. For $i = 0, \ldots, n$, let $r_i(x)$ denote the remainder of dividing x^{2^i} by $q(x)$. To satisfy equation (1) we must have

$$\sum_{i=0}^{n} t_i r_i(x) \equiv 0 \mod q(x). \tag{2}$$

Expanding (2), we obtain a set of n equations in $n + 1$ variables $t_0, \ldots, t_n$. The solution that results in a polynomial with minimum degree determines $p(x)$. Note that this method requires the determination of the irreducible polynomials corresponding to the σ-orbits. This adds to the computational complexity of this algorithm, while the irreducible polynomials are not required in the final result.

References

1. J. L. Carter and M. N. Wegman, "Universal Class of Hash Functions," *Journal of Computer and System Sciences*, vol. 18, no. 2, pp. 143–154, 1979.
2. K. Chen, "A New Identification Algorithm," in *Cryptography: Policy and Algorithms Conference* (E. Dawson and J. Golic, eds.), vol. 1029 of *Lecture Notes in Computer Science (LNCS)*, (Queensland, Australia), pp. 244–249, Springer-Verlag, July 1995.
3. E. Gabidulin, "Theory of Codes with Maximum Rank Distance," *Problems of Information Transmission*, vol. 21, no. 1, pp. 1–12, 1985.
4. T. Johansson, "Authentication Codes for Nontrusting Parties Obtained from Rank Metric Codes," *Design, Codes and Cryptography*, no. 6, pp. 205–218, 1995.
5. H. Krawczyk, "The Shrinking Generator; some practical consideration," in *Proceedings of Fast Software Encryption Workshop, FSE '93*, pp. 45–46, LNCS, Springer-Verlag, 1993.
6. H. Krawczyk, "LFSR-based Hashing and Authentication," in *Advances in Cryptology, Proceedings of CRYPTO '94* (Y. G. Desmedt, ed.), vol. 839 of *Lecture Notes in Computer Science (LNCS)*, pp. 129–139, Springer-Verlag, 1994.
7. H. Krawczyk, "New Hash Functions for Message Authentication," in *Advances in Cryptology, Proceedings of EUROCRYPT '95* (L. C. Guillou and J.-J. Quisquater, eds.), Lecture Notes in Computer Science (LNCS), (Berlin), pp. 301–310, 1995.
8. R. Lidl and H. Niederreiter, *Introduction to Finite Fields and their Applications.* Cambridge Uiversity Press, 1994.
9. U. Maurer, ed., *Advances in Cryptology, Proceedings of EUROCRYPT '96*, vol. 1070 of *Lecture Notes in Computer Science (LNCS)*, (Saragossa), Springer-Verlag, 1996.
10. B. Preneel, *Analysis and Design of Cryptographic Hash Functions.* PhD thesis, Katholieke University Leuven, Jan. 1993.

11. B. Preneel and P. C. van Oorschot, "On the Security of Two MAC Algorithms," in Maurer [9], pp. 19–32.

12. P. Rogaway, "Bucket Hashing and its Application to Fast Message Authentication," in *Advances in Cryptology, Proceedings of CRYPTO '95*, Lecture Notes in Computer Science (LNCS), pp. 30–42, Springer-Verlag, 1995.

13. V. Shoup, "On Fast and Provably Secure Message Authentication Based on Universal Hashing," in Maurer [9], pp. 321–331.

14. D. R. Stinson, "Universal Hashing and Authentication Codes," *Design, Codes and Cryptography*, vol. 4, pp. 369–380, 1994.

15. R. Taylor, "Near Optimal Unconditionally Secure Authentication," in *Advances in Cryptology, Proceedings of EUROCRYPT '94 (preprints)* (W. Wolfowicz and A. de Santis, eds.), vol. 765 of *Lecture Notes in Computer Science (LNCS)*, (Perugia, Italy), pp. 245–255, may 1994.

16. M. N. Wegman and J. L. Carter, "New Hash Functions and Their Use in Authentication and Set Equality," *Journal of Computer and System Sciences*, vol. 22, pp. 265–279, 1981.

New Constructions for Secure Hash Functions
(Extended Abstract)

William Aiello[1], Stuart Haber[2], and Ramarathnam Venkatesan[3]*

[1] BellCore (aiello@bellcore.com)
[2] Surety (stuart@surety.com)
[3] Microsoft Research (venkie@microsoft.com)

Abstract. We present new, efficient and practical schemes for construction of collision-resistant hash functions, and analyze some simple methods for combining existing hash-function designs so as to enhance their security.

In our new constructions, we first map the input to a slightly longer string using a primitive we introduce called *secure stretch functions*. These are length-increasing almost surely injective one-way functions that sufficiently randomize their inputs so that it is hard for an adversary to force the outputs to fall into a target set. Then we apply a *compression function* to the output of the stretch function. We analyze the security of these constructions under different types of assumptions on both stretch and compression functions. These assumptions combine random-function models, intractability of certain "biasing" tasks, and the degeneracy structure of compression functions. The use of stretching seems to allow reduced requirements on the compression function, and may be of independent interest.

These constructions allow one to use popular and efficient primitives such as MD5, SHA-1, and RIPEMD that may exhibit weaknesses as collision-resistant functions. But no attacks are currently known on their one-way and randomizing properties, when they are used as stretch functions as in our constructions. There are several collision-resistant hash functions based on DES for which there are no known effective attacks, but which are too slow for most practical applications. Our use of stretch functions enable us to base our compression function on DES so that the resulting hash function achieves practical speeds: a test implementation runs at 40% of the speed of MD5.

We also suggest some imperfect random-oracle models, showing how to build better primitives from given imperfect ones. In this vein, we also analyze how to defend against a collision-finding adversary for a given primitive by building "independent" primitives.

1   Introduction

In this work we present new and practical constructions for secure hash functions, and analyze their security. In addition, we present simple methods for combining existing hash-function designs so as to enhance their security. There is a compelling need for better understanding of the principles of secure hash-function design.

---

* Part of this work was done while with Bellcore and Surety.

S. Vaudenay (Ed.): Fast Software Encryption – FSE'98, LNCS 1372, pp. 150–167, 1998.
© Springer-Verlag Berlin Heidelberg 1998

Many cryptographic procedures that handle very long bit-strings make use of a hash function. The security of these procedures relies on the *collision resistance* of the hash function in use, or on the function's "randomizing" effect. A hash function f is collision-resistant if it is infeasible to find a pair of distinct arguments $x \neq x'$ such that $f(x) = f(x')$.

There are several approaches to the design of such hash functions. While it is not known whether any current designs achieve the desired properties, they generally fall into two categories: designs based on an existing block cipher (or other cryptographic primitive), and custom designs "from scratch."

Customized hash functions: There have been a number of proposals for a "practical" secure hash function, one that admits fast software implementations and for which it is hoped that the cost of computing hash collisions is infeasible in practice [Riv 90,Riv 92,NIST 94,BP 95,DBP 96,AB 96]. Several of these are in widespread use. However, the general design principles for cryptographic hash functions are not well understood. As in the case of block ciphers, in practice a good hash function is simply one that survives the current attacks. Recent collision-finding attacks due to Dobbertin, using a differential approach, have been successful against RIPEMD, MD4, and MD5 [Dob 97,Dob 96a,Dob 96b]. More recently, even the one-wayness of MD4 has been challenged [Pre 97,Dob 98].

One approach would be to try to build on existing primitives. For example, one can concatenate the outputs of two different hash functions, hoping that the two functions behave "independently" (see [Pre 93, §2.4.5] and certain commercial designs, e.g. [Sur 95]). But one's hope is weakened by a cursory look at the source code for the popular hash functions, and even more so by Dobbertin's attacks on MD4-256, which derives two 128-bit values in this manner [Dob 96a].

Current methods to extend or strengthen previous designs include the following: increase the number of rounds (as in MD5); add some coding or scrambling steps (as in SHA-1); increase the buffer size and make the mixing step vary with the round. All of these are natural attempts to increase the security of a hash-function design, but an analysis based on a set of plausible heuristic assumptions would better enhance our confidence in the result. An example of such an assumption is the ideal-cipher model for DES, discussed below.

Hash functions from (ideal) ciphers: Another well studied approach (see e.g., [MMO 85,BC$^+$ 88,Merk 89]) bases the design on an existing trusted block cipher. For security assessments of such schemes see [Merk 89,KP 97,PGV 93a]. (For this and other questions about cryptographic hash functions, [Pre 93] and [MOV 97, chap. 9] are excellent references.) Unfortunately, these designs yield implementations using DES that are slower than MD5 (for example) almost by an order of magnitude, making them unacceptable for many applications. The usual measure of the efficiency of a design based on an n-bit cipher is its *rate*, defined as the number of n-bit blocks of data compressed per applications of the cipher. (Sometimes, as in [Pre 93], "rate" is used to mean the inverse of this ratio.) One of the suggestions here enables us to increase the rate significantly, yielding practical designs.

It is common to use idealizations of block ciphers as random permutations or functions from $(n+\delta)$-bits to n-bits in the analysis. In this case one can construct n-bit valued secure hash functions (see references above). In the case of DES, where $n = 64$, this yields 64-bit hash functions, which are vulnerable to simple birthday attacks. However, it is non-trivial to construct $2n$-bit valued secure hash functions from families of n-bit valued hash functions. The $2n$-bit valued hash function must behave like a $2n$-bit valued random function for up to 2^n queries, but the n-bit primitives run into birthday collisions around $2^{n/2}$ queries, which potentially could be used in an attack against the design. A solution for this output-doubling problem was given in [AV 96]. This construction is expensive, making eight calls to the underlying random function, and hence it is not suitable for a practical $2n$-bit valued compression function.

The analysis of our construction begins by assuming that both of the two functional components are random functions. This is *not* for the purpose of proving the existence of secure hash functions, but rather to examine what security parameters can be achieved. In addition, it motivates the weaker assumptions and the analysis that follow.

1.1   New Constructions

The constructions that we propose first stretch the input string mildly, and then compress the result of this expansion. Here we briefly motivate this approach.

Expansion stage: Our first stage stretches the input mildly. We will use primitives that have reasonable one-wayness and randomizing behavior, so as to obtain an almost surely one-to-one stretch function. This trivially avoids collisions in the first stage, and allows us to analyze this stage using distributional and one-way properties of the primitives we employ. Furthermore, these properties make it infeasible for the adversary to force its outputs into a set of his choice—for example, a set of points for which he has computed collisions for the second stage. We show how to use popular hash functions like MD5 or SHA-1 to do this. We remark that in large randomness tests with MD4 and MD5, it has been observed that both functions have very good distributional properties, even when they are iterated [PV 96].

Compression stage: In our second stage we apply a compression function. This stage could simply use any candidate collision-resistant hash function such as SHA-1 or RIPEMD-160. In fact, the security of our construction does not require collision resistance from the compression stage. For example, an adversary might find collisions for the compression stage. However, the colliding strings may not be in the range of the stretch function, and even those that are will be hard to invert. On the other hand, if the adversary begins by finding many input-output pairs for the stretch function, then a successful attack on the whole construction must find compression-stage collisions from among this restricted set of fairly random points.

Constructions using existing primitives

In a practical setting this work suggests ways to use the hash functions that are currently broken or partially broken in such a way that we can depend on their one-wayness and randomness or distributional properties, rather than directly on their collision-security, which may be in doubt or already violated. In fact, there are many choices for each of the two components of our construction, and they can be combined independently.

Customized Hash Functions   Customized hash functions, for example, MD5, SHA-1, and RIPEMD-160, can be used in either or both stages of our construction.

If the hash function has N bits of output, then it can be used in the stretching stage as follows. If at most N bits are needed for the input to the compression stage, then simply feed ℓ-bit blocks ($\ell < N$) of input text to the hash function. If more than N bits are needed for the input to the compression stage, we propose the following simple chaining. Use the N bits of output above as the first N bits of output of the chaining rule. In addition, concatenate these bits to the next ℓ bits of input to the hash function to get another N bits of output, and continue this chaining rule as needed.

For the compression stage, any of these hash functions can be used directly on the fixed-length output of the expansion stage.

We remark here that MD4 may also be sufficient for both stages. For example, as noted above the stretching stage is required to be one-way. Although two rounds of MD4 have recently been inverted [Dob 98], the inverse found is of length 512. Note that there are very many inverses ($2^{512-128}$) for an average 128-bit output. However, MD4 might be used in our stretching stage to expand, for example, 80-bit inputs to 128-bit outputs. In this case, for an overwhelming fraction of outputs, an adversary would be required to find the *unique* inverse. In addition, with a sufficiently random and one-way stretch function, our analysis suggests that requirements for the compression function are considerably relaxed. For example, using a truly random stretch function the compression function need only have a "fairly uniform" preimage structure.

Subset Sum   Constructions based on the subset-sum function may be used in the stretching stage. The subset-sum function may also be used in the compression stage, because it is known to yield provably secure hash functions on the assumption that it is infeasible to find almost shortest vectors in lattices [GGH 96]. However, the best lattice-based attacks are quite powerful, forcing the lattices (and the cache needed for the implementation) to be relatively large. We suggest some constructions in the final version of this paper.

DES   Any DES-based hash function, e.g., [MMO 85,BC$^+$ 88,Merk 89], may be used in either stage of our construction in the same way as described above for customized hash functions. However, since these hash functions consume few bits of input per DES call (i.e., they have low rate), the resulting hash function will be unacceptably slow for most practical applications.

In this paper we propose a new DES-based construction for the compression stage. Because of the properties of the first stage, our construction uses only two DES calls to obtain a 128-bit output value. The construction is extremely simple. As in [Merk 89] we will use a modified form of DES called MDES, defined as follows: $\text{MDES}(K, x) = \text{DES}_K(x) \oplus x$ (where K is potentially the $16 * 48$-bit expanded key). The output of the stretching stage is split into two pieces, each of which is used separately as the key to one MDES call. The outputs of the two calls are simply concatenated.

Assuming that the stretch stage is a truly random function, and that DES has an almost regular preimage structure (i.e. all points in the range have approximately the same number of key-plaintext pairs mapping into them), we show that this construction is secure (see §3.3). This is a significant simplification on the requirements of the primitives to be used in a compression function. The same scheme without the randomizing initial stage is insecure; to achieve similar security would require more rigorously random-function like primitives, and many more calls to them (e.g. as in [AV 96]).

In many cipher-based constructions, the string to be hashed is used as a "key" to encrypt some initial or intermediate values of the hash function. The usual DES key scheduling algorithm stretches the given 56-bit key into 48*16 bits. One way to improve the rate of a DES-based hash function would be to skip the key-scheduling algorithm and feed 16*48 bits of input text directly as a key. This idea is swiftly rebuffed: one can use the invertibility of intermediate rounds of DES and mount a meet-in-the-middle attack, as follows (see [C 85] and [MH 81] for related attacks). The attacker can pick the text corresponding to the keys for all but three rounds arbitrarily. He picks the remaining round keys randomly, and expects a birthday collision between one round in the encrypting mode and the next level in the decrypting mode.

Our DES-based scheme, perhaps controversial, allows the thoroughly randomized output from the first stage to be used directly, and thus to skip the key scheduling algorithm. This considerably increases the rate of the resulting construction.

While this proposal clearly needs study, there is evidence to support the claim that keying DES in this manner is secure. We point out that this is similar to scheduling the round keys in DES with independent keys, a method whose security is closer to exhaustive search for the 56-bit key (rather than the extended key of length 16*48) in the sense that it takes about 2^{64} steps (including computational overhead) by current differential attacks, and one may expect this number to be somewhat smaller for linear attacks. Of course, the key is considered hidden for the differential attacks against a block cipher. Differential attacks are far more natural in the context of secure hash functions since the attacker can compute all the required input-output pairs by himself. In addition, the attacker could conceivably mount a meet-in-the-middle attack based on the invertibility of individual rounds of DES, as discussed above; but such attacks are not applicable to our use of DES, because the adversary has little effective control over the key bits.

Example parameters are as follows: we can securely stretch a 512-bit string to a 16*48-bit string and use the latter output as a DES key. Then using the stretched output as a DES key would effectively allow us to compress 512 bits per DES call. With well optimized assembly-language implementations, this results in implementations that are so much faster than standard DES-based constructions that they can be used in practice. See below for run-times of a preliminary implementation.

We stress that the second stage is not *required* to be derived from a cipher, and hence our first stage is not merely a key scheduling algorithm. Often, as in the case of DES or Tiger, the key scheduling in a hash-function design is reversible, but we demand one-way and randomizing properties in our first-stage functional component. Such reversibility may be more appropriate for ciphers, where the key is held secret, than it is for hash functions, where the collision adversary can choose the inputs. Our stretch functions may actually strengthen block-cipher constructions by helping to avoid weak keys and related-key attacks; we omit details here to obey space constraints.

However, there are attacks on adaptations of DES that skip the key-scheduler. Below we point out how the use of stretch functions can avoid these attacks. Briefly, the attacker must be able to choose some portions of the extended key during the attack, which is what the randomizing expansion step is designed to preclude: With overwhelming probability, the attacker's choice of extended keys will not be in the range of the first stage, and there is no easy way to take a small string and extend it to a string lying in the range of the first stage.

We believe our designs are useful in practice, and allow their security to be analyzed under explicitly stated assumptions on the cryptographic primitives that we use. Finding the weakest assumptions sufficient for the construction of collision-resistant hash functions is a fundamental unsolved problem. Our constructions raise some related issues that may be helpful both for the practical as well as the theoretical point of view. To summarize, the stretching stage simplifies the requirements on the compression function, which is arguably the crux of the task of designing secure functions. This is significant in itself, and may actually lead to faster constructions upon further research.

Performance

We performed a preliminary implementation to test the speed of one version of our construction and found it surprisingly fast compared to several other hash functions. Our test implementation on a 166Mhz Pentium Processor based laptop computer yielded a version running around 60Mbits/second. Here, as described above, we used MD5 for the stretch function, mapping 96*6 bits to 128*6=48*16 bits, and for the compression stage we used the DES-based construction producing 128-bit hash values.

We compared the speed of our construction both to MD5 and to DES-based hash functions. We tested with many variants for the DES-based compression schemes. The speed reported here for these functions is overestimated by assuming that they consume close to 56 bits of input per DES call. The speeds of MD5, our hash construction, and DES-based hashing are in the ratio 1 : 0.43 : 0.032. Our testing did not optimize for platform-dependent parameters such as cache size. The usually quoted speed ratios between MD5, SHA-1, RIPEMD-160, DES are 1 : 0.41 : 0.34 : 0.13. These ratios are at best treated as approximations, since many parameters can cause these ratios to vary among modern processors. For example, assembly coding can speed up different algorithms at different rates. While our DES code was optimized, our MD5 implementation was a straightforward one. In the final version, which will be available from the authors (or at http://research.microsoft.com/crypto and http://www.surety.com/pub/), we shall present more detailed performance analysis of more varied schemes.

1.2   Imperfect Random-Function Model Constructions

In §4 below, we suggest some imperfect random-oracle models, and show how to build better primitives from given imperfect ones. In this vein, we also analyze how to defend against a collision-finding adversary for a given primitive by building "independent" primitives.

2   Preliminaries

We will often model functions as random functions. A random function has the following property. When it is evaluated on an input (assumed to be different from all other inputs thus evaluated, since there is no need to evaluate the function more than once on the same input) the output is uniformly distributed and independent of all output values thus far.

We fix a bounding function $B(n)$ (e.g. $2^{0.5n}$) and this will correspond to our notion of an infeasible amount of resources (e.g. run-time or memory). We call functions (e.g. run-times) lower-bounded by $B(n)$ *infeasible* and functions that are smaller than $1/B(n)$ *negligible*. We call probabilities of the form $1 - 1/B(n)$ *overwhelming*.

A function f mapping n bits to m bits is said to be *one-way* if f is efficiently computable (e.g. in polynomial time), and given $y = f(x)$ where x is randomly chosen, any inverting algorithm $I(\cdot)$ with $f(I(y)) = y$ takes at least time $B(n)$ with overwhelming probability (over x). In addition, if $m < n$ and for any (adversary's collision-finding) algorithm C, a successful execution $C(n, m) = (x, x')$ satisfying $f(x) = f(x')$ takes time at least $B(m)$, then we call this function *collision-resistant*. For formal definitions and implementations based on various assumptions see [Dam 87,Merk 89,BY 90]; in addition, [Pre 93,MOV 97] are excellent references for this topic. It is not known what is the weakest assumption under which one can construct collision-resistant hash functions.

Given a fixed-length collision-resistant *compression function H* mapping L-bit inputs to N-bit outputs $(N < L)$, one can build a collision-resistant hash function G defined on arbitrary-length inputs following the construction of Merkle [Merk 90] and Damgard [Dam 89]. Assign a fixed N-bit "initial-value" string IV, and given an input $x = x_1 x_2 \cdots x_t$ (formatted, with "Merkle-Damgard strengthening," i.e. with appropriate padding to encode the length of the text, as t blocks of length $L - N$), let the value of $G(x)$ be defined as follows: $G_0 = IV$; $G_i = g(G_{i-1}, x_i)$, $1 \le i \le t$; $G(x) = G_t$. Thus, we will concentrate here on analyzing a fixed-length, collision-resistant compression function H.

3   New Constructions: Definition and Analysis

After describing our new construction (in §3.1), we proceed to analyze its security, first by assuming that its components are truly random functions of the appropriate class (§3.2), and then by weakening these assumptions (§3.3). This analysis treats the properties of our construction of a cryptographic hash function H with fixed-length inputs.

3.1   Definition

Our constructions first stretch the inputs and apply a compression function next. We describe the requirements on these functions after presenting some rationale for stretching.

Secure Stretch Functions   We introduce the use of *secure stretch functions*, which mildly increase the input lengths, for the purposes of constructing hash functions. A *stretch function* f maps ℓ-bit inputs into $2m$ bit inputs, where $2m > \ell$. The input strings to f will be denoted by t and the output strings will be denoted by the pair $K, \bar{K}$. Informally, they satisfy:

- One-wayness: given any $y = f(x)$ it is hard to find any x' such that $f(x') = y$.
- Outputs of f behave as if f is locally random (i.e., k-wise independent, for some $k >> 1$).

Under the randomizing conditions we pose on f's outputs, f is an injective function on an overwhelming fraction of the inputs if $2m - \ell$ is large enough. Our definition of one-wayness is also known as "preimage resistance."

Compression Functions   The outputs of these stretch functions (along with a $2n$-bit IV) are fed into a compression function $\mathbf{h}$ from $(2m + 2n)$ bits to $2n$ bits. We will consider the first $2m$ bits of input as a key. The remaining $2n$ bits of input will be denoted by the pair $x, \bar{x}$ and the output by $y, \bar{y}$. Our overall compression function will be denoted by $\mathbf{h}$, which compresses ℓ-bit strings to $2n$-bit strings. It is defined as follows:

$$f(t) = K, \bar{K}, \quad \mathbf{h}_{f(t)}(x, \bar{x}).$$

While there are many instantiations for $\mathbf{h}$ the one we will concentrate on is as follows. Let h denote a compression function from $m + n$ bits down to n bits. The first m bits of input of h will be considered a key. For now, we will define $\mathbf{h}_{K,\bar{K}}(x, \bar{x}) = h_K(x), h_{\bar{K}}(\bar{x})$.

In our implementations we use a modified form of DES as our function $h(\cdot)$, namely $h_K(x) = \text{MDES}(K, x) = \text{DES}_K(x) \oplus r(K, x)$, where $r(K, x)$ represents some simple function of K and x. (For example, $r(K, x) = x$ was suggested in [MMO 85,Merk 89].)

Note that in this case the hash value is 128 bits and to resist the attacks due to van Oorschot and Wiener [vOW 94] 192-bit hash values may be needed. It is easy to generalize our result to 192 bits by using three calls to the underlying cipher. This will be covered in the complete version of the paper.

Butterfly Compression: We define a variation on the compression function above, the "butterfly compression," as follows:

$$\mathbf{bh}_{K,\bar{K}}(x,\bar{x}) = h_K(x) \oplus r(\bar{K},\bar{x}), h_{\bar{K}}(\bar{x}) \oplus r(K,x),$$

where h and r are appropriately chosen and $r(\cdot)$ is very simple to compute (for example, $r(k,y) = y$). This variation appears to increase the complexity of the attacks using inversion algorithms. In the final version of this paper we present an analysis of this scheme.

3.2   Analysis Assuming f and h Are Random Functions

We begin our analysis of H by assuming that f and h are random functions.

Lemma 1. *When the stretching function f is a random function from ℓ bits to $2m$ bits and h is a random function from $m+n$ bits to n bits, an adversary making a total of q queries to f and h will find a collision with probability $\Theta(q^2/2^{2n})$.*

Proof: Any adversary which makes a total of q queries in sum to f and h can do no better than an adversary which makes q queries to both f and $\mathbf{h}$. We will thus analyze the latter type of adversary.

Due to the fact that both f and h are random functions, it is easy to show that the adversary maximizes its chances of finding a collision by using the outputs of its queries to f as the input to its queries to $\mathbf{h}$. Due to space limitations we omit this argument here. So, assume the adversary makes q queries to f to produce $\{(k_i,\bar{k}_i)\}$ as well as $\{(y_i,\bar{y}_i)\}$, $1 \leq i \leq q$, where $y_i = h_{k_i}(x)$ and $\bar{y}_i = h_{\bar{k}_i}(\bar{x})$.

We will assume that $m \geq n$. Fix a pair of queries, i and j, $1 \leq i, j \leq q$, and let us calculate the probability that this pair of queries yields a collision, i.e., that $(y_i,\bar{y}_i) = (y_j,\bar{y}_j)$. There are four disjoint cases.

Case 1 : $k_i = k_j$ and $\bar{k}_i = \bar{k}_j$. This event occurs with probability 2^{-2m}.
Case 2 : $k_i \neq k_j$ and $y_i = y_j$ and $\bar{k}_i = \bar{k}_j$. This event occurs with probability at most $2^{-(n+m)}$.
Case 3 : $k_i = k_j$ and $\bar{k}_i \neq \bar{k}_j$ and $\bar{y}_i = \bar{y}_j$. This event also occurs with probability at most $2^{-(m+n)}$.
Case 4 : $k_i \neq k_j$ and $k_i \neq k_j$ and $(y_i,\bar{y}_i) = (y_j,\bar{y}_j)$. This event occurs with probability at most 2^{-2n}.

Since we are assuming that $m \geq n$, the probability that there is a collision for this fixed pair of queries is at most $4/2^{2n}$. Hence, the probability that there is any collision is $4\binom{q}{2}/2^{2n}$.

This analysis is tight. Clearly, the probability that a collision occurs in q queries is at least $\Omega(q^2/2^{2n})$.   □

3.3   Analysis with Weaker Assumptions

We now assume that one of the two functions behaves like a random function and ask what conditions must be required of the other function. It turns out that rather surprisingly weaker conditions suffice.

Assuming f is random and h is almost regular

We now assume f behaves like a random function. Our goal is to show that it is sufficient to have some assumption on the distribution of the number of inverses belonging to a point in the range of h. Define $S_x(y) = \{K \mid h_K(x) = y\}$. Let $s_x(y) = |S_x(y)|$. For any fixed x, note that $\sum_y s_x(y) = 2^m$, so that the average value of $s_x(y)$ over all the values of y is 2^{m-n}. Define $\rho_x(y) = s_x(y)/2^{m-n}$ and observe that $\sum_y \rho_x(y) = 2^n$.

Definition 2. *We say that h is α-regular if for all x, $\sum_y \rho_x(y)^2 \le \alpha 2^n$.*

This condition is equivalent to the following: Let y be a random point in the range, chosen with probability 2^{-n}. Then, for a randomly chosen y, $\mathbf{E}(\rho_x(y)^2) < \alpha$.

For a random function, the value of α is a constant with high probability. Note that requiring a function to be α-regular is a weaker condition than requiring that each $\rho_x(y)$ be less than or equal to $\sqrt{\alpha}$ (for example, a $(1 + o(1))$-regular function might have, say, n values of y with $\rho_x(y) = n$ for some value of x).

Theorem 3. *If the function h is α-regular and f is a random function, then an adversary making q queries to f will find a collision of H with probability $\Theta(\alpha^2 q^2/2^{2n})$.*

Proof: Assume the adversary makes q queries. Fix i and j and consider the i-th and j-th queries. Then,

$$Prob[(y_i, \bar{y}_i) = (y_j, \bar{y}_j)]$$
$$= \sum_y \sum_{\bar{y}} Prob[y_i = y_j = y; \quad \bar{y}_i = \bar{y}_j = \bar{y}]$$
$$= \left(\sum_y Prob[k_i \in S_x(y); k_j \in S_x(y)] \right) \left(\sum_{\bar{y}} Prob[\bar{k}_i \in S_{\bar{x}}(\bar{y}); \bar{k}_j \in S_{\bar{x}}(\bar{y})] \right)$$
$$= \left(\sum_y s_x^2(y)/2^{2m} \right) \left(\sum_{\bar{y}} s_{\bar{x}}^2(\bar{y})/2^{2m} \right)$$
$$= \left(\sum_y \rho_x^2(y)/2^{2n} \right) \left(\sum_{\bar{y}} \rho_{\bar{x}}^2(\bar{y})/2^{2n} \right) \le \alpha^2/2^{2n}$$

where the last inequality follows from the regularity assumption. It follows that the probability of the adversary finding a collision is at most $\binom{q}{2}\alpha^2/2^{2n}$. $\square$

So it follows that randomness properties of f are sufficient to weaken the requirements on h considerably. Also, note that the outputs of f need not be completely independent, only 4-wise independent.

A word of caution is warranted here. If no one-wayness properties on h are imposed other than the degeneracy condition, then one must be careful about the fraction of easy points in the range of f (at which it is easy to invert f) in concrete implementations of f.

Assuming f is almost injective and h is random

In this section we consider the dual of the previous section: h will be considered a random function and we will impose some computational assumptions on f. We will first assume that f has high collision security as defined below. We stress here that the functions we consider here are not necessarily compressing.

Definition 4 (Collision Security of Hash functions). *A family of hash functions onto n bits has Collision security of $s(n)$ (example $s(n) = 2^n$) if any algorithm running in time $T(n)$ finds a collision pair with probability at most $(T(n)/s(n))^2$.*

Note that a more general version of the definition would allow the 2 to be replaced by any positive constant.

We stress that when the function is stretching and "randomizing", a large collision security is a mild and realistic assumption since the function is likely to be injective on overwhelming fraction of the range.

The collision security will help us characterize the strength of our scheme. But we will also need our stretch function f to be resistant to partial collisions. Recall that f maps l bits to $2m$ bits. Define a partial collision of size i to be when i inputs yield outputs which all have the same first m bits or which all have the same second m bits. Actually, we will need to account for all the partial collisions. Define $B_K = \{\bar{K}_i \mid (K_i, \bar{K}_i) \text{ is a response and } K_i = K\}$. Note that $\sum_K |B_K|$ is equal to the total number of queries to f. Define $b_K = |B_K|$ when $|B_K| \geq 2$ and 0 otherwise. $b_{\bar{K}}$ is defined similarly. Let $B^2 = \sum_K b_K^2 + \sum_{\bar{K}} b_{\bar{K}}^2$.

Definition 5 (Partial Collision Security of Hash functions). *A family of hash functions onto $2m$ bits has Partial Collision Security of $\bar{s}(m)$ if for any algorithm running in time $T(2m)$ the probability of B^2 exceeding $T(2m)$ is at most $T(2m)/\bar{s}(m)$.*

To give an example of this definition let us apply it to a random function, where $T(2m)$ just becomes the number of queries q. In such a case the partial collision security is $2^m/4$. To see this, it can first be shown that for a random function the expected value of $\sum_K b_K^2$ is at most $2^m \sum_{i \geq 2}(q/2^m)^i \leq 2 \cdot q^2/2^m$, where the last inequality follows whenever $q \leq 2^m/2$. The same holds for $\sum_{\bar{K}} b_{\bar{K}}^2$. Hence, the probability that B^2 exceeds q is at most $4 \cdot q/2^m$ which yields a partial collision security of $2^m/4$.

The following theorem shows that if the running time of an adversary is $o(2^n)$, $o(s(2m))$, and $o(\bar{s}(m))$ then it can only find a collision with probability $o(1)$.

Theorem 6. *If f has collision security $s(2m)$ and partial collision security $\bar{s}(m)$, and h is a random function then any algorithm running in time $T(2m)$ will find collisions with probability at most*

$$(T(2m)/s(2m))^2 + T(2m)/\bar{s}(m) + T(2m)/2^n + (T(2m)/2^n)^2$$

Proof: Suppose an adversary running in time T finds collisions on H. There are three cases.

Case 1: The adversary finds a collision on f. By assumption this happens with probability at most $(T(2m)/s(2m))^2$.

Case 2: The adversary finds partial collisions of f. For each K with $b_K > 1$, the adversary will get a collision on y and have a probability of a collision on $\bar{y}$ at most $2^{-n}b_K^2$. An analogous statement holds for each $\bar{K}$ with $b_{\bar{K}} > 1$. So, in this case the probability of a collision on the output is $2^{-n}B^2$. By assumption B^2 exceeds T with probability at most $T/\bar{s}$. Hence, this case occurs with probability at most $T/\bar{s} + T/2^n$.

Case 3: The adversary finds no collisions or partial collisions on f. Since h is random function, the probability of a collision is $q^2/2^{2n}$, where q is the number of input output pairs of f computed by the adversary. Since q is always bounded from above by T, this yields an upper bound on the probability for this case of $(T/2^n)^2$. □

4   Constructions Based on Imperfect Random-Function Models

Here we suggest simple constructions and heuristics to construct new hash functions using the old ones, so that the new one may be hard to break even if one or more of the old ones come under attack.

4.1   Composition Construction

Let f and g be two functions from finite binary strings to n bits. Then define

$$H(x) = [g(x), f(g(x), x)].$$

By *inversion security* of a one-way function f, we mean a lower bound $s_f(L)$ on the time to find an L-bit inverse of the function on all but a negligible fraction of the instances $y = f(x)$, where x is a randomly chosen L-bit string. The *collision security* of a hash function f is a lower bound $s'_f(n)$ on the time required to find two inputs x_1 and x_2 such that $f(x_1) = f(x_2)$. We first claim that the above construction at least preserves security. Note the symmetry: we need not know which of the functions is more secure, either with respect to collisions or with respect to inversion.

Lemma 7. *If f and g are collision-resistant hash functions, then H is collision-resistant. Moreover its inversion security and collision security are not lower than that of either f or g. I.e.,*

$$s'_H(n) \geq Max(s'_f(n), s'_g(n)), \qquad s_H(n) \geq Max(s_f(n), s_g(n))$$

Proof: Indeed, if $H(x_1) = H(x_2)$ is a collision for H, then we get a g-collision at $y = g(x_1) = g(x_2)$ and an f-collision at $z = f(y, x_1) = f(y, x_2)$. Similarly for the inversion security.   □

It seems to be essential to use x twice in this construction. The inversion security of H is never more than twice the maximum of the two. In the random-function model the composition of two functions usually causes more collisions, which makes it easy to distinguish a composition from a truly random function; however, here we are interested in the difficulty of finding collisions. Since the number of rounds in H is the sum of the rounds in f and g, one would heuristically expect the resultant function to be stronger.

Now we would like to obtain f and g that behave as if they were "independent." If the functions behaved almost like "random functions" then of course the construction would be secure. We want to provide a formal background for analyzing this. For this we use the model of *imperfect random functions* for the primitives. Any existing primitive with an estimated security (at least currently) can be thought of as an imperfect random function with appropriate parameters. For this we define two measures. First we consider a simpler (but more restrictive) bit-level parameter. We define the *bias* of a Boolean-valued random variable β to be $\Pr[\beta = 1]$.

Definition 8. *A function is called p-random if its output bits have bias p and are independent ($0 \le p \le 1$).*

Needless to say, considering the individual bits to be independent is less realistic, but it gives us a more reasonable heuristic than the perfect random function model. These functions are easily distinguished from truly random functions (for $p \ne \frac{1}{2}$) merely by observing the fraction of 1's in the output strings. Considering the output as a whole we make the following definition.

Definition 9. *A function (with n-bit values) is called h-imperfect if its output has min-entropy equal to hn.*

Here, the *min-entropy* of a source that outputs strings $x_1, \ldots, x_N$ with respective probabilities $p_1, \ldots, p_N$ is $\min\{-\lg p_i\}$. Note that the individual bits of the values of an h-imperfect function can be correlated, and they may not look at all random. However, when it comes to collision security, h-imperfect functions are good enough: High min-entropy is a necessary condition for a secure hash function; for example, if the min-entropy is low, then much of the probability may be concentrated in a small set, and it would to be easy to find a collision. But the condition is also sufficient, as shown by Lemma 11 below.

Obviously, a p-random function is $H(p)$-imperfect.

Lemma 10. *A collision adversary making a total of q queries to a p-random function will find a collision with probability at most $q^2/2^{\lambda H(p)/2}$, where $p = 1/2 + \epsilon$ and $\lambda = 1 - 2\epsilon^2 \lg_2 e$.*

Proof: Omitted due to space constraints.

Lemma 11. *A collision adversary making a total of q queries to an h-imperfect function (with n-bit values) will find a collision with probability at most $q^2/2^{hn/2}$.*

Proof: Let f be an h-imperfect function. If a string y occurs as an output of f with probability $p > 0$, then we have $-\lg p \geq hn$, or $p \leq 2^{-hn}$. And now consider a collision adversary that makes q queries to f, which we model as the random variables $Y_1, \ldots, Y_q$, taking respective values $y_1, \ldots, y_q$ in a particular run. The expected number of colliding pairs is

$$\sum_{i,j} \Pr[Y_i = Y_j] = \sum_{i,j} \Pr[Y_j = y_i] \leq \sum_{i,j} 2^{-hn} = \binom{q}{2} 2^{-hn} \leq q^2 2^{-hn}.$$

If we take $q = 2^{hn/2}$, then the expected number of collisions is at most $q^2 2^{-hn} = 1$. Thus the collision security of f is at least $2^{hn/2}$, as claimed.   □

4.2   Construction of "Independent" Primitives

One way to view the above construction is that it takes two imperfectly random functions and yields a function that is closer to being a truly random function. Towards this end, consider the following construction:

$$\bar{H}(x) = f(x) \oplus g(x)$$

It is relatively easy to analyze the construction at a bit level in the random-function model.

Lemma 12. *If f and g are p-random and independent, then the function $f \oplus g$ is q-random, with $|1 - 2q| = (1 - 2p)^2$.*

Proof: If α and β are two independent Boolean variables, both with bias p, then $\alpha \oplus \beta$ has bias $q = 2p(1 - p)$, and q satisfies $|1 - 2q| = (1 - 2p)^2$.   □

A Boolean variable with bias p has a gap between the probabilities of occurrence of 1 and 0 of $|p - (1 - p)| = |1 - 2p|$. The significance of this lemma is that the gap narrows quadratically as we pass from f or g to $f \oplus g$. Thus, for any p, after k iterations of this process with independent functions narrows the gap to $|1 - 2p|^{2^k}$.

However, this construction of $\bar{H}$ does not allow us to make a claim as in the above lemma, when we move from the idealized random-function world to a complexity-theoretic world where the functions involved are specific presumably collision-secure functions. That is, you can not show that $\bar{H}$ is collision-secure if f or g is. However in an imperfect random function model it is easy to show the following.

Given two imperfect random functions f and g, the expected number of queries to find collisions for H and $\bar{H}$ is the same in both cases. Thus the construction for H is "better" in that it allows us to analyze its security both in the complexity and in the random-function worlds. But we now show that the $\bar{H}$ construction has the surprising result of creating independence.

Intuitively, we say that f is independent of g if the provision of an oracle for finding collisions for f does not help in finding collisions for g. Note the asymmetry in the definition; for technical reasons, we must allow for the possibility that a collision-finder for g may help to find collisions for f.

Definition 13. *A collision oracle for f is a function from i to the set of finite binary strings. On input (i, k) it outputs a set $S_i = \{x_1, \ldots, x_j\} \subseteq \{0, 1\}^L$ $(2 \leq j \leq k)$ such that f has the same value on all points from the set. The set are distinct: for $i \neq j$, the sets S_i and S_j are unequal. The charge for a sequence of queries is the sum of $|S_i|$.*

We say that f is T-independent of g if, given a collision oracle for g, one incurs a charge of at least T for the calls to the oracle to find a collision for f. We say that f is independent of g if g is h-imperfect and f is T-independent of g with $T = \Omega(2^{nh/2})$

Now we claim the following. Let f and g be a-random and b-random respectively. Then $\bar{H}$ is T-independent of f, where T is the collision security of g. Similar comments apply for independence with respect to f. To see this, assume we are given an oracle for g to find collisions. $\bar{H}$ collides on the outputs of the oracle if and only if f collides on them as well. We can give similar constructions and results for imperfect random functions.

Thus in the imperfect oracle model, finding collisions for one of the functions does not help in finding collisions for the composite function at all, when the collision finder is used as a black box. Using this as a heuristic, if one takes f, g as SHA-1, MD5, and both are broken, then $\bar{H}$ would still need many calls to the collision finders, if the outputs of these functions behave approximately randomly even to a mild degree. This is helpful even if one of the hash functions is weak, for example, breakable in 2^{35} steps. In this case, finding collisions for the combination may still be infeasible.

For more detailed discussion and schemes based on these considerations, see the full version of this paper.

Acknowledgments. We thank Bart Preneel for helpful discussions. The third author thanks Yacov Yacobi, whose questions regarding a smart-card application provided the initial interest in this problem.

References

AV 96.   W. Aiello and R. Venkatesan. Foiling birthday attacks in length-doubling transformations. In *Advances in Cryptology—Eurocrypt '96*, Lecture Notes in Computer Science, Vol. 1070, ed. U.M. Maurer, pp. 307–320 (Springer-Verlag, 1996).

AB 96.   R. Anderson and E. Biham. Tiger: A Fast New Hash Function, In *Fast Software Encryption 3*, Lecture Notes in Computer Science, Vol. 1039 (Springer-Verlag, 1996).

BGV 96.   A. Bosselaers, R. Govaerts, J. Vandewalle. Fast hashing on the Pentium. In *Advances in Cryptology—Crypto '96*, ed. N. Koblitz, Lecture Notes in Computer Science, Vol. 1109, pp. 298–312 (Springer-Verlag, 1996).

BP 95.   A. Bosselaers and B. Preneel (eds.). *Integrity Primitives for secure information systems: Final report of RACE Integrity Primitives Evaluation (RIPE-RACE 1040)*, Chapter 3: RIPEMD. Lecture Notes in Computer Science, Vol. 1007 (Springer-Verlag, 1995).

BC$^+$ 88.   B. O. Brachtl, D. Coppersmith, M. M. Hyden, S. M. Matyas, Jr., C. H. W. Meyer, J. Oseas, Sh. Pilpel, and M. Shilling. Data authentication using modification detection codes based on a public one way encryption function. U.S. Patent No. 4,908,861, issued March 13, 1990. (Described in: C. H. Meyer and M. Shilling, Secure program load with modification detection code, In *Securicom 88: 6ème Congrès mondial de la protection et de la sécurité informatique et des communications*, pp. 111–130 (Paris, 1988).)

BY 90.   G. Brassard and M. Yung. One-way group actions. In *Advances in Cryptology—Crypto '90*, Lecture Notes in Computer Science, Vol. 537, pp. 94-107, (Springer-Verlag, 1991).

C 85.   D. Coppersmith. Another birthday attack. In *Advances in Cryptology—Crypto '85*, Lecture Notes in Computer Science, Vol. 218, pp. 14–17, (Springer-Verlag, 1986).

Dam 87.   I. Damgard. Collision-free hash functions and public-key signature schemes. In *Advances in Cryptology—Eurocrypt '87*, Lecture Notes in Computer Science, Vol. 304, pp. 203–217, Springer-Verlag (1988).

Dam 89.   I. Damgard. A design principle for hash functions. In *Advances in Cryptology—Crypto '89*, Lecture Notes in Computer Science, Vol. 435, pp. 416–427, Springer-Verlag (1988).

Dob 96a.   H. Dobbertin. Cryptanalysis of MD4. In *Fast Software Encryption*, Lecture Notes in Computer Science, Vol. 1039, ed. D. Gollman, pp. 53–69, Springer-Verlag (1996).

Dob 96b.   H. Dobbertin. Cryptanalysis of MD5 compress. Rump Session of Eurocrypt '96, presented by B. Preneel (May 1996). (Available at `http://www.iacr.org/conferences/ec96/rump/`.)

Dob 96c.   H. Dobbertin. The status of MD5 after a recent attack. *CrytoBytes*, Vol. 2, No. 2 (Summer 1996). (Available at `http://www.rsa.com/rsalabs/-pubs/cryptobytes/`.)

Dob 97.   H. Dobbertin. RIPEMD with two-round compress function is not collision-free. *Journal of Cryptology*, Vol. 10, No. 1, pp. 51–69 (1997).

Dob 98.   H. Dobbertin. The first two rounds of MD4 are not one-way. In *Fast Software Encryption*, Lecture Notes in Computer Science, Springer-Verlag (to appear).

DBP 96.   H. Dobbertin, A. Bosselaers, and B. Preneel. RIPEMD-160: A strengthened version of RIPEMD. In *Fast Software Encryption*, Lecture Notes in Computer Science, Vol. 1039, pp. 71–82, Springer-Verlag (1996).

GGH 96.   O. Goldreich, S. Goldwasser, and S. Halevi. Collision-free hashing from lattice problems. Theory of Cryptography Library, Record 96-09. (Available at `http://theory.lcs.mit.edu/~tcryptol/`.)

KP 97.   L. Knudsen, B. Preneel. Fast and secure hashing based on codes. In *Advances in Cryptology—Crypto '97*, Lecture Notes in Computer Science, Vol. 1294, pp. 485–498, Springer-Verlag (1997).

MMO 85.   S.M. Matyas, C.H. Meyer, and J. Oseas. Generating strong one-way func-
           tions with cryptographic algorithm. *IBM Technical Disclosure Bulletin,*
           vol. 27, pp. 5658–5659 (1985).

MOV 97.   A. Menezes, P. van Oorschot, S. Vanstone. *Handbook of Applied Cryptog-
           raphy* (CRC Press, 1997).

Merk 80.   R.C. Merkle. Protocols for public key cryptosystems. In *Proc. 1980 Sym-
           posium on Security and Privacy,* IEEE Computer Society, pp. 122–133
           (April 1980).

Merk 89.   R.C. Merkle. One-way hash functions and DES. In *Advances in
           Cryptology—Crypto '89,* Lecture Notes in Computer Science, Vol. 435,
           pp. 428–446 (Springer-Verlag, 1990).

Merk 90.   R.C. Merkle. A fast software one-way hash function. *Journal of Cryptol-
           ogy,* Vol. 3, pp. 43–58 (1990).

MH 81.   R.C. Merkle and M. Hellman. On the security of multiple encryption.
           Communications of the ACM, Vol. 24, No. 7, pp. 465–467 (July 1981).

MOI 90.   S. Miyaguchi, K. Ohta, and M. Iwata. 128-bit hash function (N-hash).
           NTT Review, vol. 2, pp. 128–132 (1990).

NY 89.   M. Naor and M. Yung. Universal one-way hash functions and their cryp-
           tographic applications. In *Proceedings of the 21st Symposium on Theory
           of Computing,* pp. 33–43 (ACM, 1989).

NIST 94.   National Institute of Standards and Technology. Secure Hash Standard.
           NIST Federal Information Processing Standard Publication 180-1 (May
           1994).

PV 96.   M. Peinado, R. Venkatesan. Highly parallel cryptographic attacks. In *Re-
           cent Advances in Parallel Virtual Machine and Message Passing Inter-
           face (EuroPVM-MPI'97),* Lecture Notes in Computer Science (Springer-
           Verlag, 1997).

Pre 93.   B. Preneel. *Analysis and Design of Cryptographic Hash Functions.*
           Ph.D. dissertation, Katholieke Universiteit Leuven (January 1993).

Pre 97.   B. Preneel, private communication (1997).

PGV 93a.   B. Preneel, R. Govaerts, J. Vandewalle. Hash functions based on block ci-
           phers: A synthetic approach. In *Advances in Cryptology—Crypto '93,* Lec-
           ture Notes in Computer Science, Vol. 773, pp. 368–378 (Springer-Verlag,
           1991).

PGV 93b.   B. Preneel, R. Govaerts, J. Vandewalle. Differential cryptanalysis of hash
           functions based on block ciphers. In *Proceedings of the 1st ACM Con-
           ference on Computer and Communications Security,* pp. 183–188 (ACM,
           1993).

R 78.   M.O. Rabin. Digitalized signatures. In *Foundations of Secure Computa-
           tion,* eds. R. Lipton, R. DeMillo, pp. 155–166 (Academic Press, 1978).

RP 95.   V. Rijmen, B. Preneel. Improved characteristics for differential cryptanal-
           ysis of hash functions based on block ciphers. In *Fast Software Encryption,*
           Lecture Notes in Computer Science, Vol. 1008, pp. 242–248 (Springer-
           Verlag, 1995).

Riv 90.   R. Rivest. The MD4 message digest algorithm. In *Advances in
           Cryptology—Crypto '90,* Lecture Notes in Computer Science, Vol. 537,
           pp. 303–311, (Springer-Verlag, 1991).

Riv 92.   R. Rivest. The MD5 Message-Digest Algorithm. Internet Network Work-
           ing Group Request for Comments 1321 (April 1992).

Sur 95.   Surety Technologies, Inc. Answers to Frequently Asked Questions about the Digital Notary™ System. http://www.surety.com (since January 1995).

vOW 94.   P. van Oorschot and M. Wiener. Parallel collision search with applications to hash functions and discrete logarithms. In *Proceedings of the 2nd ACM Conference on Computer and Communication Security*, pp. 210–218 (ACM Press, 1994).

Cryptanalytic Attacks on Pseudorandom Number Generators

John Kelsey[1], Bruce Schneier[1], David Wagner[2], and Chris Hall[1]

[1] Counterpane Systems
{kelsey,schneier,hall}@counterpane.com
[2] University of California Berkeley
daw@cs.berkeley.edu

Abstract. In this paper we discuss PRNGs: the mechanisms used by real-world secure systems to generate cryptographic keys, initialization vectors, "random" nonces, and other values assumed to be random. We argue that PRNGs are their own unique type of cryptographic primitive, and should be analyzed as such. We propose a model for PRNGs, discuss possible attacks against this model, and demonstrate the applicability of the model (and our attacks) to four real-world PRNGs. We close with a discussion of lessons learned about PRNG design and use, and a few open questions.

1 Introduction and Motivation

It is hard to imagine a well-designed cryptographic application that doesn't use random numbers. Session keys, initialization vectors, salts to be hashed with passwords, unique parameters in digital signature operations, and nonces in protocols are all assumed to be random[1] by system designers. Unfortunately, many cryptographic applications don't have a reliable source of real random bits, such as thermal noise in electrical circuits or precise timing of Geiger counter clicks [FMK85,Gud85,Agn88,Ric92]. Instead, they use a cryptographic mechanism, called a Pseudo-Random Number Generator (PRNG) to generate these values. The PRNG collects randomness from various low-entropy input streams, and tries to generate outputs that are in practice indistinguishable from truly random streams [SV86,LMS93,DIF94,ECS94,Plu94,Gut98].

In this paper, we consider PRNGs from an attacker's perspective. We discuss the requirements for PRNGs, give a basic model of how such PRNGs must work, and try to list the possible attacks against PRNGs. Specifically, we consider ways that an attacker may cause a given PRNG to fail to appear random, or ways he can use knowledge of some PRNG outputs (such as initialization vectors) to guess other PRNG outputs (such as session keys).

---

[1] Note that "random" is a word that is easily misused. In this paper, unless we say otherwise, the reader may assume that a "random value" is one sample of a random variable which is uniformly distributed over the entire set of n-bit vectors, for some n.

S. Vaudenay (Ed.): Fast Software Encryption – FSE'98, LNCS 1372, pp. 168–188, 1998.

1.1   Applications of Results

This research has important practical and theoretical implications:

1. A PRNG is its own kind of cryptographic primitive, which has not so far been examined in the literature. In particular, there doesn't seem to be any widespread understanding of the possible attacks on PRNGs, or of the limitations on the uses of different PRNG designs. A better understanding of these primitives will make it easier to design and use PRNGs securely.
2. A PRNG is a single point of failure for many real-world cryptosystems. An attack on the PRNG can make irrelevant the careful selection of good algorithms and protocols.
3. Many systems use badly-designed PRNGs, or use them in ways that make various attacks easier than they need be. We are aware of very little in the literature to help system designers choose and use these PRNGs wisely.
4. We present results on real-world PRNGs, which may have implications for the security of fielded cryptographic systems.

1.2   The Rest of This Paper

In Section 2, we define our model of a PRNG, and discuss the set of possible attacks on PRNGs that fit this model. In Section 3 discuss applications of those attacks on several real-world PRNGs. Then, in Section 4, we end with a discussion of the lessons learned, and a consideration of some related open problems.

2   Definitions

In the context of this paper, a PRNG is a cryptographic algorithm used to generate numbers that must appear random. Examples of this include the ANSI X9.17 key generation mechanism [ANSI85] and the RSAREF 2.0 PRNG [RSA94]. A PRNG has a secret state, S. Upon request, it must generate outputs that are indistinguishable from random numbers to an attacker who doesn't know and cannot guess S. In this, it is very similar to a stream cipher. Additionally, however, a PRNG must be able to alter its secret state by processing input values that may be unpredictable to an attacker. A PRNG often starts in an state that is guessable to an attacker (usually unintentionally), and must process many inputs to reach a secure state. Sometimes, the input samples are processed each time an output is generated: e.g., ANSI X9.17. Other times, the input samples are processed as they become available: e.g. RSAREF 2.0 PRNG.

Note that the inputs are intended to carry some unknown (to an attacker) information into the PRNG. These are the values typically collected from physical processes (like hard drive latencies [DIF94]), user interactions with the machine [Zim95], or other external, hard-to-predict processes. Typically, system implementers and designers will try to ensure that there is sufficient entropy in these inputs to make them unguessable by any practical attacker.

Note that the outputs are intended to stand in for random numbers in essentially any cryptographic situation. Symmetric keys, initialization vectors, random parameters in DSA signatures, and random nonces are common applications for these outputs.

See Figure 1 for a high-level view of a PRNG. Also, Figure 2 refines the terminology a bit, and Figure 3 shows a PRNG with periodic reseeding.

PRNGs are typically constructed from other cryptographic primitives, such as block ciphers, hash functions, and stream ciphers. There is a natural tendency to assume that the security of these underlying primitives will translate to security for the PRNG.

In this paper, we consider several new attacks on PRNGs. Many of these attacks may be considered somewhat academic. However, we believe there are situations that arise in practice in which these attacks are possible. Additionally, we believe that even attacks that are not *usually* practical should be brought to the attention of those who use these PRNGs, to prevent the PRNGs' use in an application that *does* allow the attacks.

Note that in principle, any method of distinguishing between PRNG outputs and random outputs is an attack; in practice, we care much more about the ability to learn the values of PRNG outputs not seen by the attacker, and to predict or control future outputs.

Fig. 1. Black-box view of a PRNG

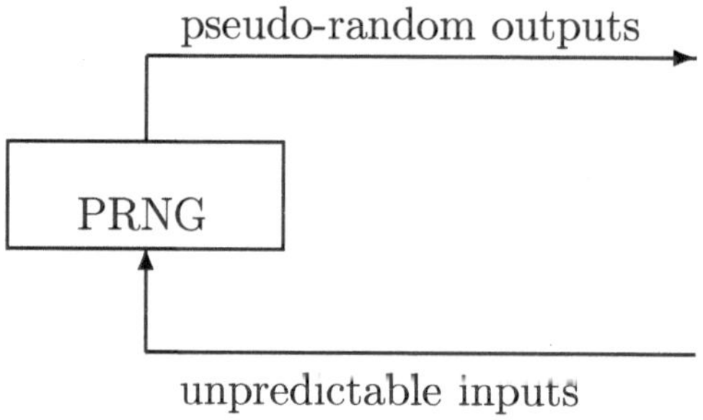

pseudo-random outputs

PRNG

unpredictable inputs

2.1   Enumerating the Classes of Attacks

1. **Direct Cryptanalytic Attack.** When an attacker is directly able to distinguish between PRNG outputs and random outputs, this is a direct cryptanalytic attack. This kind of attack is applicable to most, but not all, uses of PRNGs. For example, a PRNG used only to generate triple-DES keys may never be vulnerable to this kind of attack, since the PRNG outputs are never directly seen.

Fig. 2. View of internal operations for most PRNGs

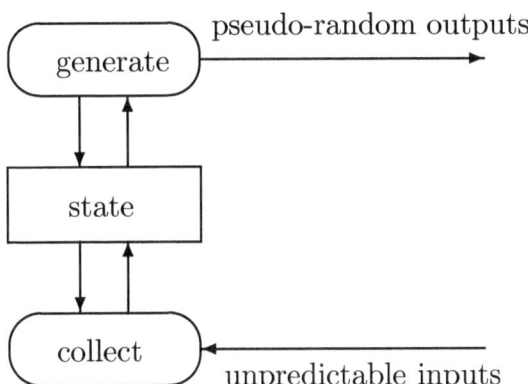

2. **Input-Based Attacks.** An input attack occurs when an attacker is able to use knowledge or control of the PRNG inputs to cryptanalyze the PRNG, i.e., to distinguish between PRNG output and random values.

   Input attacks may be further divided into *known-input*, *replayed-input*, and *chosen-input* attacks. Chosen input attacks may be practical against smart-cards and other tamper-resistant tokens under a physical/cryptanalytic attack; they may also be practical for applications that feed incoming messages, user-selected passwords, network statistics, etc., into their PRNG as entropy samples. Replayed-input attacks are likely to be practical in the same situations, but require slightly less control or sophistication on the part of the attacker. Known-input attacks may be practical in any situation in which some of the PRNG inputs, intended by the system designer to be hard to predict, turn out to be easily predicted in some special cases. (An obvious example of this is an application which uses hard-drive latency for some of its PRNG inputs, but is being run using a network drive whose timings are observable to the attacker.)

3. **State Compromise Extension Attacks.** A state compromise extension attack attempts to extend the advantages of a previously-successful effort that has recovered S as far as possible. Suppose that, for whatever reason—a temporary penetration of computer security, an inadvertent leak, a cryptanalytic success, etc.—the adversary manages to learn the internal state, S, at some point in time. A state compromise extension attack succeeds when the attacker is able to recover unknown PRNG outputs (or distinguish those PRNG outputs from random values) from before S was compromised, or recover outputs from after the PRNG has collected a sequence of inputs which the attacker cannot guess.

   State compromise extension attacks are most likely to work when a PRNG is started in an insecure (guessable) state due to insufficient starting entropy. They can also work when S has been compromised by any of the attacks in this list, or by any other method. In practice, it is prudent to assume that occasional compromises of the state S may happen; to preserve the robustness of the system, PRNGs should resist state compromise extension attacks as thoroughly as possible.

(a) **Backtracking Attacks.** A backtracking attack uses the compromise of the PRNG state S at time t to learn previous PRNG outputs.

(b) **Permanent Compromise Attacks.** A permanent compromise attack occurs if, once an attacker compromises S at time t, all future and past S values are vulnerable to attack.

(c) **Iterative Guessing Attacks.** An iterative guessing attack uses knowledge of S at time t, and the intervening PRNG outputs, to learn S at time $t + \epsilon$, when the inputs collected during this span of time are guessable (but not known) by the attacker.

(d) **Meet-in-the-Middle Attacks.** A meet in the middle attack is essentially a combination of an iterative guessing attack with a backtracking attack. Knowledge of S at times t and $t + 2\epsilon$ allow the attacker to recover S at time $t + \epsilon$.

3   Attacking Real-World PRNGs

In this section we discuss the strengths and weaknesses of four real-world PRNGs: the ANSI X9.17 PRNG, the DSA PRNG, the RSAREF PRNG, and CryptoLib.

3.1   The ANSI X9.17 PRNG

The ANSI X9.17 PRNG [ANSI85,Sch96] is intended as a mechanism to generate DES keys and IVs, using triple-DES as a primitive. (Of course, it is possible to replace triple-DES with another block cipher.) It has been used as a general-purpose PRNG in many applications.

1. K is a secret triple-DES key generated somehow at initialization time. It must be random and used only for this generator. It is part of the PRNG's secret state which is never changed by any PRNG input.

2. Each time we wish to generate an output, we do the following:
   (a) $T_i = E_K(\text{current timestamp})$.
   (b) output$[i] = E_K(T_i \oplus \text{seed}[i])$.
   (c) seed$[i + 1] = E_K(T_i \oplus \text{output}[i])$.

This generator is in widespread use in banking and other applications.

Direct Cryptanalytic Attack Direct cryptanalysis of this generator appears to require cryptanalysis of triple-DES (or whatever other block cipher is in use.) As far as we know, this has never been proven, however.

Input-Based Attacks The X9.17 PRNG has a certificational weakness (assuming a 64-bit block size) with respect to replayed-input attacks.

An attacker who can force the T values to freeze can distinguish the PRNG's outputs from random outputs after seeing about 2^{32} 64-bit outputs. In a sequence of random 64-bit numbers, we would expect to see a collision after about 2^{32} outputs. However, with T frozen, we expect a collision from X9.17 to require about 2^{63} outputs. This is a mostly academic weakness, but it may be relevant in some applications.

Otherwise, knowledge or control of inputs does not appear to weaken the PRNG against an attacker that doesn't know K.

State Compromise Extension Attacks The X9.17 PRNG does not properly recover from state compromise. That is, an attacker who compromises the X9.17 triple-DES key, K, can compromise the whole internal state of the PRNG from then on without much additional effort.

Two Design Flaws in X9.17 There are two flaws in the ANSI X9.17 PRNG that become apparent only when the PRNG is analyzed with respect to state compromise extension attacks.

1. Only 64 bits of the PRNG's state, $seed[i]$, can ever be affected by the PRNG inputs. This means that once an attacker has compromised K, the PRNG can never fully recover, even after processing a sequence of inputs the attacker could never guess.
2. The $seed[i+1]$ value is a function of the previous output, the previous T_i, and K. To an attacker who knows K from a previous state compromise, and knows the basic properties of the timestamp used to derive T_i, $seed[i+1]$ is simply not very hard to guess.

The Permanent Compromise Attack: Deriving the Internal State from Two Outputs Consider an attacker who learns K. Much later, after the seed internal variable has become totally different, he is given two successive outputs, output$[i, i+1]$. (He has not seen any intervening outputs from the PRNG.) The attacker's goal will be to learn the value of seed$[i+1]$. Of course, one can trivially mount a 64-bit search and learn the seed value.

However, there is a much more effective way to mount this attack. Suppose that each timestamp value has ten bits that aren't already known to the attacker. (This is a reasonable assumption for many systems. For example, consider a millisecond timer, and an attacker who knows to about the nearest second when an output was generated.) An attacker with two successive outputs can mount a meet-in-the-middle attack to discover the internal seed value, requiring about 2^{11} trial encryptions under the known key K. This works because we have

$$\text{seed}[i+1] = D_K(\text{output}[i+1]) \oplus T_{i+1}$$

$$\text{seed}[i+1] = E_K(\text{output}[i] \oplus T_i)$$

The attacker tries all possible values for T_i, and forms one sorted list of possible seed$[i+1]$ values. He then tries all possible values for T_{i+1}, and forms another sorted list of possible seed$[i + 1]$ values. The correct seed$[i + 1]$ value is the one that appears in both lists.

The Iterative Guessing Attack If an attacker knows seed$[i]$, and sees some function of output$[i + 1]$, he can learn seed$[i + 1]$ in almost all cases. This is true because the timestamp sample will seldom have much entropy. Using our earlier assumption of ten bits of entropy per timestamp sample, this means the attacker will need only a ten-bit guess. Note that the attacker needs only to see a *function* of the output, not the output itself. This means that a message encrypted with a key derived from the output value is sufficient to mount this attack. (Note the difference between this and the permanent compromise attack, above, in which the attacker needs raw PRNG outputs.)

Backtracking The attacker can move backwards as easily as forward with the iterative guessing attack, assuming he can find functions of the PRNG outputs. Alternatively, he may look for the successive pair of directly available PRNG outputs nearest to the unknown outputs he wants to learn, and mount the permanent compromise attack there.

Meet-in-the-Middle Attack Sometimes, a PRNG may generate a large secret value, and not directly output any bits of it. The attacker may thus know seed$[i]$ and seed$[i + 8]$, but no intervening values. Since this leaves him with (say) 80 bits of entropy, it might be naively assumed that he cannot recover these output values. However, this isn't necessarily the case, because a meet-in-the-middle attack is available. This works as follows:

1. The attacker mounts the attack described above to learn the PRNG state before and after the run of values that were used together.
2. The attacker carries out a meet-in-the-middle attack, deriving one set of possible values for seed$[i + 4]$ by guessing $T_{i+1..i+4}$ and deriving a second list by guessing $T_{i+5..i+8}$. If each sequence of four timestamps holds 40 bits of entropy, this will require 2^{41} effort. The correct value of seed$[i + 4]$ will be present in both lists, so the seed$[i + 4]$ values that match (there will be about 2^{16} of these) yield the possible sequences of timestamps, and thus, output blocks.
3. The attacker can try all these possible output sequences until he finds the right one. (For example, if the eight output blocks are used as an encryption key, 2^{16} trial decryptions will suffice to eliminate all the false alarms.)

Timer Entropy Issues In the above discussion, we have assumed that individual PRNG inputs have fixed amounts of entropy, and thus, take fixed amounts of effort to guess. In practice, this usually won't be the case. An RSA keypair generation might reasonably use two 512-bit pseudorandom starting points, thus requiring a total of sixteen PRNG output requests. However, these calls will almost certainly be made in rapid succession. Unless the timestamp on which the T_i values are based has a great deal of precision, many of these T_i values will be based on the same or very close timestamp values. This may well make meet-in-the-middle attacks practical, even though it might normally make sense to estimate at least three bits of unpredictability per timestamp.

Summary The ANSI X9.17 key generator appears to be fairly secure from all attacks that *don't* involve either stopping the timer used or compromising the internal triple-DES key. Replaying any timer input about 2^{32} times leads to a certificational weakness: a way to distinguish large numbers of X9.17 PRNG outputs from a truly random sequence of bits. Compromising the internal triple-DES key completely destroys the X9.17 PRNG: it never recovers, even after getting thousands of bits worth of entropy in its sampled timer inputs[2].

For systems that use X9.17, the most obvious way to resist this class of attack is to occasionally use the current X9.17 state to generate a whole new X9.17 state, including a new K and a new starting $seed[0]$.

3.2   The DSA PRNG

The Digital Signature Standard specification [NIST94] also describes a fairly simple PRNG based on SHA (or, alternatively, a DES construction) which was intended for generating pseudorandom parameters for the DSA signature algorithm. Since this generator appears to come with an NSA stamp of approval, it has been used and proposed for applications quite different than those for which it was originally designed.

The DSA PRNG allows an optional user input while generating keys, but not while generating DSA signature parameters. For our purposes, though, we will assume that the PRNG can be given user inputs at any time, as is true with the other PRNGs discussed in this paper. Each time the DSA PRNG generates an output, it may be provided with an optional input, W_i. Note that omitting the input from the PRNG design would guarantee that the PRNG could never recover from a state compromise.

All arithmetic in this PRNG is allowed to be done modulo 2^N, where $160 \leq N \leq 512$. In the remainder of this document, we will assume this modulus to be

---

[2] Wei Dai's Crypto++ library [Dai97] includes an implementation of a X9.17 variant with increased security against seed compromise attacks. That variant is

1. $T_i = E_K(T_{i-1} \oplus \text{current timestamp})$.
2. $\text{output}[i] = E_K(T_i \oplus \text{seed}[i])$.
3. $\text{seed}[i+1] = E_K(T_i \oplus \text{output}[i])$.

This corresponds to encrypting the timestamps in CBC mode, instead of in ECB mode as is done in the standard X9.17 generator. The timestamp is based on the program's CPU usage, and its resolution is platform-dependent; on Linux, it has a 0.01 second resolution. We have not examined this PRNG closely, but we note that our permanent compromise attack, above, can be extended to work on Crypto++'s X9.17 variant at a cost of requiring a 2^{64} search in the attack

160, since this is the weakest value (with respect to one attack) that is allowed by the design.

The DSA PRNG works as follows:

1. The PRNG maintains an ever-changing state, X_i.
2. The PRNG accepts an optional input, W_i. This may be assumed to be zero if not supplied.
3. The PRNG generates each output as follows:
   (a) output$[i]$ = hash$(W_i + X_i \bmod 2^{160})$
   (b) $X_{i+1} = X_i + \text{output}[i] + 1 \pmod{2^{160}}$

Direct Cryptanalytic Attack If the PRNG's hash function is good, then the resulting output sequence appears to be hard to distinguish from a random sequence. It would be nice, from a system designer's point of view, to have some proof of the quality of this PRNG's outputs based on the collision-resistance or one-wayness of the hash function; to our knowledge, no such proof exists.

Input-Based Attacks Consider an attacker who can control the inputs sent into W. If these inputs are sent directly in, there is a straightforward way to force the PRNG to repeat the same output forever. This has a direct relevance if this PRNG is being used in a system in which the attacker may control some of the entropy samples sent into the PRNG. To force the PRNG to repeat, the attacker forms

$$W_i = W_{i-1} - \text{output}[i-1] - 1 \pmod{2^{160}}$$

This forces the seed value to repeat, which forces the output values to repeat. Note, however, that this attack fails quickly when the user hashes his entropy samples before sending them into the PRNG. In practice, this is the natural way to process the inputs, and so we suspect that few systems are vulnerable to this attack.

State Compromise Extension Attacks The DSA PRNG doesn't handle state compromises as well as we might have liked, but it is much better in this regard than ANSI X9.17. Consider an attacker who has somehow compromised the entire internal state of the PRNG, but then lost track of its inputs and outputs for a long period. If enough entropy existed in those samples, then the DSA PRNG will become as strong as ever against attack.

Leaking Input Effects Just as with ANSI X9.17, the DSA PRNG leaks the effects of unguessable inputs in its output. Consider an attacker who has compromised the PRNG's state. The application feeds in an input that the attacker can't guess (e.g., a sample with 90 bits of entropy). If the attacker sees the next output, he doesn't need to guess the sample, because the only effect on future outputs this sample can have is through that output. Note that if the new X_{i+1} depended directly on W_i and X_i, this weakness wouldn't exist. An attacker who knew the state could still try *guessing* the entropy sample, but if he did not guess the right value, he would lose knowledge of the state.

The Iterative Guessing Attack This PRNG is vulnerable to an iterative guessing attack after the state has been compromised. That is, if an attacker knows X_i and knows that W_i has only 20 bits of entropy, he can mount a 2^{20} search, and have a list of 2^{20} 160-bit outputs, one of which is output$[i]$. Note that the attacker needs only a function of the output that he can check, such as a DSA signature made with output$[i]$ as its secret parameter value. Note also that knowledge of the correct value for output$[i]$ also uniquely determines the value of X_{i+1}.

Backtracking If an attacker knows X_i, and output$[i-1]$, then he is clearly able to backtrack to knowledge of X_{i-1}. This doesn't immediately gain him much, since he has to already know output$[i-1]$ to be able to do this. However, in some circumstances, this could turn out to be useful.

Filling in the Gaps Consider a situation in which the attacker knows X_i, X_{i+2}, and output$[i+1]$, but still needs to know output$[i]$. In this case he can solve for output$[i]$ directly:

$$\text{output}[i] = X_{i+2} - X_i - 2 - \text{output}[i+1]$$

Summary The DSA standard's PRNG appears to be quite secure when used in the application for which it was designed: DSA signature parameter generation. However, it doesn't perform well as a general-purpose cryptographic PRNG because it handles its inputs poorly, and because it recovers more slowly from state compromise than it should.

To adapt the DSA PRNG to more general use, the following measures would eliminate most of the attacks we have observed:

1. Require hashing of all PRNG inputs before applying them.
2. Update X by the following formula:

$$X_{i+1} = X_i + \text{hash}(\text{output}[i] + W_i).\ \text{modulo } 2^{160}$$

3.3   The RSAREF PRNG

The PRNG included with RSAREF 2.0 is built almost entirely around two operations: MD5 hashing and addition modulo 2^{128}. It is the most conceptually simple design of any we have analyzed. The RSAREF 2.0 PRNG consists of the following:

1. A 128 bit counter, C_i.
2. A method for processing inputs. To process input X_i, we do the following:

$$C_{i+1} = C_i + \text{MD5}(X_i)\ \text{modulo } 2^{128}.$$

3. A method for generating outputs. To generate output *output*$[i]$, we do the following:

$$\text{output}[i] = \text{MD5}(C_i)\ \text{modulo } 2^{128}$$

$$C_{i+1} = C_i + 1\ \text{modulo } 2^{128}.$$

Direct Cryptanalytic Attack We will treat MD5 as a random function. While there have been interesting cryptanalytic results on MD5 in the last several years, none of them offer an obvious way to attack the RSAREF PRNG.

Partial Precomputation Attack There is a straightforward attack on a counter-mode generator of this kind: an attacker chooses some number of successive outputs, t, that he expects to see, and then computes the hash of every tth possible counter value. He is guaranteed to see one of these hashes after t outputs; at that point, he knows the whole counter value. This attack is impractical for a 128-bit counter, but it gives an upper bound on the strength of this generator. With 2^{32} outputs, an attacker would need to do a 2^{96} precomputation to mount the attack; with 2^{48} outputs, he would need to do a 2^{80} precomputation. These attacks also require a great deal of memory, though time/memory trade-offs can reduce that.

Timing Attack The C code to add to and increment the 128-bit internal counter has the property that it will leak some information about the resulting 128-bit counter by how many 8-bit add operations the computer must execute. This opens a timing channel for an attacker.

An attacker able to observe the time taken to generate each new output can learn how many zero bytes are in the counter each time it is incremented. This is simply a matter of determining how many bytewise additions had to be done to increment the counter properly. There are two facets to this attack. First, counter values that are all-zero in their low-order few bytes leak a great deal of information through the timing channel; these can be considered a kind of weak state. Second, when combined with the partial precomputation attack discussed above, the timing information can be used to know when to bother checking the PRNG output against precomputed table. This is a small advantage.

Input-Based Attacks We note that several input-based attacks are possible against RSAREF's PRNG. In particular, chosen input attacks exist against the RSAREF PRNG. They become quite powerful when the attacker can also monitor precise timing information from the machine on which the PRNG is running.

Shortening the Cycle with a Chosen-Input Attack An attacker can force the RSAREF PRNG into a shortened cycle by choosing the input value properly. Let $input_n$ be a chosen input for the PRNG such that MD5($input_n$) has all ones in its low-order n bytes. If an attacker requests a long sequence of outputs by requesting these inputs once per output, he forces the PRNG to cycle much faster, because the low-order n bytes of the counter are fixed. Thus, for $n = 8$, the cycle length is shortened to 2^{64} outputs. Note that the attacker doesn't know what those n bytes are, but he *does* know that they are the same every time the PRNG uses them to generate another output.

A more powerful way to shorten the cycle takes advantage of the birthday paradox. Suppose x_1, x_2 are two chosen inputs such that MD5(x_1)+MD5(x_2) has all ones in its low-order n bytes. Then an attacker can feed the periodic sequence $x_1, x_2, x_1, x_2, \ldots$ as inputs to the RSAREF PRNG and observe the outputs; with this procedure, he should see a cycle after about 2^{128-8n} outputs. For example, for the case $n = 16$, it takes about 2^{64} offline work to find suitable x_1, x_2, if an attacker uses an efficient collision search algorithm (see e.g. [OW95,OW96]); this choice of chosen inputs will force the RSAREF generator to repeat immediately [3].

More generally, we can get a simple "time travel" attack: if no new inputs were mixed in during the last j outputs, then the attacker can send the RSAREF PRNG back in time j steps by finding two chosen inputs whose MD5 digests sum to $-j$ (again with the same time complexity).

A Timing + Chosen Input Attack A much more powerful attack is available if the attacker can monitor precise operation timings, and if MD5 operates in constant time. The counter increment operation in the RSAREF source code will leak how many zero bytes are in the resulting counter value by how many 8-bit additions were required, and thus, by how long the counter increment operation took. During the counter increment operation (unlike the add operation used to combine in entropy from a input), detecting n 8-bit additions means that the resulting low-order $n - 1$ bytes are zero.

The attack occurs in two stages: in the *precomputation* stage, which is done once, the attacker generates the chosen entropy values he is to use later, and also generates a table of hashed counter values. In the *Execution* stage, which is done each time he wishes to attack some RSAREF PRNG state, he uses those chosen entropy values to force the internal counter to a value that has its low-order 104 bits set to all zeros. The attack requires 2^{48} offline trial hashes and 2000 chosen-entropy requests.

The precomputation stage works as follows:

1. For $n = 1$ to 12, the attacker finds $\text{input}_{0,n}, \text{input}_{1,n}$ such that

$$\text{MD5}(\text{input}_{0,n}) + \text{MD5}(\text{input}_{1,n})$$

   is all ones in its low-order n bytes, and that its next lowest order byte is even. This is expected to take about 2^{4n} effort using a collision-search algorithm.

The stage of executing the attack works as follows:

1. The attacker watches increment timing values until he knows that the low-order byte of the counter is a zero. (He can see this, because of the extra addition operation, which alters the time taken for the input to be processed.)
2. For $n = 1$ to 12, he does the following:
   (a) He requests update with input_n. This forces the counter value to be all ones in its low n bytes.

---

[3] We note that MD5 is designed for only 64 bits of collision-resistance, and so perhaps might not be expected to provide more than 64 bits of security. However, this PRNG appears to be in use for generating 1024-bit RSA moduli and establishing triple-DES keys, so it is apparently being trusted for more than 64 bits of security.

(b) He requests an output value, and observes the time taken for the output generation, inferring how many times the PRNG executed an 8-bit add operation in the increment. He keeps requesting the update with $input_n$ and the output, until he gets $n+2$ 8-bit add operations, instead of $n+1$.

(c) At this point, he has forced the low $n+1$-bytes to zeros.

3. At the end of the above loop, the attacker has forced the low-order thirteen bytes of the counter to zero values. He now carries out a brute-force search of the remaining three bytes of C, and breaks the PRNG.

State Compromise Extension Attacks

Losing Entropy in Inputs The PRNG's input-processing mechanism has a potentially dangerous flaw: it is order-independent. That is, updating the PRNG with A, and then with B, is the same as updating it first with B, and then with A. This flaw was originally discovered by Paul Kocher [Koc95,Bal96], but it is still worth noting here. The effect of this is to make the PRNG more likely to start in an insecure state, and also to make the PRNG require considerably more entropy in its inputs before its state is unguessable.

Iterative Guessing The iterative guessing attack works here. That is, if an attacker has compromised C_i, each time the user updates his state with some X_i guessable by an attacker, and then generates an output, $output[i+1]$, which the attacker can see (even if that output is used as a symmetric encryption or authentication key, or as a key or pad encrypted under a public-key) he can maintain his knowledge of the PRNG's state. If the RSAREF PRNG manages to get updated with an unguessable input between a compromised state and a visible output, however, then he loses his knowledge of the state.

Backtracking The RSAREF PRNG is vulnerable to backtracking in a straightforward way. The iterative guessing attack works exactly as well backward as forward, and when an attacker doesn't have new entropy samples, backtracking is exactly as easy as walking the generator forward.

3.4   Summary

The RSAREF 2.0 PRNG is vulnerable to chosen-input attacks which can force it into short cycles, chosen-input timing attacks which can reveal its secret state, and iterative guessing and backtracking attacks which can allow an attacker to extend his knowledge of the secret state backward and forward through time. It also must be used very carefully, due to the fact that inputs affect it in an order-independent way.

To minimize the danger of these attacks, we make the following recommendations:

1. Guard against chosen-input attacks in the design of the system that uses the RSAREF PRNG.

2. Be careful using the RSAREF PRNG in situations where timing information might be leaked.
3. Append a current timestamp and/or a counter to all inputs before sending them into the PRNG, to eliminate the order-independence of PRNG inputs.

3.5   Cryptolib's PRNGs

Cryptolib is a cryptographic library developed primarily by Jack Lacy, Donald Mitchel, William Schnell, and Matt Blaze, and initially described in [LMS93]. The primary source of randomness in Cryptolib is TrueRand, a mechanism for pulling (hopefully) unpredictable values out of the clock skew between different timers available to the system. These values can be used directly (though the documentation warns callers not to rely on more than 16 bits of entropy per 32-bit word), or can be used to seed one of the available pseudorandom number generators, fsrRand or desRand.

fsrRand and desRand are not PRNGs by our definition, but rather are stream ciphers. That is, they do not have defined mechanisms for processing additional inputs "on the fly," but rather are seeded once and then run to generate pseudorandom numbers. This is not unreasonable, given the assumption that TrueRand delivers truly random bits as needed–the system designer can simply generate a whole new state every few minutes, and otherwise needn't worry about entropy collection. When combined, TrueRand and fsrRand or TrueRand and desRand can be analyzed in the same way as the other PRNGs in this paper. That is, we assume that the system initializes the state of either fsrRand or desRand using TrueRand, and uses one of these mechanisms to generate whatever pseudorandom values are needed, and that the whole mechanism is periodically reinitialized from TrueRand. TrueRand is thus the source of PRNG inputs, and fsrRand or desRand is the source of PRNG outputs.

Description of Algorithms

fsrRand fsrRand is described in [LMS93]. Its secret state consists of a secret DES key, K, and an array of seven 32-bit values, $X_{0..6}$, organized as a shift-register. Each time an output is required, two of the 32-bit values are taken and concatenated to form a 64-bit value. This value is encrypted with DES under the secret key. The resulting ciphertext is split into two 32-bit halves; one half is XORed back into one of the 32-bit values (in the same way a shift register value might be updated), the other half is output. The register is then shifted, so that two new values will be used to generate the next output. A more complete description can be found in [LMS93].

desRand desRand appears in the Cryptolib source code (version 1.2). Its secret state consists of a 64-bit counter, C, a secret three-key triple-DES key, K, a secret 20-byte prefix, P, and a secret 20-byte suffix, S. Each new 32-bit output is generated as follows:

1. Use the SHA1 hash function to compute $hash(P, C, S)$.
2. Use triple-DES to compute $E_K(C)$.
3. XOR together the high-order bytes of the hash value with the result from the encryption; output the high-order four bytes of this result.
4. Set $C = C + 1$.

Direct Cryptanalysis

fsrRand There is a direct cryptanalytic attack on fsrRand requiring 2^{89} effort. The attack uses the fact that, once the attacker knows K and any one PRNG output, he can build a table of the 2^{32} possible halves of the DES ciphertext that was used for feedback. For each value, he gets a whole 64-bit ciphertext, which he can decrypt into a 64-bit plaintext, yielding both 32-bit values from the array.

1. The attacker guesses the key, K.
2. The attacker gets the output when the shift register pairs used are (X_i, X_j), (X_j, U), and (V, X_i) for some other U and V. In the pair (A, B), A will be updated with the feedback.
3. For the first two output values, the attacker computes all 2^{32} possible feedback values (the 32-bit half of the DES ciphertext that was not output). This allows him to compute X_j. For each K guess, we expect there to be only one pair of feedback guesses that leads to the same X_j value.
4. The attacker uses the feedback value from the first output (learned in the previous

   step) to compute what the new X_i value should be. He then mounts another 2^{32} guess of the feedback value for the third step, and uses this to derive the current X_i and other register value. If he has the wrong K value, he expects not to find any matching value for X_i; if he has the right K value, he expects to find one value that agrees.

This demonstrates a certificational weakness in fsrRand, at most; the computational requirements are very probably outside the reach of any attacker right now[4].

desRand We are not aware of any direct cryptanalytic attacks available on desRand. The desRand design appears to us to be very conservative, and unlikely to be attacked in the future. Note that nothing like the timing attack on RSAREF's PRNG is available here, despite the use of a counter.

---

[4] It could be argued that since DES has only 56 bits of strength, this construction was intended for no more strength than that. We find this argument unconvincing. fsrRand was clearly an attempt to get more than 56 bits of strength from DES; otherwise, DES in OFB- or counter-mode would have been used.

Input-Based Attacks These systems accept input only once, and accept it directly from TrueRand or a buffer provided by the caller. This (re)initializes the PRNG. In the context of the following discussion, a known-input attack means that the attacker has learned how to predict some TrueRand values. Clearly, if the attacker can know all the TrueRand values, there is no cryptanalysis to be performed. An interesting result occurs if the PRNG becomes weak with only a small number of predictable TrueRand values.

fsrRand An attacker who knows any two X values used as a plaintext block for DES can mount a keysearch attack, and reduce the possible number of keys to about 2^{24}. He must then wait until the first of those values makes it into the DES input again, and carry out an additional 2^{32} search per candidate key; this will determine the key uniquely. This requires a total of about 2^{57} trial encryptions, and about 2^{24} blocks of memory. From this point, the attacker can quickly recover the remaining state of the PRNG. An attacker who can guess any two such X values with 2^t work can mount the same attack with 2^{57+t} trial encryption and 2^{24} blocks of memory.

An attacker who knows the key, K can recover the remaining PRNG state with a 2^{33} effort, using the same method described for direct cryptanalysis of the PRNG, above.

A more subtle concern might involve flaws in the quality of seed values from TrueRand. Consider an attacker who knows, for a given system, that only 2^8 32-bit outputs from TrueRand are possible. If fsrRand is reseeded directly from TrueRand, this leads to fairly simple attack: fsrRand's DES key must come from TrueRand, and the attacker can quickly list all possible 56-bit values that could have been generated, getting about 2^{16} of them. He can then carry out the attack described above. In general, if there are 2^m possible values for the fsrRand's DES key to get, then the attack will take 2^{m+33} trial decryptions. This is an improvement for $m < 56$, naturally.

Note that this demonstrates that fsrRand doesn't profit from the full entropy it receives during reseeding; In the example above, fsrRand would get 8 bits of entropy per 32-bit word used to reseed it, for a total of 112 bits of entropy.

desRand We are aware of no reasonable known-input attacks on desRand. An attacker with knowledge of C, P, and S, but not K, appears to have no chance to defeat the PRNG; similarly, and attacker with knowledge of C and K, but not P or S, appears to have no chance to defeat the PRNG.

State Compromise Extension Attacks The desRand and fsrRand generators don't process inputs, and so can never recover from state compromise. However, if TrueRand is used to generate a whole new state every few minutes, the scope of a state compromise is made very small. It is worth noting that both desRand and fsrRand allow an attacker in possession of their current state to go backward as well as forward, learning all values ever generated by them. That is, if the PRNG state ever is compromised, the attacker can learn every output ever generated by that state. If the system is designed to reinitialize its PRNG with TrueRand values once every hour, then this means a compromise of all PRNG outputs for that hour. If the system reinitializes the PRNG more frequently, then the attacker learns fewer outputs; if less frequently, then the attacker learns more outputs.

Summary Assuming TrueRand is a good source of unpredictable values, the PRNGs built by putting it together with either fsrRand or desRand appear to be of reasonable strength, but desRand appears to be more resistant to various attacks than fsrRand. Note, however, that nearly all of these attacks require keysearching DES or doing some similarly computationally expensive task.

We make the following recommendations:

1. System designers should verify both by statistical analysis and by an analysis of their target systems' designs whether TrueRand will reliably provide unpredictable numbers on their systems. (This holds true for every source of unpredictable inputs, for every PRNG.)
2. In environments where TrueRand's outputs may be suspect (perhaps due to malicious actions by the attacker), we recommend that desRand, rather than fsrRand, be employed.

4   Summary, Conclusions, and Open Problems

In this paper, we have argued for treating PRNGs as their own kind of cryptographic primitive, distinct from stream ciphers, hash functions, and block ciphers. We have discussed the requirements for a PRNG, developed abstract attacks against an idealized PRNG, and then demonstrated those attacks against four real-world PRNGs.

4.1   Guidelines for Using Vulnerable PRNGs

In the earlier sections, we discussed possible countermeasures for many of the attacks we had developed. Here, we propose a list of ways to protect a PRNG against each of the classes of attacks we discussed.

1. **Use a hash function to protect vulnerable PRNG outputs.** If a PRNG is suspected to be vulnerable to direct cryptanalytic attack, then outputs from the PRNG should be preprocessed with a cryptographic hash function. Note that not all possible flawed PRNGs will be secure even after hashing their outputs, so this doesn't guarantee security, but it makes security much more likely.
2. **Hash PRNG inputs with a counter or timestamp before use.** To prevent most chosen-input attacks, the inputs should be hashed with a timestamp or counter, before being sent into the PRNG. If this is too expensive to be done every time an input is processed, the system designer may want to only hash inputs that could conceivably be under an attacker's control.

Fig. 3. Generalized PRNG, with periodic reseeding

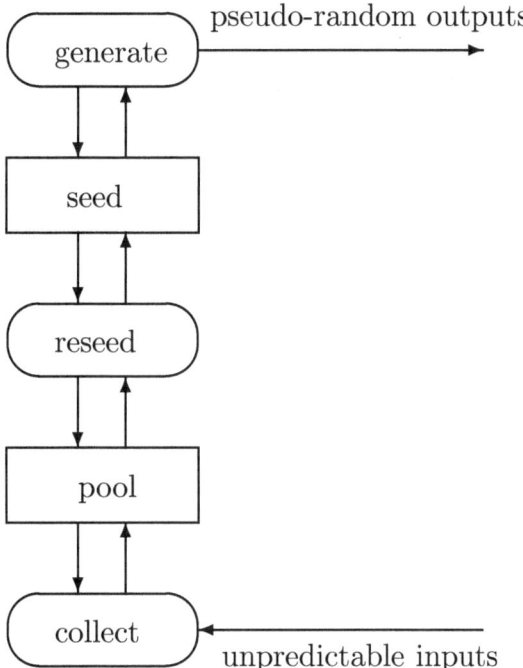

3. **Occasionally generate a new starting PRNG state.** For PRNGs like ANSI X9.17, which leave a large part of their state unchangeable once initialized, a whole new PRNG state should occasionally be generated from the current PRNG. This will ensure that any PRNG can fully reseed itself, given enough time and input entropy.
4. **Pay special attention to PRNG starting points and seed files.** The best way to resist all the state-compromise extension attacks is simply never to have the PRNG's state compromised. While it's not possible to guarantee this, system designers should spend a lot of effort on starting their PRNG from an unguessable point, handling PRNG seed files intelligently, etc. (See [Gut98] for several ways that this can be done.)

4.2   Guidelines for Designing a PRNG

Having described a set of possible attacks on PRNGs, it is reasonable to try to discuss ways to develop new PRNGs that will resist them. We propose the following guidelines for developing new PRNGs:

1. **Base the PRNG on something strong.** The PRNG should be designed so that a successful direct cryptanalytic attack implies a successful attack

on some cryptographic primitive that's believed to be strong. Ideally, this would be proven.

2. **Make sure the whole PRNG state changes over time.** The whole secret internal state should change over time. This prevents a single state compromise from being unrecoverable.

3. **Do "catastrophic reseeding" of the PRNG.** The part of the internal state that is used to generate outputs should be separate from the entropy pool. The generation state should be changed only when enough entropy has been collected to resist iterative guessing attacks, according to a conservative estimate. Figure 3 depicts a possible architecture for implementing catastrophic reseeding.

4. **Resist backtracking.** The PRNG should be designed to resist backtracking. Ideally, this would mean that output t was unguessable in practice to an attacker who compromised the PRNG state at time $t + 1$. It may also be acceptable to simply pass the PRNG's state through a one-way function every few outputs, limiting the possible scope of any backtracking attack.

5. **Resist Chosen-Input Attacks.** The inputs to the PRNG should be combined into the PRNG state in such a way that, given an unguessable sequence of inputs, an attacker who starts knowing the PRNG state but not the input sequence, and an attacker who starts knowing the input sequence but not the state, are both unable to guess the ending state. This provides some protection against both chosen-input and state compromise extension attacks.

6. **Recover from Compromises Quickly.** The PRNG should take advantage of every bit of entropy in the inputs it receives. An attacker wanting to learn the effect on the PRNG state of a sequence of inputs should have to guess the entire input sequence.

4.3   Open Problems

In this paper, we've begun the process of systematically analyzing PRNGs. However, there are several interesting areas we haven't dealt with here:

1. **Dedicated PRNG Designs.** Early in this paper, we made the assertion that PRNGs are a distinct kind of cryptographic primitive. Existing PRNGs are almost all built out of existing cryptographic primitives. This raises the question of whether it makes sense to build dedicated PRNG algorithms. Typically, the motivation for building a dedicated algorithm is to improve performance. Are there applications where the PRNG's performance is a serious enough issue to merit a new algorithm?

2. **Security Proofs.** Since most currently-fielded PRNGs are based on existing cryptographic primitives, it would be nice to see some security proofs, demonstrating that mounting some class of attack on the PRNG is equivalent to breaking an underlying block cipher, stream cipher, or hash function.

3. **Starting Points.** One likely way for an attacker to compromise the PRNG state is for the PRNG to be started in a guessable state. This raises the issue

of how a designer can ensure that his system always starts its PRNG at an unguessable state. We would like to see more discussion of these issues in the literature.

4. **Seed Compromise.** We would like to see more discussion of how to resist state compromises in fielded systems. This is an enormous practical issue, which has received relatively little attention in the literature.

5. **Analyzing Other PRNGs.** There are many PRNGs we have not discussed here, mainly due to time and space constraints. In particular, we would like to see a complete discussion of the class of PRNG used in PGP and Gutmann's Cryptlib, among other places. These PRNGs fit into our model, but look very different than any of the systems we have reviewed here: they typically maintain a considerably larger state (or "pool"), in hopes of accumulating large amounts of entropy.

6. **Developing New PRNGs.** We have discussed flaws in existing PRNGs. We are interested in seeing new designs proposed that resist our attacks. A PRNG of our own is currently under development; details will be posted to `http://www.counterpane.com` as they become available.

5   Acknowledgements

The authors would like to thank Greg Guerin, Peter Gutmann, and Adam Shostack for helpful conversations and comments on early drafts of this paper, and Ross Anderson and several anonymous referees for helpful suggestions on improving the paper's presentation.

References

Agn88.   G.B. Agnew, "Random Source for Cryptographic Systems," *Advances in Cryptology — EUROCRYPT '87 Proceedings,* Springer-Verlag, 1988, pp. 77–81.

ANSI85.   ANSI X 9.17 (Revised), "American National Standard for Financial Institution Key Management (Wholesale)," American Bankers Association, 1985.

Bal96.   R.W. Baldwin, "Proper Initialization for the BSAFE Random Number Generator," *RSA Laboratories Bulletin*, n. 3, 25 Jan 1996.

Dai97.   W. Dai, Crypto++ library,
           `http://www.eskimo.com/~weidai/cryptlib.html`.

DIF94.   D. Davis, R. Ihaka, and P. Fenstermacher, "Cryptographic Randomness from Air Turbulience in Disk Drives," *Advances in Cryptology — CRYPTO '94 Proceedings,* Springer-Verlag, 1994, pp. 114–120.

ECS94.   D. Eastlake, S.D. Crocker, and J.I. Schiller, "Randomness Requirements for Security," RFC 1750, Internet Engineering Task Force, Dec. 1994.

FMK85.   R.C. Fairchild, R.L. Mortenson, and K.B. Koulthart, "An LSI Random Number Generator (RNG)," *Advances in Cryptology: Proceedings of CRYPTO '84,* Springer-Verlag, 1985, pp. 203–230.

Gud85.   M. Gude, "Concept for a High-Performance
            Random Number Generator Based on Physical Random Noise," *Frequenz*,
            v. 39, 1985, pp. 187–190.
Gut98.   P. Gutmann, "Software Generation of Random Numbers for Cryptogra-
            phic Purposes," Proceedings of the 1998 Usenix Security Symposium,
            1998, to appear.
Koc95.   P. Kocher, post to `sci.crypt` Internet newsgroup (message-ID
            `pckDIr4Ar.L4z@netcom.com`), 4 Dec 1995.
LMS93.   J.B. Lacy, D.P. Mitchell, and W.M. Schell, "CryptoLib: Cryptography
            in Software," *USENIX Security Symposium IV Proceedings*, USENIX
            Association, 1993, pp. 237–246.
NIST92.   National Institute for Standards and Technology, "Key Management
            Using X9.17," NIST FIPS PUB 171, U.S. Department of Commerce,
            1992.
NIST93.   National Institute for Standards and Technology, "Secure Hash Stan-
            dard," NIST FIPS PUB 180, U.S. Department of Commerce, 1993.
NIST94.   National Institute for Standards and Technology, "Digital Signature Stan-
            dard," NIST FIPS PUB 186, U.S. Department of Commerce, 1994.
OW95.   P.C. van Oorschot and M.J. Wiener, "Parallel collision search with ap-
            plication to hash function and discrete logarithms," *2nd ACM Conf. on
            Computer and Communications Security*, New York, NY, ACM, 1994.
OW96.   P.C. van Oorschot and M.J. Wiener, "Improving implementable meet-
            in-the-middle attacks by orders of magnitude," *CRYPTO '96*, Springer-
            Verlag, 1996.
Plu94.   C. Plumb, "Truly Random Numbers, *Dr. Dobbs Journal*, v. 19, n. 13,
            Nov 1994, pp. 113-115.
Ric92.   M. Richterm "Ein Rauschgenerator zur Gweinnung won quasi-idealen Zu-
            fallszahlen fur die stochastische Simulation," Ph.D. dissertation, Aachen
            University of Technology, 1992. (In German.)
RSA94.   RSA Laboratories, RSAREF cryptographic library, Mar 1994,
            `ftp://ftp.funet.fi/pub/crypt/cryptography/asymmetric/rsa/`
            `rsaref2.tar.gz`.
SV86.   M. Santha and U.V. Vazirani, "Generating Quasi-Random Sequences
            from Slightly Random Sources," *Journal of Computer and System
            Sciences*, v. 33, 1986, pp. 75–87.
Sch96   B. Schneier, *Applied Cryptrography*, John Wiley & Sons, 1996.
Zim95.   P. Zimmermann, *The Official PGP User's Guide*, MIT Press, 1995.

CS-CIPHER

Jacques Stern and Serge Vaudenay *

Ecole Normale Supérieure — CNRS
{Jacques.Stern,Serge.Vaudenay}@ens.fr

Abstract. This paper presents a new block cipher which offers good encryption rate on any platform. It is particularly optimized for hardware implementation where the expected rate is several Gbps on a small dedicated chip working at 30MHz. Its design combines up to date state of the art concepts in order to make it (hopefully) secure: diffusion network based on the Fast Fourier Transform, multipermutations, highly nonlinear confusion boxes.

Recent explosion of the telecommunication marketplace motivates the research on encryption schemes. Trading security issues pushed the US government to start the development of the *Data Encryption Standard* in the 70's [1], all telecommunication devices now need to be secured by encryption. Many attacks have been proposed against DES including Biham and Shamir's differential cryptanalysis [5,6] and Matsui's linear cryptanalysis [15,16]. Still the best practical attack seems to be exhaustive search, which has become a real threat as shown by the recent success of the RSA Challenge [31]. In this paper we propose a new symmetric encryption scheme which has been designed in order to be efficient on any platform, included cheap 8-bit microprocessors (*e.g.* smart cards), modern 32-bit microprocessors (SPARC, Pentium) and dedicated chips.

Notations

- $||$ is the concatenation of two strings
- $\oplus$ is the bitwise exclusive *or* of two bitstrings (with equal lengths)
- R_l rotates a bitstring by one position to the left
- $\wedge$ is the bitwise *and* of two bitstrings (with equal lengths)
- bitstrings are written in hexadecimal by packing four bits into one digit (for instance, $\mathtt{d2}_{16}$ denotes the bitstring 11010010)
- the numbering of bits in bitstrings is from right to left starting with 0 (*i.e.* x_0 denotes the last bit in x)
- bitstring and integers are converted in such a way that $b_{n-1}||\ldots||b_0$ corresponds to an integer $b_{n-1}.2^{n-1} + \ldots + b_0$

* Part of this work has been supported by the COMPAGNIE DES SIGNAUX and may be subject to patent matter.

S. Vaudenay (Ed.): Fast Software Encryption – FSE'98, LNCS 1372, pp. 189–204, 1998.

1   Definition of CS-CIPHER

1.1   Use of CS-CIPHER

CS-CIPHER (as for the French *"Chiffrement Symétrique"*, *Symmetric Cipher*) is a symmetric block cipher which can be used in any mode to encrypt a block stream (*e.g.* the Cipher Block Chaining mode, see [2]). Basically, the CS-CIPHER encryption function maps a 64-bit plaintext block m onto a 64-bit ciphertext block m' by using a secret key k which is a bitstring with arbitrary length up to 128. The CS-CIPHER decryption function maps the ciphertext onto the plaintext by using the same secret key. We assume that m is represented by a bitstring

$$m = m_{63} \ldots m_1 m_0$$

and we similarly write

$$m' = m'_{63} \ldots m'_1 m'_0.$$

We also assume that the string k is padded with trailing zero bits to get a length of 128 bits

$$k = k_{127} \ldots k_1 k_0.$$

(A key k is therefore equivalent to another key k' which consists in padding k with a few zero bits.)

A key scheduling scheme first process the secret key k in order to obtain nine 64-bit subkeys $k^0, \ldots, k^8$ iteratively in this order. If the secret key has to be used several times, we recommend to precompute this sequence which may notably increase the encryption rate.

The encryption algorithm processes iteratively each subkey in the right order $k^0, \ldots, k^8$ whereas the decryption algorithm processes them in the reverse order $k^8, \ldots, k^0$. We thus recommend to keep a storage of all subkeys for decryption or to adapt the key scheduling scheme so that it can generate the subkeys in the reverse order.

1.2   Key scheduling scheme

Let k be the padded 128-bit secret key. We first split the bitstring into two 64-bit strings denoted k^{-2} and k^{-1} such that

$$k = k^{-1} \| k^{-2}.$$

Those strings initialize a sequence $k^{-2}, \ldots, k^8$ where $k^0, \ldots, k^8$ are the nine 64-bit subkeys to compute. The sequence comes from a Feistel scheme as

$$k^i = k^{i-2} \oplus F_{c^i}(k^{i-1})$$

for $i = 0, \ldots, 8$ where F_{c^i} is defined below (see Feistel [9]). Figure 1 illustrates the key scheduling scheme together with the encryption itself.

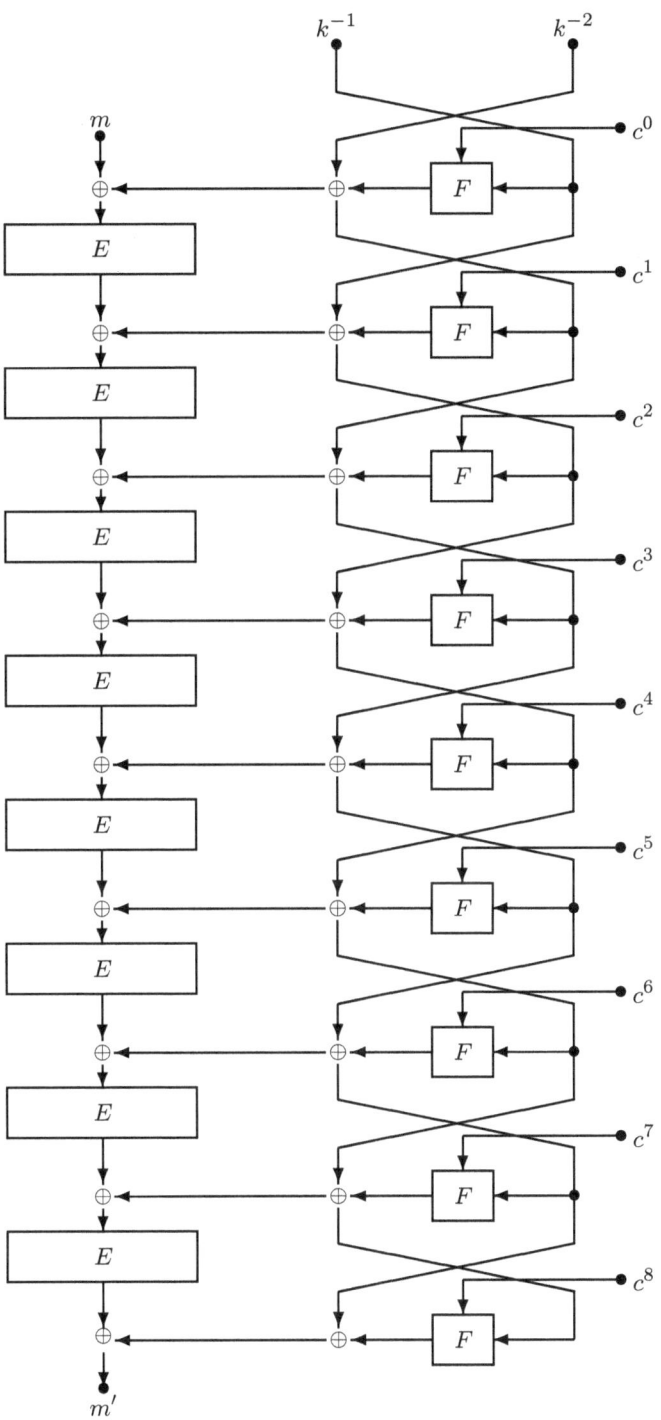

Fig. 1. Encryption process

The F_{ci} function maps a 64-bit string onto a 64-bit string by using a 64-bit constant c^i. In the definition of CS-CIPHER, $c^0, \ldots, c^8$ are defined as the first bytes of the table of a permutation P which will be defined below:

$$c^0 = \mathtt{290d61409ceb9e8f}_{16}$$
$$c^1 = \mathtt{1f855f585b013986}_{16}$$
$$c^2 = \mathtt{972ed7d635ae1716}_{16}$$
$$c^3 = \mathtt{21b6694ea5728708}_{16}$$
$$c^4 = \mathtt{3c18e6e7faadb889}_{16}$$
$$c^5 = \mathtt{b700f76f73841163}_{16}$$
$$c^6 = \mathtt{3f967f6ebf149dac}_{16}$$
$$c^7 = \mathtt{a40e7ef6204a6230}_{16}$$
$$c^8 = \mathtt{03c54b5a46a34465}_{16}.$$

F_{ci} is defined by

$$F_{ci}(x) = T(P^8(x \oplus c^i)).$$

P^8 is defined by a byte-permutation P which maps an 8-bit string onto an 8-bit string according to a table and T is a bit transposition. (Sofware implementation will use a lookup table for P whereas hardware implementation may use the inner structure of P which will be detailed below.)

Given the 64-bit string $y = x \oplus c^i$, we split it into eight 8-bit strings denoted $y_{63..56}, \ldots, y_{7..0}$ such that $y = y_{63..56} || \ldots || y_{7..0}$. We next apply the permutation P byte-wise $i.e.$ we compute

$$P^8(y_{63..56} || \ldots || y_{7..0}) = P(y_{63..56}) || \ldots || P(y_{7..0}).$$

The permutation T is the 8×8 bit-matrix transposition. More precisely, given the 64-bit string $z = P^8(x \oplus c^i)$, we first split it into eight 8-bit strings $z_{63..56}, \ldots, z_{7..0}$ as for y above and write it in a 8×8 bit-matrix fashion in such a way that the first row is $z_{63..56}$ and so on. The permutation T simply transposes the matrix so that the first eight bits of $T(z)$ are the first bits of $z_{63..56}, \ldots, z_{7..0}$ in this order, the second eight bits are the seconds bits, and so on. Thus we have

$$T(z) = z_{63} || z_{55} || \ldots || z_7 || z_{62} || z_{54} || \ldots || z_0.$$

Figure 2 illustrates how F_{ci} works in the key scheduling scheme.

1.3   Encryption scheme

The encryption process is performed through eight rounds by using a round-encryption function E which is a permutation on the set of all 64-bit strings. If m denotes the 64-bit plaintext block and $k^0, \ldots, k^8$ is the sequence of the 64-bit subkeys, the ciphertext block is

$$k^8 \oplus E(k^7 \oplus \ldots E(k^1 \oplus E(k^0 \oplus m)) \ldots)$$

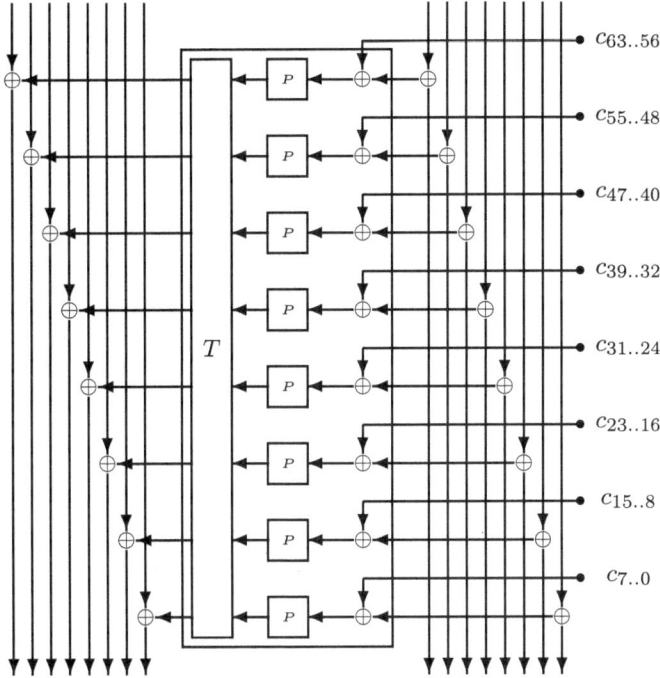

Fig. 2. The F_{c^i} function in the key scheduling scheme

as depicted on Figure 1.

The round-encryption function E is based on the Fast Fourier Transform graph and a 16-bit to 16-bit *mixing* function M as depicted on Figure 3. It also uses two 64-bit constants c and c' defined by the binary expansion of the mathematical constant

$$e = \sum_{n=0}^{\infty} \frac{1}{n!} = 2, \mathtt{b7e151628aed2a6abf7158809cf4f3c762e7160f}_{16} \ldots$$

Thus we define

$$c = \mathtt{b7e151628aed2a6a}$$
$$c' = \mathtt{bf7158809cf4f3c7}.$$

More precisely, in each encryption round, we iterate the following scheme three times

- we xor with a constant (which is successively the subkey k^i, c and c'),
- we split the 64-bit string into four 16-bit strings and we apply M to each of it, obtaining four 16-bit strings which combine into a 64-bit string,

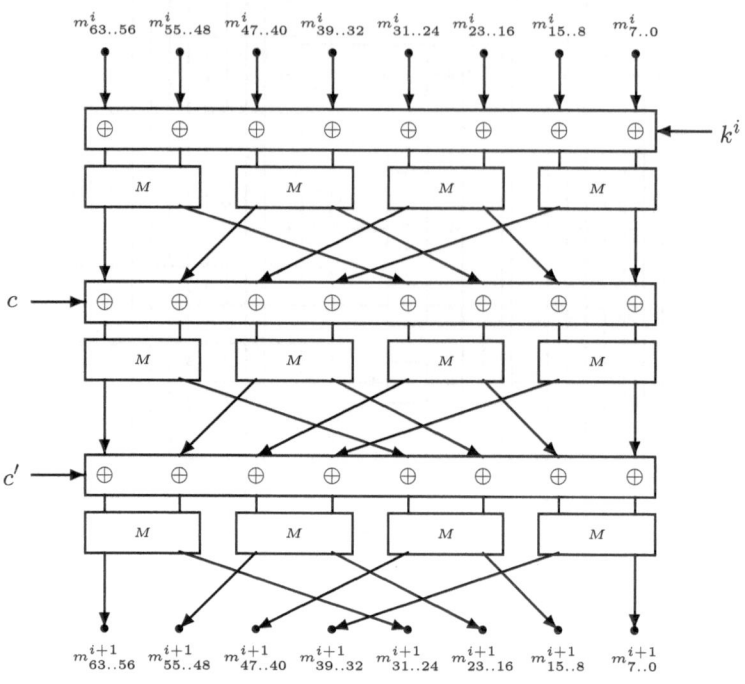

Fig. 3. One encryption round

– we split it again into eight 8-bit strings

$$r_{63..56}||r_{55..48}||r_{47..40}||r_{39..32}||r_{31..24}||r_{23..16}||r_{15..8}||r_{7..0}$$

and we change their order as

$$r_{63..56}||r_{47..40}||r_{31..24}||r_{15..8}||r_{55..48}||r_{39..32}||r_{23..16}||r_{7..0}.$$

The M function takes a 16-bit string x which is split into two 8-bit strings $x_l||x_r$ and computes $M(x) = y_l||y_r$ by

$$y_l = P(\varphi(x_l) \oplus x_r)$$
$$y_r = P(R_l(x_l) \oplus x_r)$$

where φ is defined by

$$\varphi(x_l) = (R_l(x_l) \wedge 55_{16}) \oplus x_l$$

i.e.

$$\varphi(x_7||\ldots||x_0) = x_7||(x_6 \oplus x_5)||x_5||(x_4 \oplus x_3)||x_3||(x_2 \oplus x_1)||x_1||(x_0 \oplus x_7).$$

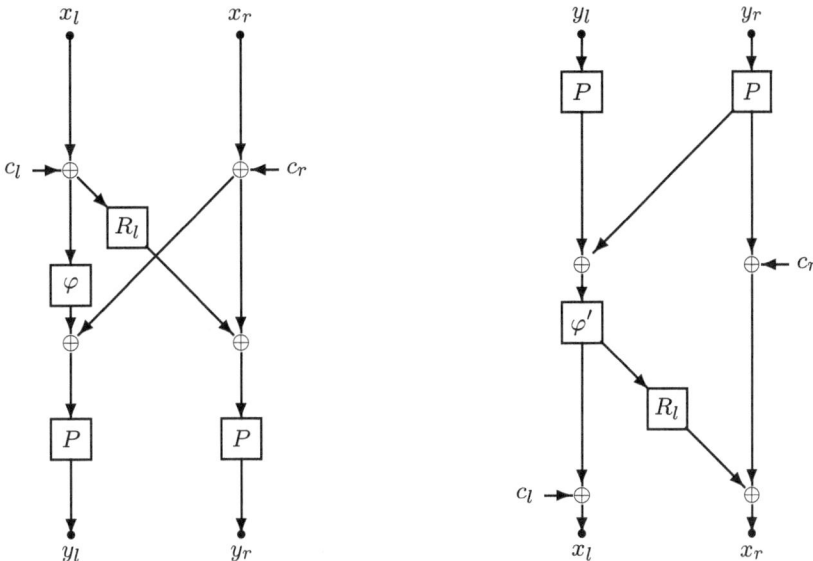

Fig. 4. Computation graph of M and M^{-1}

The M computation is depicted on Figure 4 (with the xor to its input which is always performed).

The P byte-permutation (which is also used in the key scheduling scheme) is defined by a three-round Feistel cipher represented on Figure 5: the 8-bit input x is split into two 4-bit strings $x_l\|x_r$, we compute successively

$$y = x_l \oplus f(x_r)$$
$$z_r = x_r \oplus g(y)$$
$$z_l = y \oplus f(z_r)$$

where f and g are two special functions.

The function f is defined by the table

x	0 1 2 3 4 5 6 7 8 9 a b c d e f
$f(x)$	f d b b 7 5 7 7 e d a b e d e f

which comes from

$$f(x) = \overline{x} \wedge R_l(x).$$

The function g is defined by the table

x	0 1 2 3 4 5 6 7 8 9 a b c d e f
$g(x)$	a 6 0 2 b e 1 8 d 4 5 3 f c 7 9

which does not come from a simple expression.

Finally, the value of $P(xy)$ is given as follows by the table of P.

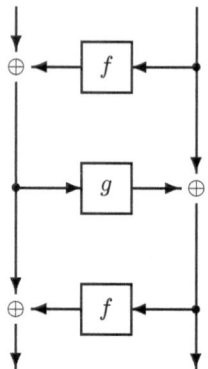

Fig. 5. The permutation P

xy	.0	.1	.2	.3	.4	.5	.6	.7	.8	.9	.a	.b	.c	.d	.e	.f
0.	29	0d	61	40	9c	eb	9e	8f	1f	85	5f	58	5b	01	39	86
1.	97	2e	d7	d6	35	ae	17	16	21	b6	69	4e	a5	72	87	08
2.	3c	18	e6	e7	fa	ad	b8	89	b7	00	f7	6f	73	84	11	63
3.	3f	96	7f	6e	bf	14	9d	ac	a4	0e	7e	f6	20	4a	62	30
4.	03	c5	4b	5a	46	a3	44	65	7d	4d	3d	42	79	49	1b	5c
5.	f5	6c	b5	94	54	ff	56	57	0b	f4	43	0c	4f	70	6d	0a
6.	e4	02	3e	2f	a2	47	e0	c1	d5	1a	95	a7	51	5e	33	2b
7.	5d	d4	1d	2c	ee	75	ec	dd	7c	4c	a6	b4	78	48	3a	32
8.	98	af	c0	e1	2d	09	0f	1e	b9	27	8a	e9	bd	e3	9f	07
9.	b1	ea	92	93	53	6a	31	10	80	f2	d8	9b	04	36	06	8e
a.	be	a9	64	45	38	1c	7a	6b	f3	a1	f0	cd	37	25	15	81
b.	fb	90	e8	d9	7b	52	19	28	26	88	fc	d1	e2	8c	a0	34
c.	82	67	da	cb	c7	41	e5	c4	c8	ef	db	c3	cc	ab	ce	ed
d.	d0	bb	d3	d2	71	68	13	12	9a	b3	c2	ca	de	77	dc	df
e.	66	83	bc	8d	60	c6	22	23	b2	8b	91	05	76	cf	74	c9
f.	aa	f1	99	a8	59	50	3b	2a	fe	f9	24	b0	ba	fd	f8	55

For instance, we have $P(26) = \text{b8}$ since $f(6) = 7$, $2 \oplus 7 = 5$, $g(5) = \text{e}$, $6 \oplus \text{e} = 8$, $f(8) = \text{e}$ and finally $5 \oplus \text{e} = \text{b}$.

1.4   Decryption scheme

Decryption is performed by iterating a decryption-round function represented on Figure 6.

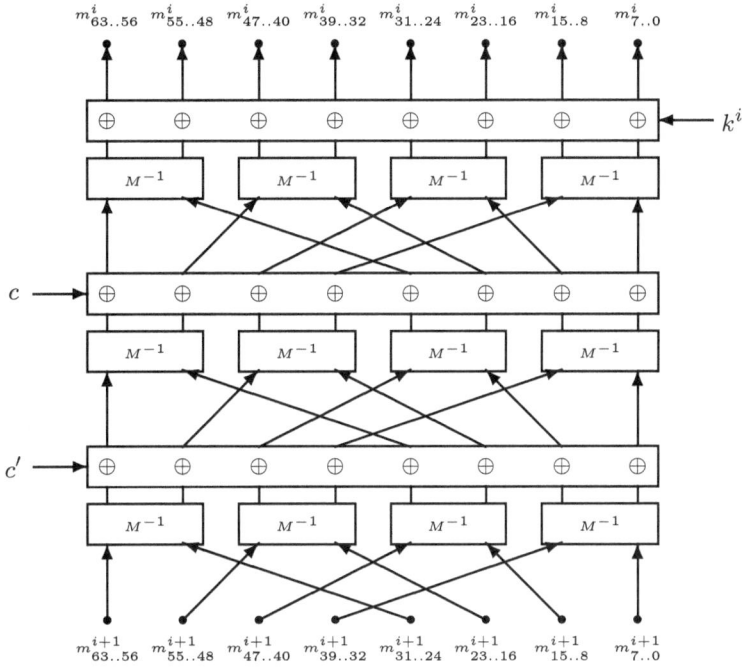

Fig. 6. One decryption round

Details of the decryption are left to the reader. We simply observe that $(x_l||x_r) = M^{-1}(y_l||y_r)$ can be computed by

$$x_l = \varphi'(P(y_l) \oplus P(y_r))$$
$$x_r = R_l(x_l) \oplus P(y_r)$$

where

$$\varphi'(x) = (R_l(x) \wedge \mathtt{aa}_{16}) \oplus x.$$

1.5   Test values

As an example we encrypt the plaintext $\mathtt{0123456789abcdef}_{16}$ with the secret key $\mathtt{0123456789abcdeffedcba9876543210}_{16}$. The subkeys sequence is

$$k^{-2} = \mathtt{fedcba9876543210}_{16}$$
$$k^{-1} = \mathtt{0123456789abcdef}_{16}$$
$$k^{0} = \mathtt{45fd137a4edf9ec4}_{16}$$
$$k^{1} = \mathtt{1dd43f03e6f7564c}_{16}$$
$$k^{2} = \mathtt{ebe26756de9937c7}_{16}$$

$$k^3 = 961704e945bad4fb_{16}$$
$$k^4 = 0b60dfe9eff473d4_{16}$$
$$k^5 = 76d3e7cf52c466cf_{16}$$
$$k^6 = 75ec8cef767d3a0d_{16}$$
$$k^7 = 82da3337b598fd6d_{16}$$
$$k^8 = fbd820da8dc8af8c_{16}$$

For instance, the first generated subkey $k^0 = 45fd137a4edf9ec4_{16}$ is

$$k^0 = k^{-2} \oplus T(P^8(k^{-1} \oplus 290d61409ceb9e8f_{16}))$$
$$= k^{-2} \oplus T(b711fa89ae0394e4_{16})$$
$$= k^{-2} \oplus bb21a9e2388bacd4_{16}.$$

The messages which enter into each round are

$$m^0 = 0123456789abcdef_{16}$$
$$m^1 = c3feb96c0cf4b649_{16}$$
$$m^2 = 3f54e0c8e61a84d1_{16}$$
$$m^3 = b15cb4af3786976e_{16}$$
$$m^4 = 76c122b7a562ac45_{16}$$
$$m^5 = 21300b6ccfaa08d8_{16}$$
$$m^6 = 99b8d8ab9034ec9a_{16}$$
$$m^7 = a2245ba3697445d2_{16}$$

and the ciphertext is $88fddfbe954479d7_{16}$. In the first round, the message m^0 is transformed though three layers into m^1. The intermediate results between the layers are $d85c19785690b0e3_{16}$ and $0f4bfb9e2f8ac7e2_{16}$. For instance, in the first layer we take m^0, xor it with k^0, apply M, permute the bytes and get $d85c19785690b0e3_{16}$.

As an implementation test, we mention that if we iterate one million times the encryption on the all-zero bitstring with the previous key, we obtain the final ciphertext $fd5c9c6889784b1c_{16}$.

2   Design arguments

The Fast Fourier Transform used in the round-encryption function E has been used in several cryptographic designs including Schnorr's FFT-Hashing [22], FFT Hash II [23], Schnorr and Vaudenay's Parallel FFT-Hashing [24], and Massey's SAFER [13,14]. This graph has been proved to have very good diffusion properties when done twice (see [25,26,30]).

The M structure implements a *multipermutation* as defined by Schnorr and Vaudenay (see [25,29,30]). In this case, it means that M is a permutation over the set of all 16-bit strings, and that fixing any of the two 8-bit inputs arbitrarily makes both 8-bit outputs be permutations of the other one. This is due to a very particular property of φ, namely that both φ and $x \mapsto \varphi(x) \oplus R_l(x)$ (which is in fact φ') are permutations. Actually, φ and φ' are linear involutions.

Those properties make E be what we call a *mixing function, i.e.* such that if we arbitrarily fix seven of the eight inputs, all outputs are permutation of the remaining free input. This performs a good diffusion.

The best general attack methods on block ciphers have been introduced by Gilbert, Chassé, Tardy-Corfdir, Biham, Shamir and Matsui (see [11,28,5,6,15,16,10]). They are now known as differential and linear cryptanalysis. We know study how CS-CIPHER has been protected against it.

The permutation P has been chosen to be an nonlinear involution in the sense that both differential and linear cryptanalysis are hard. Nonlinearity has one measure corresponding to differential cryptanalysis (which has been defined by Nyberg [19]) and one measure corresponding to linear cryptanalysis (which has been defined by Chabaud and Vaudenay [7]). Here we use the formalism introduced by Matsui [17]:

$$\mathrm{DP}_{\max}(f) = \max_{a \neq 0, b} \Pr_{X \text{ uniform}} \left[f(X \oplus a) \oplus f(X) = b \right]$$

$$\mathrm{LP}_{\max}(f) = \max_{a, b \neq 0} \left(2 \Pr_{X \text{ uniform}} [X \cdot a = f(X) \cdot b] - 1 \right)^2.$$

The functions f and g are such that $\mathrm{DP}_{\max}(f) \leq 2^{-2}$ and $\mathrm{LP}_{\max}(f) \leq 2^{-2}$. If the Theorem of Aoki and Ohta [3] (which generalizes the Theorem of Nyberg and Knudsen [20]) were applicable in this setting, we would then obtain $\mathrm{DP}_{\max}(P) \leq 2^{-4}$ and $\mathrm{LP}_{\max}(P) \leq 2^{-4}$. Both properties are however still satisfied as the experiment shows. From [19,7] it is known that for any function f on the set of all n-bit strings we have $\mathrm{DP}_{\max}(f) \geq 2^{1-n}$ and $\mathrm{LP}_{\max}(f) \geq 2^{1-n}$, but it is conjectured that 2^{2-n} is a better bound for even n (see Dobbertin [8] for instance). So our functions are reasonably nonlinear. Since it is well known that the heuristic complexity of differential or linear cryptanalysis is greater than the inverse of the product of the $\mathrm{DP}_{\max}$ or $\mathrm{LP}_{\max}$ of all active P boxes (see for instance Heys and Tavares [12]), having mixing functions makes at least five P box per round to be active, so no four rounds of CS-CIPHER have any efficient differential or linear characteristic.

3   Implementation

In any kind of implementation, the key scheduling scheme is assumed to be precomputed. This part of CS-CIPHER has not be designed to have special implementation optimization. The authors believe that every time one changes the secret key, one has to perform expensive computations (such as asymmetric cryptography, key exchange protocol or key transfer protocols) so optimizing the precomputation of the subkey sequence is meaningless. In the following Sections we only discuss implementation of the encryption (or decryption) scheme.

3.1 VLSI implementation

CS-CIPHER is highly optimized for VLSI implementations. It may be noticed that the g function has been designed to get a friendly boolean circuit implementation. Actually, Figure 7 illustrates a cheap nand-circuit with depth 4 and only 16 nand gates.

We propose two possible easy implementations. In the first one, we really implement one third of a single round encryption. It has two 64-bit input registers and one 64-bit output register. It is easy to see that an encryption can be performed by iterating this circuit 24 times and loading the subkey sequence $k^0, c, c', k^1, c, c', \ldots$ Straightforward estimates shows this circuit requires 1216 nand-gates with depth 26. This implementation can be added in any microprocessor within less than 1mm^2 in order to get a simple microcoded encryption instruction. One 30MHz-clock cycle is far enough to compute one layer, thus one 64-bit encryption requires 24 clock cycles, which leads to a 73Mbps, which is quite fast for such a cheap technology.

The second implementation consists in making a dedicated chip which consists of 24 times the previous circuit in a pipeline architecture. We estimate we need 15mm^2 in order to implement a 30000 nand-gate circuit which performs a 64-bit encryption within one 30MHz-clock cycle, which leads to an encryption rate of 2Gbps. This can be used to encrypt ATM network communications or PCI bus.

$$\begin{array}{lllll}
\text{layer4}: & g_0 = g_4 n g_5 & g_1 = g_6 n g_7 & g_2 = g_8 n g_9 & g_3 = g_{10} n g_{11} \\
\text{layer3}: & g_5 = g_6 n g_{14} & g_8 = g_{17} n g_4 & g_9 = g_7 n g_{14} & g_{10} = g_6 n g_{16} \\
\text{layer2}: & g_6 = g_{14} n g_{12} & g_7 = g_{15} n g_{16} & g_{11} = g_{13} n g_4 & g_{17} = g_{15} n g_{19} \\
\text{layer1}: & g_4 = g_{12} n g_{13} & g_{14} = g_{18} n g_{18} & g_{15} = g_{18} n g_{12} & g_{19} = g_{16} n g_{13} \\
\text{layer0}: & g_{12} & g_{13} & g_{16} & g_{18}
\end{array}$$

$$\text{Input}: \quad g_{16} g_{13} g_{18} g_{12}$$
$$\text{Output}: \quad g_0 g_1 g_2 g_3$$

Fig. 7. Implementation of g

Those results can be compared to MISTY which has been implemented by Mitsubishi. In Matsui [18], this chip is specified to require 65000 gates, working at 14MHz and encrypting at 450Mbps.

3.2 Software implementation on modern microprocessors

A straightforward non-optimized implementation of CS-CIPHER in standard C on a Pentium 133MHz (see Appendix) gives an encryption rate of 2.1Mbps which is reasonably fast compared to similar implementations of DES.

Another (non-optimized) implementation in assembly code enables the Pentium to perform a 64-bit encryption within 973 clock cycles, which leads to 8.34Mbps working at 133MHz.

An evaluation similar to the VLSI-implementation estimates shows that the number of "usual" boolean gate (xor, and, or not) required to implement 64 parallel 64-bit encryptions using Biham's bit-slice trick on a 64-bit microprocessor is 11968, which is substantially less than Biham's implementation of DES which requires about 16000 instructions (see [4]). Therefore, if we use a 300MHz Alpha microprocessor which requires .5cycles per instructions (as in [4]), we obtain an encryption rate of about 196Mbps.

3.3   Software implementation on 8-bit microprocessors

An implementation has been done for a cheap smart card platform. A compact 6805 assembly code of roughly 500 bytes can encrypt a 64-bit string in its buffer RAM by using only 6 extra byte-registers within 12633 clock cycles. This means that a cheap smart card working at 4MHz can encrypt within 3,16ms (*i.e.* at a 19,8Kbps rate), which is better than optimized implementations of DES[1]. This implementation of CS-CIPHER can still be optimized.

platform	clock frequency	encryption rate	note
VLSI 1216nand 1mm^2	30MHz	73Mbps	estimate
VLSI 30000nand 15mm^2	30MHz	2Gbps	estimate
standard C 32bits	133MHz	2Mbps	see Appendix
bit slice (Pentium)	133MHz	11Mbps	estimate
bit slice (Alpha)	300MHz	196Mbps	estimate
Pentium assembly code	133MHz	8Mbps	non-optimized
6805 assembly code	4MHz	20Kbps	non-optimized

Fig. 8. Implementations of CS-CIPHER

4   Conclusion

CS-CIPHER has been shown to offer quite fast encryption rates on several kinds of platforms, which is suitable for telecommunication applications. Figure 8 summerizes the implementation results. Its security is based on heuristic arguments.

All attacks are welcome...

Acknowledgements

We wish to thank the COMPAGNIE DES SIGNAUX for having initiated and supported this work.

References

1. Data Encryption Standard. *Federal Information Processing Standard Publication 46*, U. S. National Bureau of Standards, 1977.
2. DES Modes of Operation. *Federal Information Processing Standard Publication 81*, U. S. National Bureau of Standards, 1980.
3. K. Aoki, K. Ohta. Strict evaluation of the maximum average of differential probability and the maximum average of linear probability. *IEICE Transactions on Fundamentals*, vol. E80-A, pp. 1–8, 1997.
4. E. Biham. A fast new DES implementation in software. In *Fast Software Encryption*, Haifa, Israel, Lectures Notes in Computer Science 1267, pp. 260–272, Springer-Verlag, 1997.
5. E. Biham, A. Shamir. Differential cryptanalysis of the full 16-round DES. In *Advances in Cryptology CRYPTO'92*, Santa Barbara, California, U.S.A., Lectures Notes in Computer Science 740, pp. 487–496, Springer-Verlag, 1993.
6. E. Biham, A. Shamir. *Differential Cryptanalysis of the Data Encryption Standard*, Springer-Verlag, 1993.
7. F. Chabaud, S. Vaudenay. Links between differential and linear cryptanalysis. In *Advances in Cryptology EUROCRYPT'94*, Perugia, Italy, Lectures Notes in Computer Science 950, pp. 356–365, Springer-Verlag, 1995.
8. H. Dobbertin. Almost Perfect nonlinear power functions on $GF(2^n)$. *IEEE Trans. Inf. Theory*, submitted.
9. H. Feistel. Cryptography and computer privacy. *Scientific american*, vol. 228, pp. 15–23, 1973.
10. H. Gilbert. *Cryptanalyse Statistique des Algorithmes de Chiffrement et Sécurité des Schémas d'Authentification*, Thèse de Doctorat de l'Université de Paris 11, 1997.
11. H. Gilbert, G. Chassé. A statistical attack of the FEAL-8 cryptosystem. In *Advances in Cryptology CRYPTO'90*, Santa Barbara, California, U.S.A., Lectures Notes in Computer Science 537, pp. 22–33, Springer-Verlag, 1991.
12. H. M. Heys, S. E. Tavares. Substitution-Permutation Networks resistant to differential and linear cryptanalysis. *Journal of Cryptology*, vol. 9, pp. 1–19, 1996.
13. J. L. Massey. SAFER K-64: a byte-oriented block-ciphering algorithm. In *Fast Software Encryption*, Cambridge, United Kingdom, Lectures Notes in Computer Science 809, pp. 1–17, Springer-Verlag, 1994.
14. J. L. Massey. SAFER K-64: one year later. In *Fast Software Encryption*, Cambridge, United Kingdom, Lectures Notes in Computer Science 809, pp. 212–241, Springer-Verlag, 1994.
15. M. Matsui. Linear cryptanalysis methods for DES cipher. In *Advances in Cryptology EUROCRYPT'93*, Lofthus, Norway, Lectures Notes in Computer Science 765, pp. 386–397, Springer-Verlag, 1994.
16. M. Matsui. The first experimental cryptanalysis of the Data Encryption Standard. In *Advances in Cryptology CRYPTO'94*, Santa Barbara, California, U.S.A., Lectures Notes in Computer Science 839, pp. 1–11, Springer-Verlag, 1994.
17. M. Matsui. New structure of block ciphers with provable security against differential and linear cryptanalysis. In *Fast Software Encryption*, Cambridge, United Kingdom, Lectures Notes in Computer Science 1039, pp. 205–218, Springer-Verlag, 1996.
18. M. Matsui. New block encryption algorithm MISTY. In *Fast Software Encryption*, Haifa, Israel, Lectures Notes in Computer Science 1267, pp. 54–68, Springer-Verlag, 1997.

19. K. Nyberg. Perfect nonlinear *S*-boxes. In *Advances in Cryptology EURO-CRYPT'91*, Brighton, United Kingdom, Lectures Notes in Computer Science 547, pp. 378–385, Springer-Verlag, 1991.

20. K. Nyberg, L. R. Knudsen. Provable security against a differential cryptanalysis. *Journal of Cryptology*, vol. 8, pp. 27–37, 1995.

21. Organisation for Economic Co-operation and Development *Cryptography Policy Guidelines*, 27 March, 1997.

22. C. P. Schnorr. FFT-Hashing: an efficient cryptographic hash function. Présenté à CRYPTO'91. Non publié.

23. C. P. Schnorr. FFT-Hash II, efficient cryptographic hashing. In *Advances in Cryptology EUROCRYPT'92*, Balatonfüred, Hungary, Lectures Notes in Computer Science 658, pp. 45–54, Springer-Verlag, 1993.

24. C.-P. Schnorr, S. Vaudenay. Parallel FFT-hashing. In *Fast Software Encryption*, Cambridge, United Kingdom, Lectures Notes in Computer Science 809, pp. 149–156, Springer-Verlag, 1994.

25. C.-P. Schnorr, S. Vaudenay. Black box cryptanalysis of hash networks based on multipermutations. In *Advances in Cryptology EUROCRYPT'94*, Perugia, Italy, Lectures Notes in Computer Science 950, pp. 47–57, Springer-Verlag, 1995.

26. C. P. Schnorr, S. Vaudenay. Black box cryptanalysis of cryptographic primitives. Submitted. Early version available as LIENS Report 95–28, Laboratoire d'Informatique de l'Ecole Normale Supérieure, 1995.
`ftp://ftp.ens.fr/pub/reports/liens/liens-95-28.A4.ps.Z`

27. C. E. Shannon. Communication theory of secrecy systems. *Bell system technical journal*, vol. 28, pp. 656–715, 1949.

28. A. Tardy-Corfdir, H. Gilbert. A known plaintext attack of FEAL-4 and FEAL-6. In *Advances in Cryptology CRYPTO'91*, Santa Barbara, California, U.S.A., Lectures Notes in Computer Science 576, pp. 172–181, Springer-Verlag, 1992.

29. S. Vaudenay. On the need for multipermutations: cryptanalysis of MD4 and SA-FER. In *Fast Software Encryption*, Leuven, Belgium, Lectures Notes in Computer Science 1008, pp. 286–297, Springer-Verlag, 1995.

30. S. Vaudenay. *La Sécurité des Primitives Cryptographiques*, Thèse de Doctorat de l'Université de Paris 7, Technical Report LIENS-95-10 of the Laboratoire d'Informatique de l'Ecole Normale Supérieure, 1995.

31. Rocke Verser. Strong cryptography makes the world a safer place.
`http://www.frii.com/~rcv/deschall.htm`

Appendix

Here is a sample implementation of the heart of CS-CIPHER. The procedure takes plaintext block m and a precomputed subkey sequence k (as a 9×8 bytes array). This program is highly optimizable.

```
typedef unsigned char uint8;
#define CSC_C00 0xb7
#define CSC_C01 0xe1
#define CSC_C02 0x51
#define CSC_C03 0x62
#define CSC_C04 0x8a
#define CSC_C05 0xed
#define CSC_C06 0x2a
#define CSC_C07 0x6a
```

```
#define CSC_C10 0xbf
#define CSC_C11 0x71
#define CSC_C12 0x58
#define CSC_C13 0x80
#define CSC_C14 0x9c
#define CSC_C15 0xf4
#define CSC_C16 0xf3
#define CSC_C17 0xc7
uint8 tbp[256]={
  0x29,0x0d,0x61,0x40,0x9c,0xeb,0x9e,0x8f,
  0x1f,0x85,0x5f,0x58,0x5b,0x01,0x39,0x86,
  0x97,0x2e,0xd7,0xd6,0x35,0xae,0x17,0x16,
  0x21,0xb6,0x69,0x4e,0xa5,0x72,0x87,0x08,
  0x3c,0x18,0xe6,0xe7,0xfa,0xad,0xb8,0x89,
  0xb7,0x00,0xf7,0x6f,0x73,0x84,0x11,0x63,
  0x3f,0x96,0x7f,0x6e,0xbf,0x14,0x9d,0xac,
  0xa4,0x0e,0x7e,0xf6,0x20,0x4a,0x62,0x30,
  0x03,0xc5,0x4b,0x5a,0x46,0xa3,0x44,0x65,
  0x7d,0x4d,0x3d,0x42,0x79,0x49,0x1b,0x5c,
  0xf5,0x6c,0xb5,0x94,0x54,0xff,0x56,0x57,
  0x0b,0xf4,0x43,0x0c,0x4f,0x70,0x6d,0x0a,
  0xe4,0x02,0x3e,0x2f,0xa2,0x47,0xe0,0xc1,
  0xd5,0x1a,0x95,0xa7,0x51,0x5e,0x33,0x2b,
  0x5d,0xd4,0x1d,0x2c,0xee,0x75,0xec,0xdd,
  0x7c,0x4c,0xa6,0xb4,0x78,0x48,0x3a,0x32,
  0x98,0xaf,0xc0,0xe1,0x2d,0x09,0x0f,0x1e,
  0xb9,0x27,0x8a,0xe9,0xbd,0xe3,0x9f,0x07,
  0xb1,0xea,0x92,0x93,0x53,0x6a,0x31,0x10,
  0x80,0xf2,0xd8,0x9b,0x04,0x36,0x06,0x8e,
  0xbe,0xa9,0x64,0x45,0x38,0x1c,0x7a,0x6b,
  0xf3,0xa1,0xf0,0xcd,0x37,0x25,0x15,0x81,
  0xfb,0x90,0xe8,0xd9,0x7b,0x52,0x19,0x28,
  0x26,0x88,0xfc,0xd1,0xe2,0x8c,0xa0,0x34,
  0x82,0x67,0xda,0xcb,0xc7,0x41,0xe5,0xc4,
  0xc8,0xef,0xdb,0xc3,0xcc,0xab,0xce,0xed,
  0xd0,0xbb,0xd3,0xd2,0x71,0x68,0x13,0x12,
  0x9a,0xb3,0xc2,0xca,0xde,0x77,0xdc,0xdf,
  0x66,0x83,0xbc,0x8d,0x60,0xc6,0x22,0x23,
  0xb2,0x8b,0x91,0x05,0x76,0xcf,0x74,0xc9,
  0xaa,0xf1,0x99,0xa8,0x59,0x50,0x3b,0x2a,
  0xfe,0xf9,0x24,0xb0,0xba,0xfd,0xf8,0x55,
};
void enc_csc(uint8 m[8],uint8* k) {
  uint8 tmpx,tmprx,tmpy;
  int i;
  #define APPLY_M(cl,cr,adl,adr) \
    tmpx=m[adl]^cl; \
    tmprx=(tmpx<<1)^(tmpx>>7); \
    tmpy=m[adr]^cr; \
    m[adl]=tbp[(tmprx&0x55)^tmpx^tmpy]; \
    m[adr]=tbp[tmprx^tmpy];
  for(i=0;i<8;i++,k+=8) {
    APPLY_M(k[0],k[1],0,1)
    APPLY_M(k[2],k[3],2,3)
    APPLY_M(k[4],k[5],4,5)
    APPLY_M(k[6],k[7],6,7)
    APPLY_M(CSC_C00,CSC_C01,0,2)
    APPLY_M(CSC_C02,CSC_C03,4,6)
    APPLY_M(CSC_C04,CSC_C05,1,3)
    APPLY_M(CSC_C06,CSC_C07,5,7)
    APPLY_M(CSC_C10,CSC_C11,0,4)
    APPLY_M(CSC_C12,CSC_C13,1,5)
    APPLY_M(CSC_C14,CSC_C15,2,6)
    APPLY_M(CSC_C16,CSC_C17,3,7)
  }
  for(i=0;i<8;i++) m[i]^=k[i];
}
```

On the Design and Security of RC2

Lars R. Knudsen[1], Vincent Rijmen[2], Ronald L. Rivest[3], and Matthew J.B. Robshaw[4]

[1] Dept. of Informatics, University of Bergen, Hi-techcenter, N-5020 Bergen, Norway
`larsr@ii.uib.no`
[2] K.U. Leuven, ESAT, Kardinaal Mercierlaan 94, B-3001 Heverlee, Belgium
`vincent.rijmen@esat.kuleuven.ac.be`
[3] M.I.T. Laboratory for Computer Science, 545 Technology Square, Cambridge, MA 02139, USA
`rivest@theory.lcs.mit.edu`
[4] RSA Laboratories, 100 Marine Parkway, Redwood City, CA 94065, USA
`matt@rsa.com`

Abstract. The block cipher RC2 was designed in 1989 by Ron Rivest for RSA Data Security Inc. In this paper we describe both the cipher and preliminary attempts to use both differential and linear cryptanalysis.

1 Introduction

RC2 is a block cipher[1] that was designed in 1989 by Ron Rivest for RSA Data Security, Inc. Initially held as a confidential and proprietary algorithm, RC2 was published as an Internet Draft during 1997 [12]. RC2 has many interesting and unique design features, particularly so when one considers the style of ciphers that dominated both the literature and the market at the time of its invention. The cipher was intended to be particularly efficient on 16-bit processors and with a 64-bit block size it was intended as a drop-in replacement for DES [11]. A significant feature of RC2 is the flexibility offered to the user in terms of the effective key-size. This has now become a common feature of many block cipher proposals and it is a property that has proven to be important in commercial applications. Over the years RC2 has been deployed widely and it features prominently in the S/MIME secure messaging standard [5]. Currently there are no published results on the cryptanalytic strength of RC2. As a first step this paper sets out some details on how the basic attacks of differential [1] and linear [8] cryptanalysis might apply.

2 The Design of RC2

There are two distinct parts to using RC2. First a *key expansion* procedure takes a user-supplied key of between one and 128 bytes in length together with a

---

[1] RC2 is a registered trademark of RSA Data Security, Inc.

S. Vaudenay (Ed.): Fast Software Encryption – FSE'98, LNCS 1372, pp. 206–221, 1998.

parameter that specifies the effective key-length of encryption. From this information an array $K[\cdot]$ of 64 16-bit round keys is derived. Then a 64-bit plaintext block is encrypted using array $K[\cdot]$. Encryption consists of two styles of rounds. One is termed a MIXING round and the other a MASHING round.

Both the key expansion and encryption components rely on the use of a substitution table called PITABLE. This table specifies a random permutation on the integers $0, \ldots, 255$ and was derived from the expansion of $\pi = 3.14159\ldots$. The table itself will not concern us directly in this paper, but it is included for completeness in the Appendix. We will now describe the action of RC2 in more detail. We will use $x \lll k$ to denote the 16-bit word x rotated left by k bits, & will denote bitwise logical AND, $\oplus$ will denote bitwise exclusive-or and ~ will denote bitwise complementation. All 16-bit word addition $+$ is performed modulo 2^{16}.

2.1 Key Expansion

During the key expansion procedure both byte operations and 16-bit word operations are used. The array $K[\cdot]$ that stores the 64 16-bit round keys will be referred to in two ways.

a) For word operations the positions of the buffer will be referred to as $K[0]$, ..., $K[63]$ where each $K[i]$ is a 16-bit word.
b) For byte operations the array of round keys will be referred to as $L[0]$, ..., $L[127]$ where each $L[i]$ is an eight-bit byte. It will always be the case that $K[i] = L[2i] + 256 \times L[2i + 1]$ (That is, the lower order byte is given first).

Suppose that T bytes of key are supplied by the user with $1 \leq T \leq 128$. The key expansion procedure places the T-byte key into $L[0]$, ..., $L[T-1]$ of the key buffer. Regardless of the value of T however, the algorithm has a maximum effective key length in bits that is denoted $T1$. The effective key length in bytes $T8$ and a mask TM based on the effective key length in bits $T1$ are derived as $T8 = \lceil T1/8 \rceil$ and $TM = 255 \bmod 2^{8(1-T8)+T1}$. Key expansion consists of the following two loops and intermediate step:
   1. for $i = T, T+1, ..., 127$ do
      $L[i] = \text{PITABLE}[L[i-1] + L[i-T]]$ (addition is modulo 256)
   2. $L[128 - T8] = \text{PITABLE}[L[128 - T8] \& TM]$
   3. for $i = 127 - T8, ..., 0$ do
      $L[i] = \text{PITABLE}[L[i+1] \oplus L[i+T8]]$
At the end of this key expansion the array $K[0]$, ..., $K[63]$ contains the 64 16-bit subkey words that will be used during encryption.

2.2 Encryption and Decryption

The encryption operation is defined in terms of primitive MIX and MASH operations. An array of four 16-bit words $R[0]$, ..., $R[3]$ are used to hold the initial plaintext, the intermediate results, and the final ciphertext. Indices to this array are always given modulo 4.

MIX $R[i]$

The primitive "MIX $R[i]$" operation is defined as follows, where $s[0] = 1$, $s[1] = 2$, $s[2] = 3$, and $s[3] = 5$. Here j is a "global" variable so that $K[j]$ is always the first key word in the expanded key which has not yet been used in a MIX operation.

$$R[i] = R[i] + K[j] + (R[i-1] \,\&\, R[i-2]) + (\sim R[i-1] \,\&\, R[i-3]);$$
$$j = j + 1;$$
$$R[i] = R[i] \lll s[i];$$

MIXING round

A MIXING round consists of MIX $R[0]$, MIX $R[1]$, MIX $R[2]$, MIX $R[3]$.

MASH $R[i]$

The primitive "MASH $R[i]$" operation is defined as follows:

$$R[i] = R[i] + K[\,R[i-1] \,\&\, 003\mathtt{f}_x\,];$$

MASHING round

A MASHING round consists of MASH $R[0]$, MASH $R[1]$, MASH $R[2]$, MASH $R[3]$.

The entire encryption operation can now be described as follows. Here j is a global integer variable which is only affected by the mixing operations.

Encryption With RC2

1. Initialize words $R[0]$, ..., $R[3]$ to contain the 64-bit plaintext block.
2. Expand the key, so that words $K[0]$, ..., $K[63]$ become defined.
3. Initialize j to zero.
4. Perform five MIXING rounds.
5. Perform one MASHING round.
6. Perform six MIXING rounds.
7. Perform one MASHING round.
8. Perform five MIXING rounds.
9. The ciphertext is $R[0]$, ..., $R[3]$.

Decryption is the reverse of encryption. Since the details can easily be established they are not included here. Test vectors for encryption using RC2 are provided in the Appendix.

2.3 Features of RC2

RC2 is rather unusual in that the 64-bit plaintext block is split into four words each of 16 bits. In a style reminiscent of the hash function MD4 [13], much of the encryption process relies on one of these words being modified by a function of the other three, the four words then being swapped cyclically. This design approach was explored some seven years after the design of RC2, which now might be described as being an "unbalanced Feistel cipher" [4].

The key schedule for RC2 is also unusual. $T8$ is the number of bytes needed to contain the given $T1$ bits of key. When $T1$ is congruent to k, modulo 8, a mask TM containing ones in the low-order k bits is used to derive the correct effective key length. The first step of the key expansion expands the key to a full 128 bytes, using a non-linear byte-wide feedback shift-register approach. Step three is similar to the first, except that it starts at the high end and works towards the lower end. Steps two and three also work together to limit the effective key size to $T1$ bits. Step three corresponds to using a feedback register of only $T8$ bytes, and step two ensures that the initial state of that register has only $T1$ bits of entropy. Although the procedure limits the actually entropy of the key to $T1$ bits, it also ensures that the final key table depends upon each bit of the supplied key. If one supplies a 16-byte key, but set $T1 = 40$, then changing any bit of the supplied key should result in a different key table, although the number of possible key tables is limited to 2^{40}.

3 Differential Cryptanalysis of RC2

Differential cryptanalysis [1] can be a powerful style of attack. By choosing a pair of plaintexts with a particular difference, which can be adapted to the cipher in question, the cryptanalyst hopes that some identifiable, and unusual behavior, can be observed by processing the ciphertexts. One possible evolution of the difference between a pair of plaintexts during encryption can be described by a *characteristic*. In essence, a characteristic specifies the difference between two parallel encryptions at each stage of the encryption process and there is some associated probability that a pair being encrypted does indeed follow this description. A plaintext pair that follows the characteristic is typically called a *right pair*. A pair that does not is called a *wrong pair*.

Throughout our attack on RC2, we shall define the difference between two 16-bit words A and B to be $A \oplus B$. Furthermore in our analysis we shall be interested in how single-bit differences behave within RC2. The decision to restrict our attention to single-bit differences facilitates analysis but is also motivated by a typical assumption that characteristics involving multiple-bit differences over integer addition will generally hold with lower probability than single-bit characteristics [6]. We note that other more complex techniques [2,7] might open new avenues for the analysis of RC2.

We will use e_t to denote the 16-bit word with a single one bit in position t from the right, all other bits being set to zero. We also view the leftmost bit of a 16-bit word to be the most significant bit. Thus we shall use e_{15} to denote a 16-bit word with the only non-zero bit being the most significant bit. We will denote the word of 16 zero bits as 0000_x where the subscript x denotes hexadecimal notation and we will denote the *Hamming weight* (i.e. the number of ones in the binary expansion of some quantity x) as $\mathrm{Hwt}(x)$.

For the remainder of the paper, we shall consider MIXING and MASHING rounds in the following way. Instead of viewing the operation at each step as acting on a different word we shall consider the operations to be identical (i.e., at each MIX step $R[0] = R[0] + K[j] + (R[3] \& R[2]) + (\sim R[3] \& R[1])$) but that between steps the words are rotated cyclically (i.e., TEMP = $R[0]$; $R[0] = R[1]$; $R[1] = R[2]$; $R[2] = R[3]$; $R[3] = $ TEMP).

3.1 Some Basic Characteristics for MIX

Given an input difference $(e_t \; 0000_x \; 0000_x \; 0000_x)$ to the first MIX step in a MIXING round, the output difference before rotation will be $(e_t \; 0000_x \; 0000_x \; 0000_x)$ with probability $p \geq 1/2$. Rotation then moves this single bit difference within the word, and the four words are swapped cyclically. We can summarize the four basic characteristics which hold with probability $p \geq 1/2$ (when averaged over all plaintexts and key words) for a MIX step. The value of the rotation $s[i]$ depends on the step i in which the characteristic is applied. Note that addition within the subscript of e_t is to be performed modulo 16.

$$(e_t \; 0000_x \; 0000_x \; 0000_x) \rightarrow (0000_x \; 0000_x \; 0000_x \; e_{t+s[i]}) \tag{1}$$

$$(0000_x \; 0000_x \; 0000_x \; e_t) \rightarrow (0000_x \; 0000_x \; e_t \; 0000_x) \tag{2}$$

$$(0000_x \; 0000_x \; e_t \; 0000_x) \rightarrow (0000_x \; e_t \; 0000_x \; 0000_x) \tag{3}$$

$$(0000_x \; e_t \; 0000_x \; 0000_x) \rightarrow (e_t \; 0000_x \; 0000_x \; 0000_x) \tag{4}$$

Apart from (1) with $t = 15$ which holds with probability $p = 1$, these characteristics hold with probability $p = 1/2$ on average. There are times where the characteristics do not hold. The following are the cases where the characteristic hold with certainty:

- In (2), if $(R[2] \& e_t) = (R[1] \& e_t)$ then $p = 1$.
- In (3), if $(R[3] \& e_t) = 0000_x$ then $p = 1$.
- In (4), if $(R[3] \& e_t) = e_t$ then $p = 1$.

In the first MIXING round, the attacker chooses the plaintext and this allows the cryptanalyst to capture some of these special cases in an attack.

3.2 Some Basic Characteristics for MASH

There are two MASHING rounds in RC2 and the basic MASH step is $R[0] = R[0] + K[R[3] \& 003f_x]$. Given an input difference $(0000_x \; 0000_x \; 0000_x \; e_t)$ to a MASHING round with $(e_t \& 003f_x) = 0000_x$ the same key word $K[\cdot]$ will be added to both sets of partially encrypted data. The four basic useful characteristics for MASH are as follows:

$$(e_t \; 0000_x \; 0000_x \; 0000_x) \rightarrow (0000_x \; 0000_x \; 0000_x \; e_t) \tag{5}$$

$$(0000_x \; 0000_x \; 0000_x \; e_t) \rightarrow (0000_x \; 0000_x \; e_t \; 0000_x) \tag{6}$$

$$(0000_x \; 0000_x \; e_t \; 0000_x) \rightarrow (0000_x \; e_t \; 0000_x \; 0000_x) \tag{7}$$

$$(0000_x \; e_t \; 0000_x \; 0000_x) \rightarrow (e_t \; 0000_x \; 0000_x \; 0000_x) \tag{8}$$

Characteristic (5) holds with probability $p = 1/2$ unless $t = 15$ when it holds with probability $p = 1$, characteristics (7) and (8) hold with probability $p = 1$, and characteristic (6) holds with probability $p = 1$ if $(e_t \ \& \ \mathtt{0x3f}) = \mathtt{0000_x}$. Joining these four characteristics together to pass across a `MASHING` round with probability $p = 1$ is straightforward.

3.3   Towards a Differential Attack on RC2

In this section we combine characteristics for both `MIXING` and `MASHING` rounds while moving towards a full analysis of RC2. We will assume that the subkey words $K[0], \ldots, K[63]$ are independent and we aim to recover the expanded key table $K[\cdot]$ in our attack.

The characteristics of interest are built around single-bit differences and as noted in Section 3.1 there are advantages to having this single non-zero bit in the most significant bit of a word. Depending on which word $R[\cdot]$ is the subject of the characteristic we use, different rotation amounts feature during `MIXING`. This leads to conditions on t, the position of the single-bit difference in the plaintext, that provide some advantages in an attack. Another consideration is the presence of the `MASHING` rounds and one aim might be to nullify their action. If a one-bit characteristic specifies an input difference to a `MASHING` round of e_t in any one of the words, then provided $t = 15$ the characteristic will pass through the `MASHING` round unhindered with probability $p = 1$. If $5 < t < 15$ then there is a characteristic that holds with probability $p = 1/2$. There are six `MIXING` rounds between the two `MASHING` rounds and so with the difference $(e_t \ \mathtt{0000_x} \ \mathtt{0000_x} \ \mathtt{0000_x})$ as input to the first `MASHING` round we can establish the values of t that are useful to us.

A more accurate reflection of the success of a final attack is given by considering *differentials* [10] instead of characteristics (which provide only a lower bound to the probability of the differential). In Section 3.5 we will consider the issue of differentials in more detail but from this point on we will anticipate later analysis by referring to the use of differentials during our description of the attack. The observations provided so far allow us to present in Table 1 the differentials that are useful to us.

3.4   Recovering Key Information

In a differential cryptanalytic attack the attacker typically chooses a differential for $(n - 1)$ rounds of an n-round block cipher. The attacker then tries to deduce key information from the last round of the cipher [1]. Here, the most effective attack on RC2 appears to require that bits of the subkey $K[0]$ used in the first `MIXING` round are recovered first.

Consider a differential with input difference $(\mathtt{0000_x} \ \mathtt{0000_x} \ \mathtt{0000_x} \ e_t)$. The starting values of $R[1]$ and $R[2]$ are chosen so that $(R[1] \ \& \ e_t) = (R[2] \ \& \ e_t)$. After the first `MIX` step the difference will be $(\mathtt{0000_x} \ \mathtt{0000_x} \ e_t \ \mathtt{0000_x})$. The output difference from the second `MIX` step will depend on the value of bit t in register $R[3]$. If this bit is zero then word $R[1]$ with difference $\mathtt{0000_x}$ will be

plaintext difference	difference at start of last MIXING round	prob.	values of t
$(e_t\ 0000_x\ 0000_x\ 0000_x)$	$(e_{t+15}\ 0000_x\ 0000_x\ 0000_x)$	2^{-58}	4
$(e_t\ 0000_x\ 0000_x\ 0000_x)$	$(e_{t+15}\ 0000_x\ 0000_x\ 0000_x)$	2^{-59}	1, 2, 3
$(0000_x\ e_t\ 0000_x\ 0000_x)$	$(0000_x\ e_{t+14}\ 0000_x\ 0000_x)$	2^{-58}	5
$(0000_x\ e_t\ 0000_x\ 0000_x)$	$(0000_x\ e_{t+14}\ 0000_x\ 0000_x)$	2^{-59}	1, 3
$(0000_x\ e_t\ 0000_x\ 0000_x)$	$(0000_x\ e_{t+14}\ 0000_x\ 0000_x)$	2^{-60}	0, 2, 4
$(0000_x\ 0000_x\ e_t\ 0000_x)$	$(0000_x\ 0000_x\ e_{t+13}\ 0000_x)$	2^{-58}	14
$(0000_x\ 0000_x\ e_t\ 0000_x)$	$(0000_x\ 0000_x\ e_{t+13}\ 0000_x)$	2^{-59}	7, ..., 13
$(0000_x\ 0000_x\ 0000_x\ e_t)$	$(0000_x\ 0000_x\ 0000_x\ e_{t+11})$	2^{-58}	6
$(0000_x\ 0000_x\ 0000_x\ e_t)$	$(0000_x\ 0000_x\ 0000_x\ e_{t+11})$	2^{-59}	15, 0, ..., 5

Table 1. 26 differentials that are potentially useful in an attack on RC2. The associated probabilities are lower bounds provided by the analysis of a characteristic contained within the specified differential.

chosen. Otherwise word $R[2]$ with difference e_t will be selected and a difference will be introduced into another word. Note that the value of this bit depends on the plaintext (which we know) and on bits of the first 16-bit subkey word $K[0]$.

We can trace the output of the second MIX step to the end of the penultimate MIXING round by using the differentials in Table 1. If the pair is a right pair then we can recover one bit of information from $K[0]$ as follows. A necessary condition for a pair to be a good pair is that

$$e_t\ \&\ ((R[0] + (R[3]\ \&\ R[2])) \tag{9}$$
$$+ ({\sim}R[3]\ \&\ R[1]) + K[0])\lll 1) = 0.$$

Let $x = R[0] + (R[3]\ \&\ R[2]) + ({\sim}R[3]\ \&\ R[1])$ which we control via the choice of plaintext. Then we have the following condition for a right pair:

$$(x + K[0])\ \&\ e_{t-1} = 0. \tag{10}$$

Denote by k the value derived by setting the top $((16 - t) \bmod 16)$ bits of $K[0]$ to zero. Let $y = x\ \&\ e_{t-1}$ and let z be the quantity derived by setting the top $((16 - t + 1) \bmod 16)$ bits of x to zero. Then we have that

$$(x + K[0])\ \&\ e_{t-1} = 0 \Leftrightarrow y = (z + k)\ \&\ e_{t-1}. \tag{11}$$

To mount an attack to recover bit $(t - 1)$ of k for some given t we encrypt plaintext pairs with $z = 0$ until we obtain a right pair. Once we have a right pair we observe the value of y. From this we deduce the value of bit $(t - 1)$ in k and hence in $K[0]$. We can then repeat this approach choosing pairs with different values to z so that information on the subkey $K[0]$ is recovered bit by bit.

By using different differentials with different values of t (see Table 1) we are able to introduce some error-checking into the attack[2]. In this way the bits of

---

[2] Not all values of t are valid for use due to the two MASHING rounds.

$K[0]$ that we recover can be verified. All recovered bits of $K[0]$ have to be correct before the next bit of $K[0]$ can be correctly derived. Note that *structures* [1] can be useful in reducing the plaintext requirements for a differential attack when more than one differential is useful. With n useful differentials we can ask for a *structure* of 2^n plaintexts with specifically chosen differences. From these we derive 2^{n-1} plaintext pairs for each of the n characteristics.

There remains the issue of detecting when a data pair is a good pair. We note that the difference at the start of the final MIXING round has Hamming weight one for a good pair. We might therefore measure the Hamming weight of the ciphertext and if the weight is less than some threshold the pair can be considered a right pair. Depending on the threshold we might accept some wrong pairs as being right pairs, something that would provide a wrong answer to the bit we wish to recover with probability $1/2$. To improve the robustness of the attack one might aim to collect more right pairs. Then the value of the bit suggested most often can be assumed to be the correct value to the key bit we are trying to recover. As a demonstration we provide the success rate for different amounts of plaintext in experiments on eight-round RC2. (There are eight MIXING rounds with a MASHING inserted after round five as occurs in RC2.) A decision on whether a good pair had occurred was made according to whether the Hamming weight of the difference in the ciphertext was less than some threshold. Then, once a value for the key bit had been counted more than the other (this difference being denoted by *excess*) that value for the key bit was set. Each entry in the table was computed after 20 experiments.

	Hamming weight					
excess	11		12		13	
2	90%	2^{29}	20%	$2^{28.5}$	0%	$2^{27.5}$
4	100%	2^{30}	95%	2^{30}	20%	$2^{29.5}$
8	100%	2^{31}	100%	2^{31}	65%	2^{31}

3.5   The Effectiveness of Differential Cryptanalysis

As we previously mentioned it is differentials and their probabilities that reflect the effectiveness of a differential attack. Whereas a characteristic describes one specific evolution of differences through encryption, from a given starting difference there might well have been other "paths" through the cipher to the same target difference than the one described by one particular characteristic. With RC2 this leads to a particularly interesting interaction between the MIXING and MASHING rounds.

First we will consider in abstract terms the probability that a one-bit difference in some word a produces a one-bit difference in the word d when we define $d = a + b + c$ for unknown constants b and c. One approach might be to consider this as two separate additions and to consider the intermediate word $e = a + b$ first. Since a one-bit difference in a produces a one-bit difference in e with probability $1/2$ and a one-bit difference in e provides a one-bit difference in $d = e + c$ with probability $1/2$ we would say that the characteristic over the two additions has probability $1/4$. However it would then be misleading to use this characteristic to provide an approximation to the probability of the differential from a to d. Instead, the probability of the propagation of a one-bit difference from a to d is $1/2$ since $b + c$ is a fixed value. Consequently the probability of the differential from a to d must also be $1/2$.

Recall that the probability of the differential is given by the sum of the probabilities of all the characteristics that satisfy the differential. By looking at two successive additions in isolation we inadvertently restrict our attention to single-bit differences in the intermediate value e. Let α, $0 \le \alpha \le n - 1$, denote the position of the one bit difference in a. A one-bit difference in a will give a difference in e with Hamming weight h with probability 2^{-h}, $1 \le h < n - \alpha$, and with probability $2^{-n+\alpha+1}$ for $h = n - \alpha$. Since this h-bit difference was caused by a one-bit difference in the previous step[3] an h-bit difference in e will be transformed to a one-bit difference in d by the addition of c with probability $1/2$. Thus we get

$$p = 2^{-1}(2^{-n+\alpha} + \sum_{h=1}^{n-\alpha} 2^{-h}), \quad \text{if } \alpha < n - 1 \tag{12}$$

$$p = 1 \text{ if } \alpha = n - 1. \tag{13}$$

One place where this has an effect is when a MIXING round follows a MASHING round. Each word $R[0]$, ..., $R[3]$ is modified by a MASH step in turn. At the first subsequent MIX step $R[0]$ is modified by means of addition. By looking at the two additions in isolation one under-estimates the probability of the differential.

In the analysis of RC2 we need to take account of this effect since it applies to some extent to the MIXING rounds as well as during the transition between MIXING and MASHING rounds. Within the MIXING rounds an intermediate quantity is used as input to a multiplexor function. This reduces the probability that this particular characteristic is followed by a factor of 2^{-h} for each multiplexor when the Hamming weight of the difference is h. If we denote the number of multiplexing functions between two successive additions by k then (12) can be rewritten as follows:

$$p = \sum_{h=1}^{n-\alpha} 2^{-h} \cdot 2^{-hk} \cdot 2^{-1} + 2^{-(n-\alpha)} \cdot 2^{-(n-\alpha)k} \cdot 2^{-1} \tag{14}$$

$$= 2^{-1}\left(\sum_{h=1}^{n-\alpha} 2^{-(k+1)h} + 2^{-(k+1)(n-\alpha)}\right) \tag{15}$$

$$= 2^{-1}\left(\frac{1 - 2^{-(k+1)(n-\alpha)}}{1 - 2^{-(k+1)}}\frac{1}{2^{(k+1)}}\right) + 2^{-(k+1)(n-\alpha)} \tag{16}$$

$$\approx 2^{-1}\left(\frac{1}{2^{(k+1)} - 1}\right). \tag{17}$$

---

[3] In general it is not true that an h-bit difference goes to a one-bit difference with such a high probability.

The last approximation is reasonable for smaller α ($\alpha < n - 3$) but would need some correction for larger values of α. For $k = 0, 1, 2, 3$, (17) gives $p = 1/2, 1/6, 1/14, 1/30$, which should be compared with the respective probabilities of the characteristics we previously derived: $1/4, 1/8, 1/16, 1/32$. In the case of two consecutive MIXING rounds we have that $k = 3$ and so the probability of a one-bit to one-bit differential across two MIXING rounds is $1/30 \times 2^{-3} = 1/240$.

The effect we are using here can be extended to a series of additions whereby the intermediate values of interest have differences with a variety of Hamming weights even though the starting and ending difference have weight one. Consider three consecutive mixing rounds. Let a be a one-bit difference in the leftmost words of two inputs and let α be the position of that bit, where $0 \leq \alpha \leq n - 1$. Let d be the difference in the leftmost words after three mixing rounds and suppose that h_1 and h_2 denote the Hamming weights of the leftmost words after one, respectively two, mixing rounds. Then the probability that d is a one-bit difference can be estimated as follows, where $k = 3$ and where for simplicity we have eliminated the term for $h = n - \alpha$.

$$p \simeq \sum_{h_1=1}^{n-\alpha} \sum_{h_2=1}^{n-\alpha} 2^{-h_1} \cdot 2^{-h_1 k} \cdot 2^{-h_2} \cdot 2^{-h_2 k} \cdot 2^{-1} \tag{18}$$

$$= 2^{-1} \left(\sum_{h_1=1}^{n-\alpha} 2^{-(k+1)h_1} \cdot \sum_{h_2=1}^{n-\alpha} 2^{-(k+1)h_2} \right) \tag{19}$$

$$\approx 2^{-1} \left(\frac{1}{2^{(k+1)} - 1} \right) \left(\frac{1}{2^{(k+1)} - 1} \right). \tag{20}$$

Again the final approximation requires that α is small. For $k = 3$ p is $2^{-1}(1/15)^2$. We can now estimate the probability of the differential over three mixing rounds by $2^{-1}(1/15)^2 \times 1/8 \simeq 1/3600$. This extends easily to more rounds and in general the probability of a differential over r mixing rounds is $(1/15)^{r-1} \times 1/16$. Note that the MASHING rounds can be passed with probability one.

For a more accurate assessment a slight correction should be applied for rounds where the difference is close to the most significant bit, but experimental evidence given below suggests that the expressions derived are reasonable to use. The number of rounds in the table refers to the number of MIXING rounds used. After five MIXING rounds an additional MASHING round is inserted as occurs when encrypting with RC2. The final column is derived as an average over at least five sets of experiments for each row.

# rounds	# pairs/test	# right pairs expected	# right pairs obtained
3	2^{19}	146	146
4	2^{22}	78	79
6	2^{29}	44	47
7	2^{31}	12	13

Note that the probability of the differential obtained in this section does not take into account text pairs which have internal differences in more than one word before they resynchronize. This was observed occasionally during experiments but cases where differences in more than one word resynchronize are rare and we ignore their impact on our estimates.

3.6   Differential Cryptanalysis of RC2

We arrive at the following estimates for the data required to recover information about the subkey $K[0]$. Once this subkey word has been recovered the attack is repeated on what would now become a reduced version of RC2. When we take into account the key-recovery techniques of Section 3.4 we estimate that a differential attack on RC2 with r MIXING rounds (including the MASHING rounds) requires at most 2^{4r} chosen plaintexts. An attack on RC2 with 16 MIXING rounds requires use of a differential with probability at least $2^{-58.7}$ ("at least" since we have not yet accounted for such phenomena as a one-bit difference in the most significant bit at a MIXING round). In this regard RC2 with 16 MIXING rounds compares favorably to DES (2^{47} pairs [1]) and 12-round RC5 (2^{44} pairs [3]). It is fair to observe, however, that RC2 is not a fast cipher and an optimized version of DES and 12-round RC5 are both likely to be faster than RC2.

4   Linear Cryptanalysis of RC2

Linear cryptanalysis has provided the best theoretical attack on DES in terms of data requirements [9]. However, its usefulness on other ciphers is often limited. The aim of such an attack is to relate bits of the plaintext and ciphertext to bits of the key via a linear equation which holds with some probability p. Such an approximation can generally be used to provide an estimate for one bit of the key and more advanced techniques are available to extract more key information [9]. If an approximation holds with probability p then the important quantity for the cryptanalyst is the absolute value of the bias of the approximation $b = |p - 1/2|$. Typically the data required to use such an approximation is given by $c \times b^{-2}$ known plaintexts for some small constant c [9].

   The MIX step in RC2 is $R[0] = R[0] + K[j] + (R[3]\ \&\ R[2]) + (\sim R[3]\ \&\ R[1])$. Across integer addition the best linear approximation involves the least significant bit of each quantity, and will hold with probability one. The multiplexor function $x = (R[3]\ \&\ R[2]) + (\sim R[3]\ \&\ R[1])$ has linear approximations of varying usefulness. The absolute value of the highest non-trivial bias is $1/4$ when averaged over all plaintexts. In a slight abuse of notation we will consider a 16-bit word x as a vector in Z_2^{16} and we will use the 16-bit quantity α to indicate the bits of x that are to be used in a linear approximation. This is most conveniently described by means of the scalar product of two vectors. Thus the $\{0,1\}$-vector α will be used to denote the specific bits of x to be used in an approximation and $\alpha \cdot x$ is the value of these bits combined using exclusive-or. Useful linear approximations across the multiplexor are of the form

$$\alpha \cdot x = \alpha \cdot R[1] \qquad \alpha \cdot x = \alpha \cdot R[1] \oplus \alpha \cdot R[3]$$
$$\alpha \cdot x = \alpha \cdot R[2] \qquad \alpha \cdot x = \alpha \cdot R[2] \oplus \alpha \cdot R[3]$$

where $\mathrm{Hwt}(\alpha) = 1$. More generally approximations to the multiplexor function with non-zero bias have the form

$$\delta \cdot x = \alpha \cdot R[1] \oplus \beta \cdot R[2] \oplus \gamma \cdot R[3] \tag{21}$$

where δ is the bitwise inclusive-or of α and β and γ is either 0 or it consists of ones in positions where either α or β have ones. The greater the value of $\mathrm{Hwt}(\gamma)$ the lower the absolute value of the bias of the approximation.

The following approximation to the first MIX step (which includes the cyclic swap of the $R[\cdot]$ words) might be useful

$$e_1 \cdot (R[3]^{\mathrm{new}}) = e_0 \cdot (R[0]^{\mathrm{old}}) \oplus e_0 \cdot (K[j]) \oplus e_0 \cdot R[2]^{\mathrm{old}}.$$

This has a bias of absolute value $1/4$. The following steps require no approximation and there appears to be no better non-trivial linear approximations for a complete MIXING round. We might illustrate this approximation in the following way:

step	$R[0]$	$R[1]$	$R[2]$	$R[3]$	round 1
	e_0	—	e_0	—	start
1	—	—	—	e_1	$\|b\| = 1/4$
2	—	—	e_1	—	$\|b\| = 1/2$
3	—	e_1	—	—	$\|b\| = 1/2$
4	e_1	—	—	—	$\|b\| = 1/2$

In continuing this approximation into the next MIXING round we would be forced to approximate the bit $e_1 \& R[0]$. One integer addition involves the sub-key word $K[4]$ and depending on this value the bias of the approximation will vary[4]. The second integer addition involves the output from the multiplexor function. By the conditions given above this approximation must involve $e_1 \& R[1]$ or $e_1 \& R[2]$ and we can construct the following approximations for the second and third MIXING rounds. Here we assume that the bias of the approximation across the multiplexor function is at most $1/4$. Similarly, we assume that the bias of the approximation across the integer addition is at most $1/4$. This occurs in approximating steps 1 and 3 and the value of $|b|$ is given for those steps individually.

step	$R[0]$	$R[1]$	$R[2]$	$R[3]$	round 2
	e_1	—	—	—	start
1	—	e_1	—	e_2	$\|b\| \le 1/8$
2	e_1	—	e_2	—	$\|b\| = 1/2$
3	—	$e_1 \oplus e_2$	—	e_4	$\|b\| \le 1/8$
4	$e_1 \oplus e_2$	—	e_4	—	$\|b\| = 1/2$

4.1   The Effectiveness of Linear Cryptanalysis

The typical way to measure the effectiveness of linear cryptanalysis is to appeal to the so-called *piling-up lemma* [8]. By doing this, we are lead to estimate a bias

---

[4] Note that the whole issue of key-dependence in linear cryptanalysis is a complex one that is rarely addressed in detail.

of $\leq 2^{-2} \times 2^{-3} \times 2^{-3} \times 2^{2} = 2^{-6}$ for our approximation to the first two `MIXING`
rounds of RC2. In the case of RC2, however, routine use of the piling-up lemma
can lead to misleading results.

As an example, suppose that the two subkeys used in steps one and three of
round two are zero. In isolation the approximation to step one (A_1, say) holds
with probability 5/8. In step three we find that the second approximation (A_2,
say) involves bits that previously determined whether A_1 held. Analysis shows
that the probability that A_2 holds given that A_1 held is 13/20 and not 5/8 when
A_2 is considered in isolation. Furthermore, the probability that A_2 doesn't hold
when A_1 doesn't hold is 5/12 instead of 3/8. So when the two approximations
are combined the probability that the combined approximation to round two
holds is $(5/8 \times 13/20) + (3/8 \times 5/12) = 9/16$ which leads to a bias of 1/16. This
is greater than the 1/32 predicted by use of the piling-up lemma.

Much of the complicated interaction between the two approximations is due
to the role of addition in the cipher. As an example, if we suppose that approximation A_1 holds, then it can be shown that the probability that the least
significant bit of $R[2]$ is equal to zero is 11/20. Since this bit plays a pivotal role
in determining whether A_2 holds it is no surprise that the piling-up lemma gives
misleading results.

For the user of RC2 there is circumstantial evidence that linear cryptanalysis is unlikely to pose a threat to RC2. Such attacks appear to be ineffective
for ciphers that mix integer addition and bitwise operations unless the approximation can be limited to the least significant bits across an addition [6]. Such a
restriction appears unlikely as an extension of the current approximation into a
third `MIXING` round illustrates:

step	$R[0]$	$R[1]$	$R[2]$	$R[3]$	round 3
	$e_1 \oplus e_2$	—	e_4	—	start
1	—	$e_1 \oplus e_2 \oplus e_4$	—	$e_2 \oplus e_3$	$|b| \leq 1/16$
2	$e_1 \oplus e_2 \oplus e_4$	—	$e_2 \oplus e_3$	—	$|b| = 1/2$
3	—	$e_1 \oplus e_3 \oplus e_4$	—	$e_4 \oplus e_5 \oplus e_7$	$|b| \leq 1/128$
4	$e_1 \oplus e_3 \oplus e_4$	—	$e_4 \oplus e_5 \oplus e_7$	—	$|b| = 1/2$

Nevertheless, there are complex interactions between the individual steps of
RC2 and these often provide unintuitive results. In particular we have discovered
cases where adding a non-trivial approximation to an existing approximation
actually boosts the absolute value of the bias. (Such an example can be found
in step 3 above when the subkeys in all rounds are set to zero.) Under such
circumstances the true effectiveness of linear cryptanalysis in attacking RC2 has
to remain an open problem.

5   Conclusions

In this paper we have described the block cipher RC2. While the cipher is perhaps
slower than other alternatives available today, it does appear to offer effective
resistance to differential cryptanalysis. Our attempts to apply linear cryptanalysis to RC2 have provided some intriguing insights, but are as yet insufficient to

determine the actual resistance of RC2 to linear cryptanalysis; this remains an open problem. It is important that RC2 continues to come under close scrutiny from the cryptanalytic community.

References

1. E. Biham and A. Shamir. *Differential Cryptanalysis of the Data Encryption Standard*. Springer-Verlag, New York, 1993.
2. J. Borst, L.R. Knudsen, and V. Rijmen. Two attacks on reduced IDEA. In W. Fumy, editor, *Advances in Cryptology — Eurocrypt '97*, volume 1233 of *Lecture Notes in Computer Science*, pages 1–13, 1997. Springer Verlag.
3. A. Biryukov and E. Kushilevitz. Improved cryptanalysis of RC5. Preprint.
4. M. Blaze and B. Schneier. The MacGuffin block cipher algorithm. In B. Preneel, editor, *Fast Software Encryption*, volume 1008 of *Lecture Notes in Computer Science*, pages 97–110, 1995. Springer Verlag.
5. S. Dusse, P. Hoffman, B. Ramsdell, L. Lundblade, and L. Repka. S/MIME Message Specification. September 23, 1997. Available from http://www.imc.org/draft-dusse-smime-msg.
6. B.S. Kaliski and Y.L. Yin. On differential and linear cryptanalysis of the RC5 encryption algorithm. In D. Coppersmith, editor, *Advances in Cryptology — Crypto '95*, volume 963 of *Lecture Notes in Computer Science*, pages 171–184, 1995. Springer Verlag.
7. L.R. Knudsen and W. Meier. Improved differential attacks on RC5. In N. Koblitz, editor, *Advances in Cryptology — Crypto '96*, volume 1109 of *Lecture Notes in Computer Science*, pages 216–228, 1996. Springer Verlag.
8. M. Matsui. Linear cryptanalysis method for DES cipher. In T. Helleseth, editor, *Advances in Cryptology — Eurocrypt '93*, volume 765 of *Lecture Notes in Computer Science*, pages 386–397, 1994. Springer-Verlag.
9. M. Matsui. The first experimental cryptanalysis of the Data Encryption Standard. In Y. Desmedt, editor, *Advances in Cryptology — Crypto '94*, volume 839 of *Lecture Notes in Computer Science*, pages 1–11, 1994. Springer-Verlag.
10. X. Lai, J. Massey and S. Murphy. Markov ciphers and differential cryptanalysis. In D. Davies, editor, *Advances in Cryptology — Eurocrypt '91*, volume 547 of *Lecture Notes in Computer Science*, pages 17–38, 1991. Springer Verlag.
11. National Institute of Standards and Technology (NIST). *FIPS Publication 46-2: Data Encryption Standard*. December 30, 1993.
12. R.L. Rivest. A Description of the RC2^{TM} Encryption Algorithm. File `draft-rivest-rc2desc-00.txt` available from `ftp://ftp.ietf.org/internet-drafts/`.
13. R.L. Rivest. The MD4 message digest algorithm. In A.J. Menezes and S.A. Vanstone, editors, *Advances in Cryptology — Crypto '90*, volume 537 of *Lecture Notes in Computer Science*, pages 303–311, 1991. Springer-Verlag.
14. R.L. Rivest. The RC5 encryption algorithm. In B. Preneel, editor, *Fast Software Encryption*, volume 1008 of *Lecture Notes in Computer Science*, pages 86–96, 1995. Springer Verlag.

Appendix

The substitution table **PITABLE** specified in
hexadecimal notation for input byte **ab**.

a*										*b						
	0	1	2	3	4	5	6	7	8	9	a	b	c	d	e	f
00:	d9	78	f9	c4	19	dd	b5	ed	28	e9	fd	79	4a	a0	d8	9d
10:	c6	7e	37	83	2b	76	53	8e	62	4c	64	88	44	8b	fb	a2
20:	17	9a	59	f5	87	b3	4f	13	61	45	6d	8d	09	81	7d	32
30:	bd	8f	40	eb	86	b7	7b	0b	f0	95	21	22	5c	6b	4e	82
40:	54	d6	65	93	ce	60	b2	1c	73	56	c0	14	a7	8c	f1	dc
50:	12	75	ca	1f	3b	be	e4	d1	42	3d	d4	30	a3	3c	b6	26
60:	6f	bf	0e	da	46	69	07	57	27	f2	1d	9b	bc	94	43	03
70:	f8	11	c7	f6	90	ef	3e	e7	06	c3	d5	2f	c8	66	1e	d7
80:	08	e8	ea	de	80	52	ee	f7	84	aa	72	ac	35	4d	6a	2a
90:	96	1a	d2	71	5a	15	49	74	4b	9f	d0	5e	04	18	a4	ec
a0:	c2	e0	41	6e	0f	51	cb	cc	24	91	af	50	a1	f4	70	39
b0:	99	7c	3a	85	23	b8	b4	7a	fc	02	36	5b	25	55	97	31
c0:	2d	5d	fa	98	e3	8a	92	ae	05	df	29	10	67	6c	ba	c9
d0:	d3	00	e6	cf	e1	9e	a8	2c	63	16	01	3f	58	e2	89	a9
e0:	0d	38	34	1b	ab	33	ff	b0	bb	48	0c	5f	b9	b1	cd	2e
f0:	c5	f3	db	47	e5	a5	9c	77	0a	a6	20	68	fe	7f	c1	ad

Test vectors for the RC2 encryption algorithm.

key length (bytes)	effective key length (bits)	key	plaintext	ciphertext
8	63	00000000 00000000	00000000 00000000	ebb773f9 93278eff
8	64	ffffffff ffffffff	ffffffff ffffffff	278b27e4 2e2f0d49
8	64	30000000 00000000	10000000 00000001	30649edf 9be7d2c2
1	64	88	00000000 00000000	61a8a244 adacccf0
7	64	88bca90e 90875a	00000000 00000000	6ccf4308 974c267f
16	64	88bca90e 90875a7f 0f79c384 627bafb2	00000000 00000000	1a807d27 2bbe5db1
16	128	88bca90e 90875a7f 0f79c384 627bafb2	00000000 00000000	2269552a b0f85ca6
33	129	88bca90e 90875a7f 0f79c384 627bafb2 16f80a6f 85920584 c42fceb0 be255daf 1e	00000000 00000000	5b78d3a4 3dfff1f1

Serpent: A New Block Cipher Proposal

Eli Biham[1], Ross Anderson[2], and Lars Knudsen[3]

[1] Technion, Haifa, Israel; biham@cs.technion.ac.il
[2] Cambridge University, England; rja14@cl.cam.ac.uk
[3] University of Bergen, Norway; lars.knudsen@ii.uib.no

Abstract. We propose a new block cipher as a candidate for the Advanced Encryption Standard. Its design is highly conservative, yet still allows a very efficient implementation. It uses the well-understood DES S-boxes in a new structure that simultaneously allows a more rapid avalanche, a more efficient bitslice implementation, and an easy analysis that enables us to demonstrate its security against all known types of attack. With a 128-bit block size and a 256-bit key, it is almost as fast as DES on a wide range of platforms, yet conjectured to be at least as secure as three-key triple-DES.

1 Introduction

For many applications, the Data Encryption Standard algorithm is nearing the end of its useful life. Its 56-bit key is too small, as shown by a recent distributed key search exercise [21]. Although triple-DES can solve the key length problem, the DES algorithm was also designed primarily for hardware encryption, yet the great majority of applications that use it today implement it in software, where it is relatively inefficient.

For these reasons, the US National Institute of Standards and Technology has issued a call for a successor algorithm, to be called the *Advanced Encryption Standard* or *AES*. The essential requirement is that AES should be both faster and more secure than triple DES; specifically, it should have a 128 bit block length and a key length of 256 bits (though keys of 128 and 192 bits must also be supported).

In this paper, we present a candidate for AES. Our design philosophy has been highly conservative; we did not feel it appropriate to use novel and untested ideas in a cipher which, if accepted after a short review period, will be used to protect enormous volumes of financial transactions, health records and government information over a period of decades.

We therefore decided to use the S-boxes from DES, which have been subjected to intense study over many years and whose properties are thus well understood, in a new structure which is optimized for efficient implementation on modern processors while simultaneously allowing us to apply the extensive analysis already done on DES. As a result, we can show that our design resists all known attacks, including those based on both differential [7] and linear [20] techniques.

S. Vaudenay (Ed.): Fast Software Encryption – FSE'98, LNCS 1372, pp. 222–238, 1998.
© Springer-Verlag Berlin Heidelberg 1998

We propose several variants of the cipher, which we have tentatively named Serpent. The primary variant is a 32-round cipher which we believe to be as secure as three-key triple-DES, but which is only slightly slower than DES when implemented in C on a Pentium (in some assembly languages it may be faster than DES). It is an SP-network operating on four 32-bit words, thus giving a block size of 128 bits.

The additional variants have increased block sizes. The block size can be doubled to 256 bits either by increasing the word size from 32 to 64 bits (which will be well suited to the new generation of 64-bit processors), or by using the round function in a Feistel construction. These two variants can be combined to give a cipher with 512-bit blocks.

At this stage all the variants are still tentative. We are still working on improvements and analysis. As usual in this field, we encourage interested parties to analyze the cipher, inform us of any weakness, and pass on any remarks or suggestions for improvements.

All values used in the ciphers are represented in little-endian, including the bit order (0–31 in 32-bit words, or 0–127 in the full 128-bit blocks), and the order of words in the block. Thus bit 0 is the least significant bit, and word 0 is the least significant word. The notation is important, as there are two equivalent representations of Serpent: a standard representation and a bitslice representation.

2 The Cipher

The main variant of our cipher encrypts a 128-bit plaintext P to a 128-bit ciphertext C in r rounds under the control of $r + 1$ 128-bit subkeys $\hat{K}_0, \ldots, \hat{K}_r$. (We have chosen $r = 32$ as the default, and will henceforth replace r by 32 in order to make the description of the cipher more readable.)

The cipher is an SP-network and consists of:

- an initial permutation IP;
- 32 rounds, each consisting of a key mixing operation, a pass through S-boxes, and (in all but the last round) a linear transformation. In the last round, this linear transformation is replaced by an additional key mixing operation;
- a final permutation FP.

The initial and final permutations do not have any cryptographic significance. They are used to simplify an optimized implementation of the cipher, which is described in the next section, and to improve its computational efficiency. Both these two permutations and the linear transformation are specified in the appendix; their design principles will be made clear in the next section.

We use the following notation. The initial permutation IP is applied to the plaintext P giving $\hat{B}_0$, which is the input to the first round. The rounds are numbered from 0 to 31, where the first round is round 0 and the last is round 31. The output of the first round (round 0) is $\hat{B}_1$, the output of the second round (round 1) is $\hat{B}_2$, the output of round i is $\hat{B}_{i+1}$, and so on, until the output of the last round (in which the linear transformation is replaced by an additional key mixing) is denoted by $\hat{B}_{32}$. The final permutation FP is now applied to give the ciphertext C.

Each round function R_i ($i \in \{0, \ldots, 31\}$ uses only a single replicated S-box. For example, R_0 uses S_0, 32 copies of which are applied in parallel. Thus the first copy of S_0 takes bits 0,1,2 and 3 of $\hat{B}_0 \oplus \hat{K}_0$ as its input and returns as output the first four bits of an intermediate vector; the next copy of S_0 inputs bits 4–7 of $\hat{B}_0 \oplus \hat{K}_0$ and returns the next four bits of the intermediate vector, and so on. The intermediate vector is then transformed using the linear transformation, giving $\hat{B}_1$. Similarly, R_1 uses 32 copies of S_1 in parallel on $\hat{B}_1 \oplus \hat{K}_1$ and transforms their output using the linear transformation, giving $\hat{B}_2$.

In the last round R_{31}, we apply S_{31} on $\hat{B}_{31} \oplus \hat{K}_{31}$, and XOR the result with $\hat{K}_{32}$ rather than applying the linear transformation. The result $\hat{B}_{32}$ is then permuted by FP, giving the ciphertext.

Thus the 32 rounds use 32 different S-boxes each of which maps four input bits to four output bits. Each S-box is used only in one round, in which it is used 32 times in parallel. The 32 S-boxes are chosen as the 32 separate lines of the eight DES S-boxes; thus our S_0 (used in round 0) is the first line of the DES $S1$, our S_1 (used in round 1) is the second line of the DES $S1$, our S_4 (used in round 4) is the first line of the DES $S2$ and so on.

As with DES, the initial permutation is the inverse of the final permutation. Thus the cipher may be formally described by the following equations:

$$\hat{B}_0 = IP(P)$$
$$\hat{B}_{i+1} = R_i(\hat{B}_i)$$
$$C = IP^{-1}(\hat{B}_r)$$

where

$$R_i(X) = L(\hat{\mathcal{S}}_i(X \oplus \hat{K}_i)) \qquad i = 0, \ldots, r - 2$$
$$R_i(X) = \hat{\mathcal{S}}_i(X \oplus \hat{K}_i) \oplus \hat{K}_r \quad i = r - 1$$

where $\hat{\mathcal{S}}_i$ is the application of the S-box S_i 32 times in parallel, and L is the linear transformation.

Although each round of the proposed cipher might seem weaker than a round of DES, we shall see below that their combination overcomes the weakness. The greater speed of each round, and the increased number of rounds, make the cipher both almost as fast as DES and much more secure.

2.1   Decryption

Decryption is different from encryption in that the inverse of the S-boxes must be used, as well as the inverse linear transformation and reverse order of the subkeys.

3 An Efficient Implementation

Much of the motivation for the above design will become clear as we consider how to implement the algorithm efficiently. We do this in bitslice mode. For a full description of a bitslice implementation of DES, see [9]; the basic idea is that just as one can use a 1-bit processor to implement an algorithm such as DES by executing a hardware description of it, using a logical instruction to emulate each gate, so one can also use a 32-bit processor to compute 32 different DES blocks in parallel — in effect, using the CPU as a 32-way SIMD machine.

This is much more efficient than the conventional implementation, in which a 32-bit processor is mostly idle as it computes operations on 6 bits, 4 bits, or even single bits. The bitslice approach was used in the recent successful DES key search, in which spare CPU cycles from thousands of machines were volunteered to solve a challenge posed by RSADSI. However the problem with using bitslice techniques for DES encryption (as opposed to keysearch) is that one has to process many blocks in parallel, and although special modes of operation can be designed for this, they are not the modes in common use.

Our cipher has therefore been designed so that all operations can be executed using 32-fold parallelism during the encryption or decryption of a single block. Indeed the bitslice description of the algorithm is much simpler than its conventional description. No initial and final permutations are required, since the initial and final permutations described in the standard implementation above are just the conversions of the data from and to the bitslice representation. We will now present an equivalent description of the algorithm for bitslice implementation.

The cipher consists simply of 32 rounds. The plaintext becomes the first intermediate data $B_0 = P$, after which the 32 rounds are applied, where each round $i \in \{0, \ldots, 31\}$ consists of three operations:

1. Key Mixing: At each round, a 128-bit subkey K_i is exclusive or'ed with the current intermediate data B_i
2. S Boxes: The 128-bit combination of input and key is considered as four 32-bit words. The S-box, which is implemented as a sequence of logical operations (as it would be in hardware) is applied to these four words, and the result is four output words. The CPU is thus employed to execute the 32 copies of the S-box simultaneously, resulting with $\mathcal{S}_i(B_i \oplus K_i)$
3. Linear Transformation: The 32 bits in each of the output words are linearly mixed, by

$$X_0, X_1, X_2, X_3 = \mathcal{S}_i(B_i \oplus K_i)$$
$$X_0 = X_0 <\!\!<\!\!< 13$$
$$X_2 = X_2 <\!\!<\!\!< 3$$
$$X_1 = X_1 \oplus X_0 \oplus X_2$$
$$X_3 = X_3 \oplus X_2 \oplus (X_0 <\!\!< 3)$$

$$X_1 = X_1 <<< 1$$
$$X_3 = X_3 <<< 7$$
$$X_0 = X_0 \oplus X_1 \oplus X_3$$
$$X_2 = X_2 \oplus X_3 \oplus (X_1 << 7)$$
$$X_0 = X_0 <<< 5$$
$$X_2 = X_2 <<< 22$$
$$B_{i+1} = X_0, X_1, X_2, X_3$$

where $<<<$ denotes rotation, and $<<$ denotes shift. In the last round, this linear transformation is replaced by an additional key mixing: $B_r = S_{r-1}(B_{r-1} \oplus K_{r-1}) \oplus K_r$. Note that at each stage $IP(B_i) = \hat{B}_i$, and $IP(K_i) = \hat{K}_i$.

The first reason for the choice of linear transformation is to maximize the avalanche effect. The DES S-boxes have the property that a single input bit change will cause two output bits to change; as the difference sets of {0, 1, 3, 5, 7, 13, 22} modulo 32 have no common member (except one), it follows that a single input bit change will cause a maximal number of bit changes after two and more rounds. The effect is that each plaintext bit, and each round key bit, affect all the data bits after three rounds. Even if an opponent chooses some subkeys and works backwards, it is still the case that each key bit affects each data bit over six rounds. (Some historical information on the design of the above linear transformation is given in the appendix.) The second reason is that it is simple, and can be used in a pipelined processor with a minimum number of pipeline stalls. The third reason is that it was analyzed by programs we developed for investigating block ciphers, and we found bounds on the probabilities of the differential and linear characteristics. These bounds show that this choice suits our needs, although we would like to improve on it.

So we are still considering other, simpler, choices for the linear transformation. One possibility is to adapt an LFSR-like transform of the form $X_i = X_i \oplus \mathrm{ROL}(X_{i-1}, r_i)$ for $i = 1, \ldots, 6$, where the four data words are $X_0, \ldots, X_3$, the indices of X are taken modulo 4, and the r_i's are fixed. The problems with such a scheme are that it is hard to pipeline, and that every characteristic can be rotated in all its words and still remain with the same probability. We are still working on other possible linear transformations.

4   The Key Schedule

As with the description of the cipher, we can describe the key schedule in either standard or bitslice mode. For reasons of space, we will give the substantive description for the latter case.

Our cipher requires 132 32-bit words of key material. We first expand the user supplied 256 bit key K to 33 128-bit subkeys $K_0, \ldots, K_{32}$, in the following way. We write the key K as eight 32-bit words $w_{-8}, \ldots, w_{-1}$ and expand these to an intermediate key (which we call prekey) $w_0, \ldots, w_{131}$ by the following affine recurrence:

$$w_i = (w_{i-8} \oplus w_{i-5} \oplus w_{i-3} \oplus w_{i-1} \oplus \phi \oplus i) <<< 11$$

where ϕ is the fractional part of the golden ratio $(\sqrt{5}+1)/2$ or $\texttt{0x9e3779b9}$ in hexadecimal. The underlying polynomial $x^8 + x^7 + x^5 + x^3 + 1$ is primitive, which together with the addition of the round index is chosen to ensure an even distribution of key bits throughout the rounds, and to eliminate weak keys and related keys.

The round keys are now calculated from the prekeys using the S-boxes, again in bitslice mode. The S-box inputs and outputs are taken at a distance of 33 words apart, in order to minimize the key leakage in the event of a differential attack on the last few rounds of the cipher. We use the S-boxes to transform the prekeys w_i into words k_i of round key by dividing the vector of prekeys into four sections and transforming the i'th words of each of the four sections using $S_{(r+3-i) \bmod r}$. This can be seen simply for the default case $r = 32$ as follows:

$$\{k_0, k_{33}, k_{66}, k_{99}\} = S_3(w_0, w_{33}, w_{66}, w_{99})$$
$$\{k_1, k_{34}, k_{67}, k_{100}\} = S_2(w_1, w_{34}, w_{67}, w_{100})$$
$$\cdots$$
$$\{k_{31}, k_{64}, k_{97}, k_{130}\} = S_4(w_{31}, w_{64}, w_{97}, w_{130})$$
$$\{k_{32}, k_{65}, k_{98}, k_{131}\} = S_3(w_{32}, w_{65}, w_{98}, w_{131})$$

We then renumber the 32-bit values k_j as 128-bit subkeys K_i (for i $\in \{0, \ldots,$ r$\}$) as follows:

$$K_i = \{k_{4i}, k_{4i+1}, k_{4i+2}, k_{4i+3}\} \tag{1}$$

Where we are implementing the algorithm in the form initially described in section 2 above rather than using bitslice operations, we now apply IP to the round key in order to place the key bits in the correct column, i.e., $\hat{K}_i = IP(K_i)$.

5   Security

As mentioned above, the differential and linear properties of the DES S-boxes are well understood. Our preliminary estimates indicate that the number of known/chosen plaintexts required for either type of attack would exceed 2^{128} (they are certainly well over 2^{100} and we are working on more accurate bounds). There is thus no indication of any useful shortcut attack; we believe that such an attack would require a new theoretical breakthrough. In any case, it should be noted that regardless of the design of a 128 bit block cipher, it is normally prudent to change keys well before 2^{64} blocks have been encrypted, in order to avoid the collision attack of section 5.2 below. This would easily prevent all known shortcut attacks.

We designed the cipher with a view to reducing or avoiding vulnerabilities arising from the following possible weaknesses and attacks. In our analysis, we use conservative bounds to enable our claims to resist reasonable improvements in the studied attacks. For example, we analyze the cipher using 24-round and 28-round characteristics, shorter by 8 and 4 rounds than the cipher, while the best attack on DES uses characteristics that are shorter by only three rounds. Our estimates of the probabilities of the best characteristics are also very conservative; in practice they should be considerably lower. Therefore, our complexity claims are probably much lower than the real values, and Serpent is expected to be much more secure than we actually claim.

5.1   Dictionary Attacks

As the block size is 128 bits, a dictionary attack will require 2^{128} different plaintexts to allow the attacker to encrypt or decrypt arbitrary messages under an unknown key. This attack applies to any deterministic block cipher with 128-bit blocks regardless of its design.

5.2   Modes of Operation

After encrypting about 2^{64} plaintext blocks in the CBC or CFB mode, one can expect to find two equal ciphertext blocks. This enables an attacker to compute the exclusive-or of the two corresponding plaintext blocks [18]. With progressively more plaintext blocks, plaintext relationships can be discovered with progressively higher probability. This attack applies to any deterministic block cipher with 128-bit blocks regardless of its design.

5.3   Key-Collision Attacks

For key size k, key collision attacks can be used to forge messages with complexity only $2^{k/2}$ [5]. Thus, the complexity of forging messages under 128-bit keys is only 2^{64}, under 192-bit keys it is 2^{96}, and under 256-bit keys it is 2^{128}. This attack applies to any deterministic block cipher, and depends only on its key size, regardless of its design.

5.4   Differential Cryptanalysis

An important fact about Serpent is that any characteristic must have at least one active S-box in each round. At least two active S-boxes are required on average, due to the property that a difference in only one bit in the input causes a difference of at least two bits in the output of each S-box. Therefore, if only one bit differs in the input of some round, then at least two differ in the output, and these two bits affect two distinct S-boxes in the following round, whose output differences affect at least four S-boxes in the following round.

We searched for the best characteristics of this cipher. For this, we made a worst case assumption that all the entries in the difference distribution tables have probability $1/2$, except the few entries which have only one bit input difference and one bit output difference, which are assumed impossible (probability zero). These bounds are satisfied by all the S-boxes, except for one entry of S_{30}, where the maximal value is $10/16$: the highest probabilities in the various S-boxes are $6/16$ and $8/16$, except in S_{21} in which it is $4/16$ and in S_{30} in which it is $10/16$. We assume later that round 30 is not approximated by the characteristic anyway. Thus the following results hold independently of the order of the S-boxes used in the cipher, and independently of the choice of the S-boxes, so long as they satisfy these minimal conditions. We searched for the best characteristics with up to seven rounds, and the ones with the highest probabilities are given in Table 1.

Rounds	Differential Probability	Linear Probability $(1/2 \pm p)$	p^{-2}
1	2^{-1}	$1/2 \pm 6/16 = 1/2 \pm 2^{-1.4}$	$2^{2.8}$
2	2^{-3}	$1/2 \pm (6/16)^3 = 1/2 \pm 2^{-2.2}$	$2^{4.4}$
3	2^{-7}	$1/2 \pm (6/16)^8 = 1/2 \pm 2^{-4.3}$	$2^{8.6}$
4	2^{-13}	$1/2 \pm (6/16)^{14} = 1/2 \pm 2^{-6.8}$	$2^{13.6}$
5	2^{-21}	$1/2 \pm (6/16)^{20} = 1/2 \pm 2^{-9.3}$	$2^{18.6}$
6	2^{-29}	$1/2 \pm (6/16)^{27} = 1/2 \pm 2^{-12.2}$	$2^{24.4}$
7	$< 2^{-35}$	$1/2 \pm (6/16)^{33} = 1/2 \pm 2^{-13.8}$	$> 2^{27.6}$

Table 1. Bounds on the Probabilities of Differential and Linear Characteristics

We can see that the probability of a 6-round characteristic is bounded by 2^{-29}. Thus, the probability of a 24-round characteristic is bounded by $2^{-4 \cdot 29} = 2^{-116}$. In practice, the probability of the best 24-round characteristic is expected to be much lower than this. Thus, even if an attacker can implement an 8R-attack, still the attack requires more than 2^{117} chosen plaintexts (and again, this is a very conservative estimate). If the attacker can implement only a 4R-attack, using a 28-round characteristic, the probability of the characteristic is bounded by $2^{-4 \cdot 35} = 2^{-140}$, and the attack requires more plaintexts than are available. A 3R-attack would require even more plaintexts.

Notice that if the linear transformation had used only rotates, then every characteristic could have 32 equiprobable rotated variants, with all the data words rotated by the same number of bits. This is the reason that we also use shift instructions, which avoid most of these rotated characteristics.

We have bounded the probabilities of characteristics. However, it is both much more important and much more difficult to bound the probabilities of differentials. In order to reduce the probabilities of differentials we have (1) reduced the probabilities of the characteristics, (2) ensured that there are few characteristics with the highest possible probability, and that they cannot be rotated and still remain valid, (3) arranged for characteristics to affect many different bits, so that they cannot easily be unified into differentials.

We conjecture that the probability of the best 28-round differential is not higher than 2^{-120}, and that such a differential if it exists would be very hard to find. (Note that for any fixed key there expected to be differentials with probability 2^{-120}, but averaging over all possible keys reduces this average probability to about 2^{-128}.)

5.5 Linear Cryptanalysis

In linear cryptanalysis, it is possible to find one-bit to one-bit relations of the S-boxes. The probability of these relations are $1/2 \pm 2/16$. Thus, a 28-round linear characteristic with only one active S-box in each round would have probability $1/2 \pm 2^{27}(2/16)^{28} = 1/2 \pm 2^{-57}$, and that an attack based on such relations would require about 2^{114} known plaintexts, if it were possible at all (as the linear transformation assures that in the round following a round with only one active S-box, at least two are active).

More general attacks can use linear characteristics with more than one active S-box in some of the rounds. In this case the probabilities of the S-boxes are bounded by $1/2 \pm 6/16$. As with differential cryptanalysis, we can bound the probability of characteristics. We searched for the best linear characteristic of this cipher under the assumptions that a probability of any entry is not further from $1/2$ than $6/16$ and that the probability of a characteristic which relates one bit to one bit is not further from $1/2$ than $2/16$. Note that due to the relation between linear and differential characteristics, the searches are very similar; we actually modified the search program used in the differential case to search for the best linear characteristics with up to seven rounds, and those with the highest probabilities are given in Table 1.

We can see that the probability of a 6-round characteristic is bounded by $1/2 \pm 2^{-12.2}$, from which we can conclude that the probability of a 24-round characteristic is bounded by $1/2 \pm 2^{-45.8}$. The probability of a 28-round characteristic is bounded by $1/2 \pm 2^{-52.2}$, and an attack based on it would require at least 2^{104} known plaintexts. Again, we wish to emphasize that all these figures are conservative lower bounds, and that the actual complexities of attacks are expected to be substantially higher.

Based on these figures we conjecture that the probability of the best 28-round linear differential is bounded by $1/2 \pm 2^{-50}$, so an attack would need at least 2^{100} blocks. Again, this is a very conservative estimate; we believe the real figure is over 2^{128} and that linear attacks are thus infeasible. We are working on more accurate figures; meantime the normal prudent practice of changing keys well before 2^{64} blocks have been encrypted will prevent linear attacks.

5.6 Higher Order Differential Cryptanalysis

It is well known that a dth order differential of a function of nonlinear order d is constant, and this can be exploited in higher order differential attacks [4,17,19]. The DES S-boxes all have nonlinear order 5 [18]. From this one would expect that the nonlinear order of the output bits after r rounds is about 3^r, with the maximum value of 127 reachable after five rounds. Therefore we are convinced that higher order differential attacks are not applicable to Serpent.

5.7   Truncated Differential Cryptanalysis

For some ciphers it is possible and advantageous to predict only the values of parts of the differences after each round. This notion, of truncated differential attacks, was introduced by Knudsen in [17]. However, the method seems best applicable to ciphers where all operations are done on larger blocks of bits. Because of the strong diffusion over many rounds, we believe that truncated differential attacks are not applicable to Serpent.

5.8   Related Keys

As the key schedule uses rotations and S-boxes, it is highly unlikely that keys can be found that allow related key attacks [8,15,16]. Moreover, different rounds of Serpent use different S-boxes, so even if related keys were found, related-key attacks would not be applicable.

Serpent has none of the simpler vulnerabilities that can result from exploitable symmetries in the key schedule: there are no weak keys, semi-weak keys, equivalent keys, or complementation properties.

5.9   Other Attacks

Davies' attack [12,13] and the Improved Davies attack [6] are not applicable, since the S-boxes are invertible, and no duplications of data bits are applied.

As far as we know, neither statistical cryptanalysis [22] nor partitioning cryptanalysis [14] provides a less complex attack than differential or linear cryptanalysis.

5.10   Fault Analysis

We have not been concerned in this design to build in any particular protection against attacks based on induced faults [3,10,11]. If an attacker can progressively remove the machine instructions by which this cipher is implemented, or progressively destroy selected gates, or progressively modify the bits of the key register, then he can clearly extract the key. We tend to the view that an attacker with the ability to inspect or modify the implementation detail will have many attacks based not just on compromising keys but on subverting protocols, extracting plaintext directly and so on [2]. The mechanisms required to protect against such attacks are largely independent of the design of any block cipher used [1], and are thus beyond the scope of this work.

6   Performance

We implemented this cipher on a 133MHz Pentium/MMX processor. A 32-round bitslice (unoptimized) implementation (available online from the authors' web pages) gave speeds which are only slightly slower than DES: it encrypted 8,976,157 bits per second, while the best optimized DES implementation (Eric Young's Libdes) encrypts 9,824,864 bits per second on the same machine.

The performance of the cipher on other processors in bitslice mode should be only slightly slower than the standard implementation of DES. When coded in assembly language this cipher might be even faster than DES. It takes somewhat over 2000 instructions to encrypt 128 bits versus typically 700 instructions to encrypt 64 bits in DES. The reason our cipher is not 50% slower is that it has been designed to make good use of pipelining.

The instruction count is based on the observation that a gate circuit of any of the 4x4 S-boxes requires between 19 and 28 gates on the Pentium, between 18 and 28 on MMX (using only MMX instructions), and between 18 and 25 on the DEC Alpha (the numbers vary due to the different sets of instructions, which are detailed in the appendix). MMX has the additional advantage that it can operate on 64-bit words, or alternatively on two 32-bit words at once (so two encryptions can be done in parallel using the same or different keys). It is also implemented with greater parallelism on some recent chips (e.g. the Pentium II). On the other hand, it does not have rotate operations, so rotates require four instructions (copy, shift left, shift right, and OR).

It is also worth remarking that if this cipher is adopted as the Advanced Encryption Standard, and chip makers wish to support high speed implementation, then it may not be necessary to add a hardware encryption circuit to the CPU. It would be sufficient to add what we call the 'BITSLICE instruction'. This executes an arbitrary boolean function on four registers under the control of a truth table encoded in a (64-bit) fifth register. We estimate that the cost of implementing this on an n-bit processor will be only about $100n$ gates, and it would have many uses other than cryptography (an example would be image processing). If supported, one BITSLICE instruction would replace most of the instructions in each round, and Serpent would become two or three times faster than DES.

It is also worth noting that hardware implementations of the cipher can iteratively apply one round at a time, although the S-boxes in each round are different. The trick is similar to the BITSLICE instruction: the designers of the hardware can design the round function to get a description of the S-boxes as a parameter in some register, and compute the S-boxes according to this description. This trick crucially reduces the number of gates required for the hardware implementation of the cipher. An estimate of the gate count will be provided in the full AES submission.

7   Other Variants

As we remarked above, there are two ways in which the block size can be doubled:

1. increase the word length (in the bitslice implementation) from 32 to 64 bits (or more);
2. Use the round-function as the F-function in a Feistel construction.

If both of these are done, then the block size will be quadrupled.

These variants might require other modifications of the cipher, such as modifications in the rotation constants. We believe that these variants are secure (or can easily be made so). Work on them is ongoing.

8   Conclusion

We have presented a cipher which we have engineered to satisfy the AES requirements. It is about as fast as DES, and conjectured to be as secure as three-key triple DES. Its security is partially based on the reuse of the thoroughly studied components of DES, and thus can draw on the wide literature of block cipher cryptanalysis published in the last decade. Its performance comes from allowing an efficient bitslice implementation on a range of processors, including the market leading Intel/MMX and compatible chips.

This is still a preliminary design and may change between the time of writing and the final AES submission. Readers are invited to attack the cipher, to test implementations in various environments, and to report any interesting findings to the authors. A patent application has been filed, but it is our intention to grant a worldwide royalty-free license for conforming implementations in the event that this cipher is adopted as the Advanced Encryption Standard.

Finally, up to date information on Serpent, including the latest revision of the paper, can be found on the authors' home pages:

http://www.cs.technion.ac.il/~biham/
http://www.cl.cam.ac.uk/~rja14/
http://www.ii.uib.no/~larsr/

Acknowledgments

The first author was supported by Intel Corporation during a visit to Cambridge in September 1997 while much of this work was done; and the name of the cipher was suggested by Gideon Yuval (see Amos 5.19).

References

1. DG Abraham, GM Dolan, GP Double, JV Stevens, "Transaction Security System", in *IBM Systems Journal* v 30 no 2 (1991) pp 206–229

2. RJ Anderson, MG Kuhn, "Tamper Resistance — a Cautionary Note", in *The Second USENIX Workshop on Electronic Commerce Proceedings* (Nov 1996) pp 1–11

3. RJ Anderson, MG Kuhn, "Low Cost Attacks on Tamper Resistant Devices", *to appear in proceedings of Security Protocols 97*

4. E Biham, *'Higher Order Differential Cryptanalysis'*, unpublished paper, 1994

5. E Biham, *How to Forge DES-Encrypted Messages in 2^{28} Steps*, Technical Report CS884, Technion, August 1996

6. E Biham, A Biryukov, "An Improvement of Davies' Attack on DES", in *Journal of Cryptology* v 10 no 3 (Summer 97) pp 195–205

7. E Biham, A Shamir, *'Differential Cryptanalysis of the Data Encryption Standard'* (Springer 1993)

8. E Biham, "New Types of Cryptanalytic Attacks Using Related Keys", in *Journal of Cryptology* v 7 (1994) no 4 pp 229–246

9. E Biham, "A Fast New DES Implementation in Software", in *Fast Software Encryption — 4th International Workshop, FSE '97*, Springer LNCS v 1267 pp 260–271

10. E Biham, A Shamir, "Differential Fault Analysis of Secret Key Cryptosystems", in *Advances in Cryptology — Crypto 97*, Springer LNCS v 1294 pp 513–525

11. D Boneh, RA DeMillo, RJ Lipton, "On the Importance of Checking Cryptographic Protocols for Faults", in *Advances in Cryptology — Eurocrypt 97*, Springer LNCS v 1233 pp 37–51

12. DW Davies, *'Investigation of a Potential Weakness in the DES Algorithm'*, private communication (1987)

13. D Davies, Murphy, "Pairs and Triplets of DES S Boxes", in *Journal of Cryptology* v 8 no 1 (1995) pp 1–25

14. C Harpes, JL Massey, "Partitioning Cryptanalysis", in *Fast Software Encryption — 4th International Workshop, FSE '97*, Springer LNCS v 1267 pp 13–27

15. J Kelsey, B Schneier, D Wagner, "Key-Schedule Cryptanalysis of IDEA, GDES, GOST, SAFER and Triple-DES", in *Advances in Cryptology — Crypto 96*, Springer LNCS v 1109 pp 237–251

16. LR Knudsen, "Cryptanalysis of LOKI91", in *Advances in Cryptology — Auscrypt'92* Springer LNCS

17. LR Knudsen, "Truncated and Higher-Order Differentials", in *Fast Software Encryption — 2nd International Workshop, FSE '94*, Springer LNCS v 1008 pp 196–211

18. L.R. Knudsen, *Block Ciphers – Analysis, Design and Applications*, Ph.D. Thesis, Aarhus University, Denmark, 1994.

19. X.J. Lai, *'Higher Order Derivative and Differential Cryptanalysis'*, in *Communication and Cryptography, Two Sides of one tapestry*, R. Blahut (editor), Kluwer Academic Publishers, 1994
communication, September 30, 1993.

20. M Matsui, "Linear Cryptanalysis Method for DES Cipher", in *Advances in Cryptology — Eurocrypt 93*, Springer LNCS v 765 pp 386–397

21. RSA Data Security Inc., www.rsa.com

22. S Vaudenay, "An Experiment on DES Statistical Cryptanalysis", in *3rd ACM Conference on Computer and Communications Security, March 14-16, 96, New Delhi, India; proceedings published by ACM* pp 139–147

A   Appendix

A.1   The Initial Permutation IP:

```
 0  32  64  96   1  33  65  97   2  34  66  98   3  35  67  99
 4  36  68 100   5  37  69 101   6  38  70 102   7  39  71 103
 8  40  72 104   9  41  73 105  10  42  74 106  11  43  75 107
12  44  76 108  13  45  77 109  14  46  78 110  15  47  79 111
16  48  80 112  17  49  81 113  18  50  82 114  19  51  83 115
20  52  84 116  21  53  85 117  22  54  86 118  23  55  87 119
24  56  88 120  25  57  89 121  26  58  90 122  27  59  91 123
28  60  92 124  29  61  93 125  30  62  94 126  31  63  95 127
```

A.2   The Final Permutation FP:

```
 0   4   8  12  16  20  24  28  32  36  40  44  48  52  56  60
64  68  72  76  80  84  88  92  96 100 104 108 112 116 120 124
 1   5   9  13  17  21  25  29  33  37  41  45  49  53  57  61
65  69  73  77  81  85  89  93  97 101 105 109 113 117 121 125
 2   6  10  14  18  22  26  30  34  38  42  46  50  54  58  62
66  70  74  78  82  86  90  94  98 102 106 110 114 118 122 126
 3   7  11  15  19  23  27  31  35  39  43  47  51  55  59  63
67  71  75  79  83  87  91  95  99 103 107 111 115 119 123 127
```

A.3   The Linear Transformation:

For each output bit of this transformation, we describe the list of input bits whose parity becomes the output bit. In each row we describe four output bits, which later enter the same S-box in the next round. The bits are listed from 0 to 127.

```
{16 52 56   70  83  94 105} {72 114 125} { 2   9 15   30  76  84 126} {36  90 103}
{20 56 60   74  87  98 109} { 1  76 118} { 2   6 13   19  34  80  88} {40  94 107}
{24 60 64   78  91 102 113} { 5  80 122} { 6  10 17   23  38  84  92} {44  98 111}
{28 64 68   82  95 106 117} { 9  84 126} {10  14 21   27  42  88  96} {48 102 115}
{32 68 72   86  99 110 121} { 2  13  88} {14  18 25   31  46  92 100} {52 106 119}
{36 72 76   90 103 114 125} { 6  17  92} {18  22 29   35  50  96 104} {56 110 123}
{ 1 40 76   80  94 107 118} {10  21  96} {22  26 33   39  54 100 108} {60 114 127}
{ 5 44 80   84  98 111 122} {14  25 100} {26  30 37   43  58 104 112} { 3 118    }
{ 9 48 84   88 102 115 126} {18  29 104} {30  34 41   47  62 108 116} { 7 122    }
{ 2 13 52   88  92 106 119} {22  33 108} {34  38 45   51  66 112 120} {11 126    }
{ 6 17 56   92  96 110 123} {26  37 112} {38  42 49   55  70 116 124} { 2  15  76}
{10 21 60   96 100 114 127} {30  41 116} { 0  42 46   53  59  74 120} { 6  19  80}
{ 3 14 25  100 104 118    } {34  45 120} { 4  46 50   57  63  78 124} {10  23  84}
{ 7 18 29  104 108 122    } {38  49 124} { 0   8 50   54  61  67  82} {14  27  88}
{11 22 33  108 112 126    } { 0  42  53} { 4  12 54   58  65  71  86} {18  31  92}
{ 2 15 26   37  76 112 116} { 4  46  57} { 8  16 58   62  69  75  90} {22  35  96}
{ 6 19 30   41  80 116 120} { 8  50  61} {12  20 62   66  73  79  94} {26  39 100}
{10 23 34   45  84 120 124} {12  54  65} {16  24 66   70  77  83  98} {30  43 104}
{ 0 14 27   38  49  88 124} {16  58  69} {20  28 70   74  81  87 102} {34  47 108}
{ 0  4 18   31  42  53  92} {20  62  73} {24  32 74   78  85  91 106} {38  51 112}
{ 4  8 22   35  46  57  96} {24  66  77} {28  36 78   82  89  95 110} {42  55 116}
```

```
{ 8 12 26  39  50  61 100} {28  70  81} {32 40 82  86  93  99 114} {46  59 120}
{12 16 30  43  54  65 104} {32  74  85} {36 90 103 118           }.{50  63 124}
{16 20 34  47  58  69 108} {36  78  89} {40 94 107 122           } { 0  54  67}
{20 24 38  51  62  73 112} {40  82  93} {44 98 111 126           } { 4  58  71}
{24 28 42  55  66  77 116} {44  86  97} { 2 48 102 115           } { 8  62  75}
{28 32 46  59  70  81 120} {48  90 101} { 6 52 106 119           } {12  66  79}
{32 36 50  63  74  85 124} {52  94 105} {10 56 110 123           } {16  70  83}
{ 0 36 40  54  67  78  89} {56  98 109} {14 60 114 127           } {20  74  87}
{ 4 40 44  58  71  82  93} {60 102 113} { 3 18 72 114 118 125    } {24  78  91}
{ 8 44 48  62  75  86  97} {64 106 117} { 1  7 22  76 118 122    } {28  82  95}
{12 48 52  66  79  90 101} {68 110 121} { 5 11 26  80 122 126    } {32  86  99}
```

A.4 S-Boxes

Here are the S-boxes S_0 through S_{31} (each on a separate line):

```
14 4 13 1 2 15 11 8 3 10 6 12 5 9 0 7
0 15 7 4 14 2 13 1 10 6 12 11 9 5 3 8
4 1 14 8 13 6 2 11 15 12 9 7 3 10 5 0
15 12 8 2 4 9 1 7 5 11 3 14 10 0 6 13
15 1 8 14 6 11 3 4 9 7 2 13 12 0 5 10
3 13 4 7 15 2 8 14 12 0 1 10 6 9 11 5
0 14 7 11 10 4 13 1 5 8 12 6 9 3 2 15
13 8 10 1 3 15 4 2 11 6 7 12 0 5 14 9
10 0 9 14 6 3 15 5 1 13 12 7 11 4 2 8
13 7 0 9 3 4 6 10 2 8 5 14 12 11 15 1
13 6 4 9 8 15 3 0 11 1 2 12 5 10 14 7
1 10 13 0 6 9 8 7 4 15 14 3 11 5 2 12
7 13 14 3 0 6 9 10 1 2 8 5 11 12 4 15
13 8 11 5 6 15 0 3 4 7 2 12 1 10 14 9
10 6 9 0 12 11 7 13 15 1 3 14 5 2 8 4
3 15 0 6 10 1 13 8 9 4 5 11 12 7 2 14
2 12 4 1 7 10 11 6 8 5 3 15 13 0 14 9
14 11 2 12 4 7 13 1 5 0 15 10 3 9 8 6
4 2 1 11 10 13 7 8 15 9 12 5 6 3 0 14
11 8 12 7 1 14 2 13 6 15 0 9 10 4 5 3
12 1 10 15 9 2 6 8 0 13 3 4 14 7 5 11
10 15 4 2 7 12 9 5 6 1 13 14 0 11 3 8
9 14 15 5 2 8 12 3 7 0 4 10 1 13 11 6
4 3 2 12 9 5 15 10 11 14 1 7 6 0 8 13
4 11 2 14 15 0 8 13 3 12 9 7 5 10 6 1
13 0 11 7 4 9 1 10 14 3 5 12 2 15 8 6
1 4 11 13 12 3 7 14 10 15 6 8 0 5 9 2
6 11 13 8 1 4 10 7 9 5 0 15 14 2 3 12
13 2 8 4 6 15 11 1 10 9 3 14 5 0 12 7
1 15 13 8 10 3 7 4 12 5 6 11 0 14 9 2
7 11 4 1 9 12 14 2 0 6 10 13 15 3 5 8
2 1 14 7 4 10 8 13 15 12 9 0 3 5 6 11
```

A.5   Lists of Relevant Instructions on Various Processors

The relevant instructions on the following processors are:

Pentium: AND, OR, XOR, NOT, rotate
MMX: AND, OR, XOR, NOT, ANDN, only shifts
Alpha: AND, OR, XOR, NOT, ANDN, ORN, XORN, only shifts

where the ANDN operation on x and y is $x \wedge (\neg y)$, the ORN operation is $x \vee (\neg y)$, and the XORN operation is $x \oplus (\neg y)$ (or equivalently $\neg(x \oplus y)$).

On MMX a rotate takes four instructions, while on an Alpha it takes three. On Pentium and MMX it might be necessary to copy some of the registers before use, as instructions have only two arguments; but some instructions can refer directly to memory. The Alpha instructions have 3 arguments (src1, src2 and destination), but cannot refer directly to memory.

A.6   Historical Remarks

Here we describe some design history. In our first design, the linear transformations were just bit permutations, which were applied as rotations of the 32-bit words in the bitslice implementation. In order to ensure maximal avalanche, the idea was to choose these rotations in a way that ensured maximal avalanche in the fewest number of rounds. Thus, we chose three rotations at each round: we used (0, 1, 3, 7) for the even rounds and (0, 5, 13, 22) for the odd rounds. The reason for this was that (a) rotating all four words is of course useless (b) a single set of rotations did not suffice for full avalanche (c) these sets of rotations have the property that no difference of pairs in either of them coincides with a difference either in the same set or the other set.

However, we felt that the avalanche was still slow, as each bit affected only one bit in the next round, and thus one active S-box affected only 2–4 out of the 32 S-boxes in the next round. As a result, we had to use 64 rounds, and the cipher was only slightly faster than triple-DES. So we moved to a more complex linear transformation; this improved the avalanche, and analysis showed that we could now reduce the number of rounds to 32. We believe that the final result is a faster and yet more secure cipher.

We also considered "improving" the cipher by replacing the XOR operations by seemingly more complex operations, such as additions. We did not do this due to two major reasons: (1) Our analysis takes advantage of the independence between the bits in the XOR operation, as it allows us to describe the cipher in a standard way, and use the known kinds of security analysis. This analysis would not hold if the XOR operations were replaced; (2) in some other ciphers the replacement of XORs by additions (or other operations) has turned out to weaken the cipher, rather than strengthening it.

A.7   Reference Implementation

An unoptimized reference C implementation is available from the authors' home pages. Note however that the cipher may still be modified in the future as it progresses through the AES selection process.

Attacking Triple Encryption

Stefan Lucks[*]

Theoretische Informatik
Universität Mannheim
68131 Mannheim A5, Germany
lucks@pi3.informatik.uni-mannheim.de

Abstract. The standard technique to attack triple encryption is the meet-in-the-middle attack which requires 2^{112} encryption steps. In this paper, more efficient attacks are presented. One of our attacks reduces the overall number of steps to roughly 2^{108}. Other attacks optimize the number of encryptions at the cost of increasing the number of other operations. It is possible to break triple DES doing 2^{90} single encryptions and no more than 2^{113} faster operations.

1   Introduction

The most well-known symmetric encryption algorithm is the Data Encryption Standard (DES). It defines a block cipher with 64-bit blocks and 56-bit keys. Due to questions raised regarding the small key size, several varieties of multiple encryption have been considered for the DES, including double and triple DES.

Fig. 1. Double encryption (top) and triple encryption (bottom)

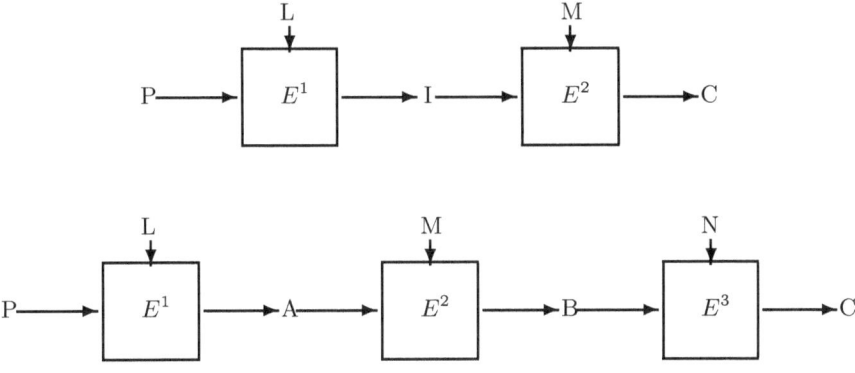

In this paper, we consider arbitrary single encryption functions $E\{0,1\}^k \times \{0,1\}^s \longrightarrow \{0,1\}^s$ with k-bit keys and a block size of s bits, and in particular

---

[*] A part of this research was done while the author was at the University of Göttingen.

S. Vaudenay (Ed.): Fast Software Encryption – FSE'98, LNCS 1372, pp. 239–253, 1998.
© Springer-Verlag Berlin Heidelberg 1998

point out the consequences of our findings for triple DES. Since multiple encryption is mainly of relevance to strengthen block ciphers with a small key space, we concentrate on $k \leq s$. With two k-bit keys L and M and two encryption functions E^1 and E^2, double encryption is defined by $C = E^2_M(E^1_L(P))$. Here, C denotes the ciphertext and P the plaintext. Similarly, triple encryption is defined by $C = E^3_N(E^2_M(E^1_L(P)))$. Figure 1 describes double and triple encryption. If $L = N$, this defines the special case of two-key triple encryption. In this paper, we concentrate on the case of general (three-key) triple encryption.

Double DES is double encryption with $E^1 = E^2 = E$. Triple DES is usually defined by $E^1 = E^3 = E$, $E^2 = D$, where E denotes the (single) DES encryption function and D its decryption counterpart.

In general, we assume the functions E^i and D^i to behave like a set of 2^k random permutations E^i_K with $K \in \{0,1\}^k$, chosen according to the uniform probability distribution. Usually, nonrandom statistical properties are considered to be weaknesses of block ciphers. In the special case of the DES, two important statistical weaknesses are known, the complementation property, which is exploited in Section 6 of this paper, and a small number of weak keys.

All attacks considered in this paper are key-search attacks and exploit known (or chosen) pairs of plaintext and ciphertext. To measure the complexity of an attack, we consider four values:

1. The number of known plaintext-ciphertext pairs.
2. The storage space required for the attack.
3. The number of single encryptions $y := E^i_K(x)$ or $x := D^i_K(y)$ to mount the attack.
4. The overall number of operations ("steps") to mount the attack.

The third value demands some explanation: Clearly, given a key K and a plaintext x (or a ciphertext y) the attacker can compute the corresponding ciphertext $y := E^i_K(x)$ (or the corresponding plaintext). A good block cipher behaves like a random permutation, hence given some triples (plaintext,ciphertext,key) one can't find other triples more efficiently than by encrypting/decrypting again.

Attacking multiple encryption without breaking the underlying encryption function can be described as attacking multiple encryption in the presence of encryption/decryption oracles. Figure 2 visualizes such an oracle. The underlying cipher is treated as a black box. We simply write "single encryption" for accessing the encryption/decryption oracle. Much work has been done with respect to this model.

This view also motivates to specifically count the single encryptions, in addition to counting all steps. Note that such a single encryption counts as one step, but in practice is an exceptionally complex step by itself, compared to common operations like comparisons and table look-ups.

One may as well concentrate on the number of single encryptions and disregard the number of steps and the amount of space required. This is an approved method for estimating the minimum strength of a composed cipher, in order to demonstrate the soundness of the composition technique. In this context,

Fig. 2. An encryption/decryption oracle

Key K

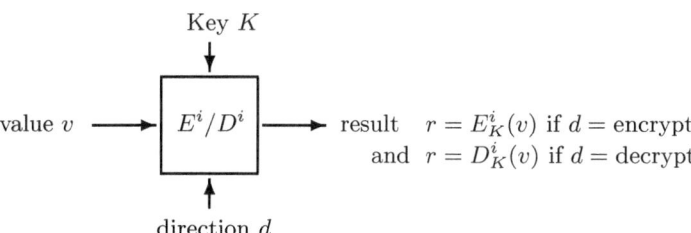

value v ⟶ E^i/D^i ⟶ result $r = E^i_K(v)$ if d = encrypt
and $r = D^i_K(v)$ if d = decrypt

direction d

one neglects possible weaknesses of the underlying encryption functions. Idealizations of the underlying encryption functions are accessible to the attacker by querying encryption/decryption oracles, but the attacker has no knowledge about the oracles' internals. In the sequel, we refer to this point of view as the "black-box-only" model.

The rest of this paper is organized as follows. Section 2 describes previously known attacks, concentrating on the meet-in-the-middle attack. In Section 3, we introduce the notion of "t-collisions" and use it for a technique to reduce the number of steps. In Sections 4 and 5 we consider that single encryptions are much slower than each of the other steps, and we design attacks optimized to save single encryptions (but not the total number of steps). In Section 6 we exploit the complementation property of DES and triple DES to further improve our attacks. Finally, in Section 7 we concentrate on the consequences of our findings for the security of triple DES.

2   Previous Work

Double encryption can be broken with meet-in-the-middle (MITM). This attack requires $\lceil 2k/s \rceil$ known plaintext/ciphertext pairs on the average, about 2^k units of storage, about 2^k single encryptions, and about as much steps. For a plaintext p and a corresponding ciphertext c, compute all values $I_L = E^1_L(p)$ and store all pairs (I_L, L) in a table, indexed by I_L. Since there are 2^k keys L, this requires 2^k units of storage, 2^k steps and 2^k single encryptions. Now, all values $I'_M = D^2_M(c)$ are computed. For the correct key pair (L, M) the equation $I_L = I'_M$ must hold. Thus the attacker needs to look up I'_M in the previously computed table of pairs (I_L, L).

Two-key triple encryption can be broken by a chosen plaintext attack using about 2^k units of everything: 2^k plaintext/ciphertext pairs, 2^k units of storage, 2^k single encryptions, and 2^k steps, see [5].

The best known way to attack general triple encryption is also by MITM [4, Section 7.2.3]. Let a plaintext/ciphertext pair (p, c) be given. Proceed as follows:

1. Compute all values $b_N = D_N^3(c)$, $N \in \{0,1\}^k$, and store the pairs (b_N, N) in a table, indexed by b_N.
2. Compute all values $b_{L,M} = E_M^2(E_L^1(p))$ with $L, M \in \{0,1\}^k$, and look for $(b_{L,M}, N)$ in the previously computed table of pairs (b_N, N).[1]
3. Test all key triples (L, M, N) with $b_{L,M} = b_N$ until only one such triple remains.

The first stage requires about 2^k steps and single encryptions and as much units of storage. The second requires about 2^{2k} steps and single encryptions. The third stage is cheap. Note that we need at least $l \geq \lceil 3k/s \rceil$ pairs of plaintext and ciphertext for the attack. In the case of triple DES, we need $l \geq 3 = \lceil 3*56/64 \rceil$ such pairs, about 2^{56} units of storage, about 2^{112} single encryptions and the same number of steps (mainly table look-ups). (The *expected* number of steps and single encryptions needed for the MITM attack is 2^{111}. This is the number we use when comparing the MITM attack with our probabilistic attacks.)

Advanced MITM techniques for attacking two-key triple encryption have been studied by van Oorschot and Wiener [6]. The same authors also proposed advanced MITM techniques for attacking double encryption [7].

Kelsey, Schneier, and Wagner [2] demonstrated how to attack three-key triple DES using related-key techniques. Let a plaintext p and a corresponding ciphertext c be known to the attacker. Assume the attacker to be able to change the first subkey from L to $L + \Delta$ (both L and $L + \Delta$ unknown to the her, but Δ known). If the attacker receives the decryption of c under the modified key, then she can find the subkey L using only 2^k steps (and the same number of single encryptions). The second and third subkeys M and N can be found as in the case of double encryption.

If the same plaintext is encrypted 2^{28} times using triple DES under 2^{28} different keys, an attacker can recover one of the 2^{28} keys using 2^{84} steps (and the same number of single encryptions). This result is due to Biham [1].

DESX is a variant of DES, where encrypting and decrypting requires to compute one single encryption and two XORs of s-bit blocks. Kilian and Rogaway [3] describe the security of DESX in the black-box-only model, concentrating on finding a lower bound for the number of single encryptions every black box attack needs.

3   How to Save Steps

In this section, we describe an "operation optimized" attack to save some steps of computation, compared to MITM.

Consider a function $f : \{0,1\}^* \longrightarrow \{0,1\}^s$. A *collision* is a pair x, y of inputs with $x \neq y$ and $f(x) = f(y)$. The value $v \in \{0,1\}^s$ is associated with a *t-collision*, if there exists a set S with $|S| \geq t$ inputs and $f(x) = v$ for all

---

[1] When computing the complexity of this stage, the operation of computing a value $b_{L,M} = E_M^2(E_L^1(p))$ *and* looking up the pair $(b_{L,M}, N)$ in a table *and* the operations to maintain the loop together count as *one step*.

$x \in S$. Assuming the function $f : \{0,1\}^* \longrightarrow \{0,1\}^s$ to behave like a random function, and that $1 \leq w \leq 2^t$ inputs are randomly chosen, the expected number of values $v \in \{0,1\}^s$ associated with a t-collision[2] is about $w^t * 2^{-s(t-1)}$. Given plaintext/ciphertext pairs (p_i, c_i), our attack depends on finding t-collisions for the functions $f_{p_i} : \{0,1\}^k \longrightarrow \{0,1\}^s$, $f_{p_i}(L) = E_L^1(p_i)$. We consider all keys $L \in \{0,1\}^k$, hence the number of inputs for the function f_{p_i} is $w = 2^k$.

We write $K_1(a, i)$ for the set of all keys which encrypt the plaintext p_i to the ciphertext a using E^1. Similarly, we write $K_3(b, i)$ for all keys which decrypt the ciphertext c_i to b, using E^3. I.e.,

$$K_1(a, i) = \{ L \in \{0,1\}^k \mid E_L^1(p_i) = a \} \quad \text{and}$$
$$K_3(b, i) = \{ N \in \{0,1\}^k \mid E_L^3(b) = c_i \}.$$

If $|K_1(a, i)| \geq t$, the value a is associated with a t-collision. Given a pair (p_i, c_i), we choose set $S_A(i) \subseteq \{0,1\}^s$ of values associated with t-collisions:

$$S_A(i) = \{ a \in \{0,1\}^s \mid \text{there exists a } t\text{-collision } K_1(a, i). \}.$$

Our attack works like this:

$i := 1$;
Repeat:
  let $(p_i, c_i) \in (\{0,1\}^s)^2$ be a known pair of plaintext and ciphertext;
  initialize the sets $K_1(\cdot, i)$, $K_3(\cdot, i)$, and $S_A(i)$ to be empty;
  A. for $L \in \{0,1\}^k$: $a := E_L^1(p_i)$;
              $K_1(a, i) := K_1(a, i) + \{L\}$;
              if $|K_1(a, i)| \geq t$ then $S_A(i) := S_A(i) + \{a\}$;
    (* Now $S_A(i)$ is the set of all values a associated with a t-collision. *)
    for $N \in \{0,1\}^k$: $b := D_N^3(c_i)$;
              $K_3(b, i) := K_3(b, i) + \{N\}$;
  B. for $a \in S_A(i)$:
      for $M \in \{0,1\}^k$: $b := E_M^2(a)$;
              for $N \in K_3(b, i)$:
                  for $L \in K_1(a, i)$: $\mathtt{tripletest}(i, L, M, N)$;
  $i := i + 1$;
until $\mathtt{tripletest}$ accepts.

The procedure "$\mathtt{tripletest}$" can be realized like this:

$\mathtt{tripletest}(i, L, M, N)$ is
  $S_I := \{1, \ldots, l\} - \{i\}$;
  $d := 3k - s + \delta$;
  repeat: choose $j \in S_I$ at random;

---

[2] Cf. Rivest and Shamir [8], who exploit this for one of their micropayment schemes. To verify this estimation, one can use a well known special case, the "birthday paradox": *The expected number of inputs for f until the first 2-collision occurs is $c * 2^{s/2}$ with a small constant c.* (Actually, $c = \sqrt{\pi/2} \approx 1.25$, cf. [4, Section 2.1.5]).

$$S_I := S_I - \{j\};$$
$$c := E_N^3(E_M^2(E_L^1(p_j)));$$
$$d := d - s;$$
until $(d \leq 0)$ or $(c_j \neq c)$;
if $(c_j = c)$ then accept (L, M, N) as the correct key-triple and stop
    else reject (L, M, N) and continue.

When "`tripletest`" is called, the equation

$$E_N^3(E_M^2(E_L^1(p_i))) = c_i$$

holds. In the procedure, we are looking for $j \neq i$ such that $E_N^3(E_M^2(E_L^1(p_j))) \neq c_j$. If we fail often enough, i.e., $\lceil \frac{3k-s+\delta}{s} \rceil$ times, we accept the key-triple (L, M, N) as correct. The value δ serves as a security parameter, the risk to accept an incorrect key-triple is no more that $2^{-\delta}$.

On the average, a wrong key-triple requires insignificantly more than three single encryptions, i.e., one computation of $c := \ldots$, since for $j \in \{1, \ldots, l\} - \{i\}$, the equation $E_N^3(E_M^2(E_L^1(p_j))) = c_j$ holds one out of 2^s times. The correct triple is always accepted after $\lceil \frac{3k-s+\delta}{s} \rceil$ rounds. E.g., two rounds are sufficient for triple DES ($k = 56$ and $s = 64$) if $\delta = 20$. In the sequel, we assume δ to be "large enough" and ignore the risk of accepting an incorrect key-triple.

Let t be chosen such that $w^t * 2^{-s(t-1)} \ll 2^k$.

Theorem 1. *The expected number of pairs (p_i, c_i) for the operation optimized attack to succeed is $2^k/(w^t * 2^{-s(t-1)}t)$ (except for a small factor).*

Sketch of proof. Every t-collision $K_1(a, i) \in S_A(i)$ consists of at least t keys and hence has at least a $t*2^{-k}$ chance to contain the correct first key L. Every index i corresponds to a pair (p_i, c_i) of plaintext and ciphertext. For every i, we expect to find about $w^t * 2^{-s(t-1)}$ values a to be associated with a t-collision $K_1(a, i)$. Thus the expected number of (plaintext,ciphertext)-pairs we need to consider in order to find the correct first key L is $2^k/(w^t * 2^{-s(t-1)}t)$.

It is easy to verify the following: If (L, M, N) is the correct key triple, $K_1(a, i) \in S_A(i)$ and $L \in K_1(a, i)$, then the procedure `tripletest`(i, L, M, N) is executed in stage B with the index i and the keys (L, M, N) as parameters. $\quad\square$

Theorem 2. *With l pairs (p_i, c_i), the operation optimized attack requires the following resources:*

- $\Theta(2^k)$ *units of storage space, and*
- $\Theta(w^t * 2^{-s(t-1)} * 2^k * l + w^t * 2^{-s(t-1)} * l * t * 2^{2k-s})$ *steps (and as much single encryptions).*

Proof. Both stages are to be executed l times. During every iteration of the "Repeat" loop, no results of previous iterations are needed. Hence, the amount of storage for the attack can be estimated by the storage space during one iteration, and the required number of steps is l times the average number of steps during one iteration. Below, we estimate the storage space and the number of steps for one such iteration.

Both loops of stage A are iterated 2^k times, hence the number of steps is about $2*2^k$. When the first loop is finished, the storage space for the sets $K_1(a, i)$ (i.e., 2^k units) is no longer needed, and can be reused, with the exception of the sets $K_1(a, i) \in S_A(i)$. For the second loop, 2^k units of storages space are needed for the sets $K_3(\cdot, i)$. Since $w^t * 2^{-s(t-1)} \ll 2^k$, we expect the probability for $|S_A(i)| > 2^k$ to be negligible, and we approximate the storage space for stage A by 2^k.

Now, we consider stage B. For every pair (p_i, c_i) we expect the existence of $w^t * 2^{-s(t-1)}$ t-collisions; thus the loop "for $a \in S_A(i)$" is iterated $w^t * 2^{-s(t-1)}$ times on the average. The loop "for M ..." is iterated 2^k times, hence, $w^t * 2^{-s(t-1)} * 2^k$ single encryptions $b := E_M^2(a)$ are done. So far, we need $w^t * 2^{-s(t-1)} * 2^k$ steps. The expected size of a set $K_3(b, i)$ is $2^{k-s} \le 1$. $K_1(a, i)$ is a t-collision, thus it contains about t keys L, and the procedure `tripletest` is to be called $w^t * 2^{-s(t-1)} * 2^k * 2^{k-s} * t$ times.

During each of the iterations of stage A and B and hence during the complete algorithm $\Theta(2^k)$ units of storage space are needed, and the number of steps is $l\Theta(2^k + w^t * 2^{-s(t-1)} * 2^k + w^t * 2^{-s(t-1)} * 2^k * 2^{k-s} * t) = \Theta(w^t * 2^{-s(t-1)} * 2^k + w^t * 2^{-s(t-1)} * 2^k * 2^{k-s} * t)$, similarly to the number of single encryptions.   □

The constants hidden by the asymptotics are small. Given l pairs of plaintext and ciphertext, we need about 2^k units of storage space—as is the case for the MITM attack. The number of steps is $\text{STEPS}_A + \text{STEPS}_B^o + \text{STEPS}_B^i$. Here $\text{STEPS}_A \approx 2 * l * 2^k$ is the number of steps steps for stage A, $\text{STEPS}_B^o \approx l * w^t * 2^{-s(t-1)} * 2^k$ is the number of steps for the outer loops of stage B (i.e., the number of times the operation $b := E_M^2(a)$ is executed), and $\text{STEPS}_B^i \approx 3 * l * w^t * 2^{-s(t-1)} * 2^{2k-s}$ is the number of steps for all loops of stage B and for `tripletest`.

For triple DES ($k = 56$ and $s = 64$) expected number of t-collisions is about $2^{kt} * 2^{-s(t-1)}$. If $t = 8$, then $2^{kt} * 2^{-s(t-1)} = 2^0 = 1$, we expect one 8-collision, the attack requires about 2^{56} units of storage space, and for $l = 2^k/(w^t * 2^{-s(t-1)}t) = 2^{53}$ we need about $\text{STEPS}_A \approx 2^{110}$ steps (mainly table look-ups and single encryptions) for the attack—instead of 2^{111} similar steps for MITM.

We can improve this by choosing $t = 7$: The expected number of 7-collisions is $2^{kt} * 2^{-s(t-1)} = 2^8 = 256$. Again, the attack requires about 2^{56} units of storage space. For $l = 2^k/(w^t * 2^{-s(t-1)}t) = 2^{56}/(256 * 8) = 2^{45}$, we get $\text{STEPS}_A \approx 2^{102}$, $\text{STEPS}_B^i \approx 3 * 2^{104}$, and

$$\begin{aligned}
\text{STEPS}_B^o &\approx l * w^t * 2^{-s(t-1)} * 2^k \\
&= (2^k/(w^t * 2^{-s(t-1)}t)) * w^t * 2^{-s(t-1)} * 2^k \\
&= 2^{2k}/7 = 2^{112}/7 \\
&\approx 2^{109.2},
\end{aligned}$$

thus we only need slightly more than 2^{109} steps (and as much single encryptions).

From a practical point of view, the operation optimized attack is not very useful for breaking triple DES. It is faster than MITM, but requires much more pairs of known plaintexts/ciphertext (e.g., about 2^{45} if $t = 7$, compared to 3 for MITM). But from a theoretical point of view, the attack's performance clearly indicates triple DES to be weaker than widely believed.

4   How to Save Single Encryptions

The previous section's technique to reduce the number of steps seems to be at a dead end. So in the next two sections, we concentrate on reducing the number of single encryptions instead the number of steps. This section deals with an "encryption optimized" attack. Instead of l sets $S_A(i)$ depending on p_i, we choose one fixed set S_A and no longer exploit the occurances of t-collisions.

Let there be l plaintext/ciphertext pairs (p_1, c_1), ..., (p_l, c_l) known to the attacker. In the previous section, we computed a set $S_A(i) \subseteq \{0,1\}^s$ for every index $i \in \{1, \ldots, l\}$. Now instead, we choose one set $S_A = S_A(1) = \cdots = S_A(l)$. The size $|S_A|$ of S_A is fixed and $|S_A| \ll 2^s$. We assume the $a \in S_A$ to be chosen randomly. (We use the independence of the sets S_A and $\{a \in \{0,1\}^s \mid a = E_L^1(p_i)\}$, where L denotes the correct first key.) Our attack consists of three stages:

1. For $a \in S_A$: compute the sets

$$S_1(a) = \{ \ (i, L) \in \{1, \ldots, l\} \times \{0,1\}^k \mid E_L^1(p_i) = a \ \}.$$

2. For $b \in \{0,1\}^s$ and $i \in \{1, \ldots, l\}$: compute the sets

$$K_3(b, i) = \{ \ N \in \{0,1\}^k \mid E_N^3(b) = c_i \ \}.$$

3. For $M \in \{0,1\}^k$ and $a \in S_A$:
   $b := E_M^2(a)$;
   for $(i, L) \in S_1(a)$:
      for $N \in K_3(b, i)$:
         `tripletest`(i, L, M, N).

What is the chance to find the correct key by using the algorithm?

Theorem 3. *If $|S_A| \approx 2^s/l$, the encryption optimized attack's probability to find the correct key-triple is close to $1/2$.*

Sketch of proof. The attack succeeds in finding the correct key triple (L, M, N), if for any $i \in \{1, \ldots, l\}$ the operation "`tripletest`(i, L, M, N)" is executed, i.e., if a pair (i, a) exists in $\{1, \ldots, l\} \times S_A$ with $E_L^1(p_i) = a$. The existence of such a pair can be expected if $l * |S_A| \approx 2^s$ (due to the birthday paradox).   □

One of the resources required to mount the attack is the number l of known plaintext/ciphertext pairs. What about the other resources?

Theorem 4. *If $|S_A| \approx 2^s/l$, $s \geq k$ and $2^{2k-s} \geq l \geq 2^{2s-2k}$, the encryption optimized attack requires the following resources:*

  – *Exactly l known plaintext/ciphertext pairs,*

- $\Theta(l * 2^k)$ *units of storage space,*
- $\Theta(2^{2k})$ *steps,*
- *and* $\Theta(2^{3k-s})$ *single encryptions.*

Proof. Let the sets $S_1(\cdot)$ and $K_3(\cdot, \cdot)$ be initialized to be empty. The first two stages can be realized like this:

1. For $i \in \{1, \ldots, l\}$ and $L \in \{0, 1\}^k$: $a := E_L^1(p_i)$;
$$S_1(a) := S_1(a) + \{(i, L)\}.$$
$$(* \text{ Or: If } a \in S_A, \text{ then } S_1(a) := \ldots *)$$
2. For $i \in \{1, \ldots, l\}$ and $N \in \{0, 1\}^k$: $b := D_N^3(c_i)$;
$$K_3(b, i) := K_3(b, i) + \{N\}.$$

Each loop is iterated $l * 2^k$ times, hence the overall number of elements in the sets $K_3(\cdot, \cdot)$ are $l * 2^k$, the overall number of elements in the sets $S_1(\cdot)$ is the same (but we only need the sets $S_1(a)$ for $a \in S_A$), and the number of steps and single encryptions for the first two stages is $\Theta(l * 2^k)$.

Stage 3 requires much less storage than the first two stages. Its outer loop "For $M \in \{0, 1\}^k$ and $a \in S_A$" is iterated $2^k * |S_A| \approx 2^{k+s}/l$ times. On the average, the middle loop "for $(i, L) \in S_1(a)$" is iterated $l * 2^k/2^s$ times, and the inner loop "for $N \in K_3(b, i)$" is iterated 2^{k-s} times. Since $2^{k-s} < 1$, the outer and the middle loop determine the number $(2^k * |S_A|)(l * 2^k/2^s) \approx 2^{2k}$ of steps for stage 3. But for the single encryptions, we count how often the operation "$b := E_M^2(a)$" is executed in the outer loop (i.e., $2^k * |S_A| \approx 2^{k+s}/l$) and add three times the number the operation "$c := E_N^3(E_M^2(E_L^1(p_j)))$" is executed within the procedure `tripletest` (i.e., $3(2^k * |S_A|)(l * 2^k/2^s)(2^{k-s}) \approx 3 * 2^{3k-s}$). If $l \geq 2^{2s-2k}$, as required, the `tripletest` part dominates the sum, i.e., the number of single encryptions in stage 3 is about $3 * 2^{3k-s} = \Theta(2^{3k-s})$.

Thus the storage requirement for the attack is dominated by stage 2, the number of steps and the number of single encryptions are dominated by stage 3. Hence we need $\Theta(l * 2^k)$ units of storage space, $\Theta(2^{2k})$ steps and especially $\Theta(2^{3k-s})$ single encryptions.   □

As one can easily deduce from the proof, the constants hidden by the asymptotics are small. We need about $l * 2^k$ units of storage, about 2^{2k} steps, and about $3 * 2^{3k-s} + 2^{k+s}/l$ single encryptions. For triple DES, we may choose l within the range $2^{16} \leq l \leq 2^{48}$. Given, say, $l = 2^{16}$ known pairs of plaintext and ciphertext, we need roughly 2^{72} units of storage (mainly for the elements of the sets $K_3(\cdot, \cdot)$), cycle through a loop for about 2^{112} times, and have to encrypt/decrypt about

$$3 * 2^{3k-s} + 2^{k+s}/l + l * 2^{k+1}$$
$$\approx 3 * 2^{104} \quad + \quad 2^{104} \quad + \quad 2^{73}$$
$$\approx 2^{106}$$

times.

Unlike the operation optimized attack, in which we decreased the number of steps, this section's encryption optimized attack reduces the number of single encryptions but not the number of steps. This optimization reduces the time of the attack as single encryptions are considerably slower than other operations.

5   How to Save more Single Encryptions

The encryption optimized attack's efficiency is limited, since `tripletest` is executed 2^{3k-s} times, which induces $3 * 2^{3k-s}$ single encryptions. As we argued in the sketch of proof of theorem 3, the correct key triple (L, M, N) is found "if a pair (i, a) exists in $\{1, \ldots, l\} \times S_A$ with $E_L^1(p_i) = a$." In this section, we modify the attack; we only execute `tripletest` if there exist *two* pairs $(i, a), (j, a') \in \{1, \ldots, l\} \times S_A$ with $E_L^1(p_i) = a$ and $E_L^1(p_j) = a'$. This idea leads to the "advanced attack". (More generally, we execute `tripletest` if r pairs $(i_1, a_1), \ldots, (i_r, a_r)$ with $E_L^1(p_j) = a_j$ exist in $\{1, \ldots, l\} \times S_A$. In this paper, we concentrate on $r \in \{1, 2\}$.) On one hand, this forces us to increase the number of known plaintext/ciphertext pairs $(p., c.)$ in order to succeed. On the other hand, we need to execute the `tripletest` much less frequently.

The first two stages are the same as before, for stage 3 we do the following:

3. For $M \in \{0, 1\}^k$:
    $S := \{\}$;
    for $a \in S_A$:
        $b := E_M^2(a)$;
        for $(i, L) \in S_1(a)$:
            for $N \in K_3(b, i)$:
                if $(L, N) \in S$ then `tripletest`(i, L, M, N)
                else $S := S + \{(L, N)\}$.

Theorem 5. *If $|S_A| \approx 2 * 2^s/l$, the advanced attack's probability to find the correct key-triple is close to $1/2$.*

Sketch of proof.   Let (L, M^*, N) denote the correct key triple. We consider the iteration of the loop "For $M \in \{0, 1\}^k$:" with $M = M^*$, all other iterations cannot succeed anyway. If $|S_A| \approx 2 * 2^s/l$, the expected number r of pairs $(i_1, a_1), \ldots, (i_r, a_r) \in \{1, \ldots, l\} \times S_A$ with $E_L^1(p_j) = a_j$ is $r = 2$. If there actually exist two such pairs (i_1, a_1) and (i_2, a_2) in $\{1, \ldots, l\} \times S_A$, then the following inclusions hold

$$(i_1, L) \in S_1(a_1), \qquad N \in K_3(E_M^2(a_1), i_1),$$
$$(i_2, L) \in S_1(a_2), \quad \text{and} \quad N \in K_3(E_M^2(a_1), i_2).$$

In this case, the key pair (L, N) is found twice within the execution of the algorithm. At first "$(L, N) \in S$" is wrong and (L, N) is inserted into the set S. The second time "$(L, N) \in S$" is true, `tripletest`(i, L, M, N) is executed (with $i \in \{i_1, i_2\}$) and accepts because $(L, M, N) = (L, M^*, N)$ is the correct key triple.   □

Theorem 6. *If $|S_A| \approx 2^{s+1}/l$, the advanced attack requires the following resources:*

- *Exactly l known plaintext/ciphertext pairs,*

  - $\Theta(l * 2^k)$ *units of storage space,*
  - $\Theta(2^{2k})$ *steps,*
  - *and* $\Theta(l * 2^k + 2^{k+s}/l)$ *single encryptions.*

Proof. The resource requirements for the first two stages of the advanced attack are the same as for the encryption optimized attack.

In the third stage, and for fixed M and a, the loop "for $(i, L) \in S_1(a)$" is iterated about $l * 2^{k-s}$ times, the inner loop "for $(L, N) \in S$" is iterated about 2^{k-s} times. Hence the size of the set S is roughly $|S| \approx l * 2^{2k-2s} \leq l$. The sets $K_3(\cdot, \cdot)$ require $l \times 2^k = \Theta(l * 2^k)$ units of storage and thus dominate the advanced attack's storage requirements.

Similarly to the proof of theorem 4, the number of steps is $(2^k * |S_A|)(l * 2^k/2^s) \approx 2^{2k+1} = \Theta(2^{2k})$.

The first two stages together require $l*2^{k+1}$ single encryptions. The operation $c := E_N^3(\ldots)$ in the procedure `tripletest` is to be executed about $2^{3k-2s+1}$ times, inducing $3 * 2^{3k-2s+1}$ single encryptions. The operation $b := E_M^2(a)$ is to be executed $2^k * |S_A| \approx 2^{k+s+1}/l$ times. Since $l * 2^{k+1} \gg 3 * 2^{3k-2s+1}$, the number of single encryptions is about

$$l * 2^{k+1} + 3 * 2^{3k-2s+1} + 2^{k+s+1}/l \approx l * 2^{k+1} + 2^{k+s+1}/l,$$

i.e., $\Theta(l * 2^k + 2^{k+s}/l)$. $\square$

In practice, we need about $l * 2^k$ units of storage, about 2^{2k+1} steps, and about $2^{k+s+1}/l + l * 2^{k+1}$ encryptions/decryptions. If we fix $l = 2^{s/2}$, we need

$$\text{about} \quad 2^{k+(s/2)+2} \quad \text{single encryptions.} \tag{1}$$

For attacking triple DES, given $l = 2^{32}$ known pairs of plaintext and ciphertext, we need 2^{88} units of storage space and 2^{113} steps, but only 2^{90} single encryptions. In comparison to the operation optimized attack, the advanced attack allows us to drastically reduce the amount of single encryptions at the cost of doubling the number of steps. So what is our gain? As we mentioned in the introduction, a single encryption is a very complex operation, compared to, say, table look-ups. If we assume one implementation of DES to require 8 table look-ups per round, i.e., $8 * 16 = 2^7$ table look-ups per encryption, our speed-up can be estimated like this:

  - The expected number of 2^{111} steps and as much single encryptions of the MITM attack actually correspond to about $1.3 * 2^{109}$ triple encryptions.
  - The operation optimized attack of section 3 needed 2^{109} steps and single encryptions. These correspond to about $1.3 * 2^{107}$ triple encryptions.
  - The encryption optimized attack's 2^{112} steps (mostly table look-ups) and 2^{106} single encryptions. This is equivalent to about 2^{105} triple encryptions.
  - This section's attack requires 2^{113} steps (mostly table look-ups) and 2^{90} single encryptions. This corresponds to about $1.3 * 2^{104}$ triple encryptions.

Our result 1 for triple encryption (i.e. 2^{90} single encryptions to break triple DES) is very close to Kilian's and Rogaway's *lower bound* [3] for the number of single encryptions required to break DESX. For details, see appendix A.

6   A Special Variant for Triple DES

So far, we pretended the underlying single block cipher to be ideal, i.e., to behave like a random permutation. But DES is not an ideal block cipher. Most important in this context is the complementation property: If $\bar{x}$ denotes the complement of the bit-string x, then for every plaintext $p \in \{0,1\}^s$ and every key $K \in \{0,1\}^k$:

$$\mathrm{DES}_K(p) \;=\; \overline{\mathrm{DES}_{\overline{K}}(\overline{p})}.$$

How does the complementation property affect the efficiency of our attacks?

First, we note there is not much harm for the attacker. The encryption optimized attack succeeds, if the sets $\{p_1, \ldots, p_l\}$ and S_A are chosen such that there exists a $(i, a) \in \{1, \ldots, l\} \times S_A$ with $E_L^1(p_i) = a$, L the correct first subkey, cf. proof of theorem 3. This probability is not at all affected by the complementation property $E_{\overline{L}}^1(\overline{p_i}) = \overline{a}$. We may argue similarly for the advanced attack. The success rate of the operation optimized attack depends on the probability that for a plaintext p_i the correct first subkey L participates in a t-collision $K(a, i) = \{L, L_2, \ldots, L_t\}$, i.e., $E_L^1(p_i) = E_{L_2}^1(p_i) = \ldots = E_{L_t}^1(p_i) = a$. Again, this probability is not affected by the complementation property $E_{\overline{L}}^1(\overline{p_i}) = \overline{a}$.

Second, there are many ways for the attacker to exploit the complementation property for a small improvement of an attack. For the sake of shortness, we concentrate on one example. Recall the attack in section 3. Let S_A be chosen such that for all $a \in \{0,1\}^s$ the equivalence $a \in S_A \iff \bar{a} \in S_A$ holds. The attack is unchanged, except for stage B:

> B. for $a \in S_A(i)$:
>> for $M \in \{0\} \times \{0,1\}^{k-1}$:
>>> $b := E_M^2(a)$;
>>> for $N \in K_3(b, i)$:
>>>> for $L \in K_1(a, i)$: $\mathtt{tripletest}(i, L, M, N)$;
>>> ($*$ Next, we exploit $\bar{b} = E_{\overline{M}}^2(\bar{a})$. $*$)
>>> for $N \in K_3(\bar{b}, i)$:
>>>> for $L \in K_1(\bar{a}, i)$: $\mathtt{tripletest}(i, L, \overline{M}, N)$;

The analysis in section 3 is not much affected. Neither the expected number of pairs of plaintext and ciphertext changes, nor the complexity STEPS_A of stage A, nor the attack's storage requirements.

With respect to stage B, the loop "for $a \in S_A(i)$" is iterated $w^t * 2^{-s(t-1)}$ times on the average. The loop "for $M \ldots$" is only iterated 2^{k-1} times, hence, $w^t * 2^{-s(t-1)} * 2^{k-1}$ single encryptions $b := E_M^2(a)$ are done. So far, we need $\mathrm{STEPS}_B^o = w^t * 2^{-s(t-1)}2^{k-1}$ steps for stage B. Together, the two loops "for $N \in \ldots$" need as much time as before: $\mathrm{STEPS}_B^i = w^t * 2^{-s(t-1)} * 2^k * 2^{k-s} * t$.

If we choose the parameters were $t = 7$, $u \approx 2.2$, and $l = 2^{45}$, the operation optimized attack's complexity is the sum of three numbers $\mathrm{STEPS}_A \approx 2^{102}$, $\mathrm{STEPS}_B^i \approx 3 * 2^{104}$, and $\mathrm{STEPS}_B^o \approx 2^{109.2}$.

This section's variant does not affect STEPS_A and STEPS_B^i, hence

$$\text{STEPS}_B^o \approx 2^{108.2}$$

approximates the overall number of steps and single encryptions.

7   Comparison and Conclusion

Based on today's technology, neither MITM nor any of our attacks constitutes a practical way to break triple DES. If in the future an attack like MITM will be considered practical for doing this, certainly some of the required resources will be more valuable than others. This paper provides a variety of options how to possibly save such a bottleneck resource. A comparison is given in table 1.

attack	sect.	l	memory	steps	single encryptions
MITM	2	3	2^{56}	2^{111}	2^{111}
op. optim. (variant)	3 6	2^{45} (2^{45})	2^{56} 2^{56}	$2^{109.2}$ $2^{108.2}$	$2^{109.2}$ $2^{108.2}$
encr. optim.	4	l 2^{16} 2^{24} 2^{32}	$l*2^k$ 2^{72} 2^{80} 2^{88}	2^{2k} 2^{112} 2^{112} 2^{112}	$3*2^{3k-s}+2^{k+s}/l$ 2^{106} $3*2^{104}$ $3*2^{104}$
advanced	5	l 2^{16} 2^{24} 2^{32}	$l*2^k$ 2^{72} 2^{80} 2^{88}	2^{2k+1} 2^{113} 2^{113} 2^{113}	$l*2^{k+1}+2^{k+s+1}/l$ 2^{105} 2^{97} 2^{90}

Table 1. Attacking triple DES with l known (chosen) pairs of plaintext and ciphertext and the expected number of resources required.

Van Oorschot and Wiener [6,7] considered attacks with decreased memory requirements at the cost of increased running times. Usually, reducing storage requirements is seen as the main goal of improving an attack like MITM. The approach in sections 4 and 5 is to decrease the running time at the cost of storage. As an anonymous referee criticized, this seems to make our attacks less realistic. The current author's reply is that the basic MITM attacks on double encryption and two-key triple encryption both have balanced time-memory characteristics, i.e., require roughly one step of computation per unit of memory. In this case, trading away storage space at the cost of additional computational steps, as van Oorschot and Wiener did, certainly makes such attacks more realistic. On the other hand, the MITM attack on general (three-key) triple encryption has a highly unbalanced time-memory characteristic: 2^k units of memory and 2^{2k} steps are needed, i.e., 2^k steps per unit of memory. If k is reasonably large, e.g., $k = 56$, decreasing the running time at the cost of additional memory requirements actually appears to make such attacks *more* realistic. (Today though, our attacks are far from being practical, as is the MITM attack. It is quite difficult to reasonably estimate the economically best time-memory characteristic of a future technology for which such attacks are practical.)

Even though our attacks are far from being practical today, this paper demonstrates that *it is too optimistic to identify the complexity of breaking triple DES and similar block ciphers with the complexity of the MITM attack.* Also, this paper alludes that the ability to quickly perform many single DES operations is not crucial for breaking triple DES (though even the required number of single DES operations is too large to be considered feasible today). The number of memory accesses, i.e., table look-ups, appears to be dominating—with great consequences on the difficulty of massively parallel triple DES cracking.

8   Acknowledgements

The author is thankful to Rüdiger Weis for discussing DESX and very much appreciates referees' aid in improving the presentation of this material.

References

1. E. Biham, *How to forge DES-Encrypted Messages in 2^{28} steps*, Technical report CS0884, Computer Science department, Technion, 1996, found in the www[3].
2. J. Kelsey, B. Schneier, D. Wagner, "Key-Schedule Cryptanalysis of 3-WAY, IDEA, G-DES, RC4, SAFER, and Triple-DES", *Crypto '96*, Springer LNCS 1109, 237–251.
3. J. Kilian, P. Rogaway, *How to protect DES against exhaustive key search*, Crypto '96, Springer LNCS 1109, 252–267, full version found in the www[4].
4. A.J. Menezes, P.C. van Oorschot, S.A. Vanstone, *Handbook of applied cryptography*, CRC Press, 1997.
5. R.C. Merkle, M.E. Hellman, *On the security of multiple encryption* Communications of the ACM, Vol. 24, No. 7 (1981).
6. P.C. van Oorschot, M.J. Wiener, *A known-plaintext attack on two-key triple encryption*, Eurocrypt '90, Springer LNCS 473, 318–325.
7. P.C. van Oorschot, M.J. Wiener, *Improving implementable meet-in-the-middle attacks by orders of magnitude* Crypto '96, Springer LNCS 1109, 229–236.
8. R.L. Rivest, A. Shamir, *PayWord and MicroMint, two simple micropayment schemes*, CryptoBytes, Vol. 2, No. 1 (1996), 7–11.

[3] http://www.cs.technion.ac.il/Reports/

[4] http://wwwcsif.cs.ucdavis.edu/~rogaway/papers/list.html

A   On Triple DES and DESX

As mentioned in the introduction, the black-box-only model provides a proven environment to demonstrate the soundness of a composed cipher. Kilian and Rogaway [3] analyze the DESX block cipher and its security in this model. Note that in the black-box-only model, one concentrates on the number of encryptions and disregards all other operations.

A generalized variant of DESX is EX, based on the encryption function $E : \{0,1\}^k \times \{0,1\}^s \longrightarrow \{0,1\}^s$. An EX key is a triple $(L, M, N) \in \{0,1\}^k \times \{0,1\}^s \times \{0,1\}^s$. The encryption function is $EX_{(L,M,N)}(p) = N \oplus E_L(M \oplus p)$, where "$\oplus$" denotes the bit-wise XOR. Compared to triple DES, DESX is amazingly elegant and efficient.

Let l denote the number of known (or chosen) pairs of plaintext and ciphertext. Kilian and Rogaway prove for EX that the attacker's advantage in distinguishing between random nonsense unrelated to E, and EX encryptions using a key-triple (L, M, N) unknown to the attacker, is $\epsilon \leq l * x * 2^{-k-s+1}$. Here, x denotes the number of single encryptions. If $\epsilon = 1/2$ and $l = s/2$, this requires

$$\text{about} \quad x \geq 2^{k+(s/2)-2} \quad \text{single encryptions,} \tag{2}$$

e.g., about $x \geq 2^{85}$ for DESX. (Note that Kilian and Rogaway consider $k = 55$ and ignore the additional key bit of DES. This is necessary for lower bounds in the black-box-only model due to the DES complementation property.) By presenting a chosen plaintext attack, Kilian and Rogaway also demonstrate that the above bound is tight, except for a small factor.

Our result (1) in section 5 for breaking triple encryption is surprisingly close to Kilian's and Rogaway's lower bound (2) for EX. We conclude, in order to find a combined cipher *provably much more secure* than EX (or DESX), one has to abstain from triple encryption (triple DES) or to forego the black-box-only model. In other words, this paper gives evidence that it will be difficult to prove triple DES to be much stronger than the more efficient DESX construction.

Cryptanalysis of Some Recently-Proposed Multiple Modes of Operation

David Wagner

University of California, Berkeley
daw@cs.berkeley.edu

Abstract. In a paper cryptanalyzing many triple modes of operation, Biham proposed four new triple modes and five new quadruple modes of operation for DES. It was conjectured that the complexity (in a particular threat model) of breaking the triple modes is at least 2^{112} and that the quadruple modes are more secure than any triple mode.

We present new attacks on all but one of the proposed modes. We can break all but two of Biham's proposed modes with at most 2^{56} off-line trial encryptions and between 2 and 2^{32} (depending upon the mode) chosen-IV chosen texts; another mode can be broken with somewhat more work. This raises questions about the suitability of the proposed modes, and provides further evidence for the fragility of inner chaining; however, we emphasize that our results do not disprove Biham's conjectures, as we rely on an extended attack model which admits more powerful adversaries who can mount chosen-IV queries, a capability denied to them in Biham's model.

1 Introduction

DES is the most thoroughly-analyzed cipher in the open literature, but after more than two decades, it is reaching the end of its useful lifetime: the DES 56-bit key-length is simply too short to be secure against serious keysearch efforts. Therefore, there is great interest in the search for a multiple mode of operation for DES which provides increased strength against exhaustive keysearch while retaining the high level of analysis and confidence that single-DES currently offers.

Biham [Bih96] analyzed a great many triple modes of operation, and broke every mode considered except the commonly-used triple-DES-ECB mode (when used with some outer chaining technique). Unfortunately, due to its short 64-bit block length, triple-DES-ECB has some shortcomings: it is susceptible to dictionary attacks (when 2^{64} known texts are available) and matching-ciphertext attacks (where partial information about the plaintext is recovered by using the birthday paradox, when 2^{32} known texts are available).

To improve this state of affairs, Biham proposed 9 new block modes and 2 new stream modes of operation for DES. The complexity of attacking these new modes is conjectured to be at least 2^{112}. The quadruple modes were conjectured

S. Vaudenay (Ed.): Fast Software Encryption – FSE'98, LNCS 1372, pp. 254–269, 1998.

to be more secure than any triple mode; furthermore, the complexity of attacking two of the quadruple modes was conjectured to be at least 2^{128}.

This paper shows that, when we allow chosen-IV chosen-text attacks, most of the proposed modes are not significantly more secure than single-DES. We provide new attacks against all but one of the modes.

Note that Biham's studies were premised on a more restrictive threat model that did not admit chosen-IV attacks, so our results do not disprove Biham's conjectures; but our position is that these new results raise questions about the security of Biham's proposed modes and illustrate the application of general techniques for cryptanalysis of multiple modes of operation. See Section 3 for more discussion on this point.

The paper is organized as follows. Section 2 establishes some notation and other background, and Section 3 discusses our threat model. Section 4 shows how to attack two important classes of modes using a divide-and-conquer strategy, and applies this result to attack six of Biham's proposed modes. Section 5 shows how to attack four more of Biham's modes using narrow-pipe attacks. Finally, Section 6 discusses some implications of our results, and Section 7 wraps up the paper with some concluding remarks.

2   Preliminaries

Biham developed a concise notation for multiple modes which is worth summarizing here. All of his new modes are derived from the standard DES modes of operation—ECB, CBC, CFB, and OFB—as well as their corresponding decryption modes—ECB^{-1}, etc. The notation CBC|CFB refers to the mode where the output of DES-CBC encryption is fed to the input of DES-CFB encryption; the | operator can be extended to triple and higher-order modes. The notation OFB[CBC] refers to a mode which applies OFB to its input, then encrypts with CBC mode, and finally applies the same OFB keystream to that result. (Note: the streams xored into the input and the output of CBC are generated from a single DES key, and therefore are *the same*.) This can be generalized to modes such as OFB[CBC,CFB], where we apply OFB, then CBC, then OFB again, then CFB, and then OFB once more. (Again, all the OFB output streams are the same!) The notation OFB→CBC refers to a stream mode which applies CBC encryption to the keystream generated by OFB mode, and xors the result to the plaintext. We can of course use the → operator to define triple and higher-order modes, too.

For clarity, we will attempt to use the same notation for plaintext, ciphertext, etc. throughout this note. We write $P_0, P_1, \ldots$ (respectively $C_0, C_1, \ldots$) for the blocks of the plaintext (resp. ciphertext). We let $K0, K1, \ldots$ denote the 56-bit DES keys, and write $IV0, IV1, \ldots$ for the corresponding IVs. We number the keys $K0, K1, \ldots$ according to the order that the single-mode appears in this notation: for instance, in OFB[CBC,CBC^{-1}], the OFB-mode is keyed with $K0$, the CBC with $K1$, and the CBC^{-1} with $K2$. When multiple plaintext/ciphertext pairs are obtained in an attack run, we write $P[j]$ for the full plaintext of the

j-th message, write $P_i[j]$ for the i-th block of $P[j]$, and so on. We let $E_k(x)$ stand for the single-DES encryption of the input block x under the key k.

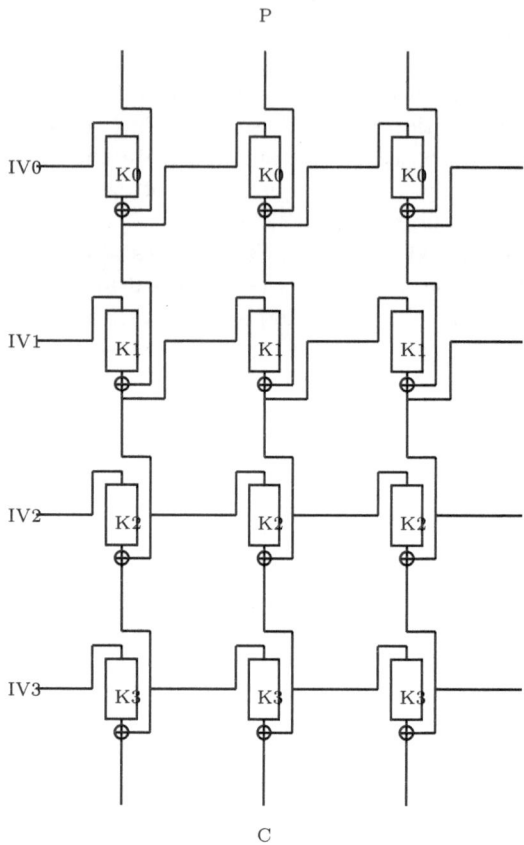

Fig. 1. The CFB|CFB|CFB^{-1}|CFB^{-1} mode.

As an example of this notation, we depict the the CFB|CFB|CFB^{-1}|CFB^{-1} mode in Figure 1.

All of our attacks recover the secret DES keys. The basic ideas behind the attacks are not entirely novel; many of them are applications of the general tools worked out by Coppersmith, Johnson, and Matyas [CJM97] and Biham [Bih94a,Bih94b,Bih96].

The nine new block modes which Biham proposed are

1. OFB[CBC,CBC^{-1}],
2. OFB[CFB,CFB^{-1}],
3. OFB[CBC,CBC],

4. OFB[CFB,CFB],
5. CBC|CBC|CBC^{-1} |CBC^{-1},
6. CFB|CFB|CFB^{-1} |CFB^{-1},
7. OFB[CBC,CBC,CBC^{-1}],
8. OFB[CBC,CBC,CBC], and
9. OFB[CFB,CFB,CFB].

The proposed stream modes are

1. OFB→CBC→CBC, and
2. OFB→CFB→CFB.

In the following sections, we find new attacks on all of these except OFB[CBC, CBC^{-1}].

3   Security model

In this section, we examine the attack model. The opponent is assumed to have the necessary computational power to perform 2^{56} off-line trial encryptions. We assume (as is standard) that the adversary can perform known-plaintext, chosen-plaintext, and chosen-ciphertext attacks.

So far we have not deviated from Biham's model. We list below three important differences.

3.1   Controlling IVs

The most important difference between the two models comes when we examine the treatment of IVs.

In our model, a mode is essentially a mini-protocol specifying how to perform secure message transport. To send the secret message P, one enciphers P under the appropriate multiple mode with key K and with randomly chosen IVs $IV0, \ldots, IVn$, transmitting the bundle $IV0, \ldots, IVn, C$ over the insecure medium; the receiver decrypts C with the specified IVs under the shared key K and recovers the decrypted message P'. The subtlety comes when we introduce active attackers with the ability to perform chosen-ciphertext attacks: such adversaries are free to specify any ciphertext C along with any set of IVs they wish, and they will receive the decryption P' of that ciphertext.

Our attack model captures this notion. Because we allow chosen-ciphertext attacks, we also allow (as a natural consequence) chosen-IV chosen-ciphertext attacks. It is worth noting that this choice induces a slight assymetry between chosen-plaintext and chosen-ciphertext attacks: adversaries may control the IV in chosen-ciphertext attacks, but not in chosen-plaintext attacks.

In contrast, Biham did not consider chosen-IV attacks; even known-IV attacks were mentioned only in a few special cases. His model is more elegant and cleaner for analysis; for instance, the symmetry ensures that the security factor for a mode is the same as for its inverse. Also, attacks are all the more compelling when they are performed in Biham's more restrictive model. Finally, Biham's attacks remain applicable even when special measures to protect the IV are taken, whereas our attacks may be stopped by such measures.

We take the conservative philosophy that our model should allow adversaries considerable leeway; if the cryptosystem can stand up to attack in such a model, our assurance of security will be all the greater. Part of our justification for this approach is that triple DES with outer chaining already offers pretty good security, with only a few shortcomings: if we want to do better, our threshold should be quite high.

Our attacks will take advantage of this ability to control the IV, so they are not directly comparable to Biham's results. However, a number of the chosen-IV attacks can be converted to known-IV attacks with only a minor increase in the complexity of cryptanalysis, so some comparisons may be possible. See Section 6. Even where we are not aware of known-IV attacks, we view our chosen-IV attacks as certificational weaknesses that should at the very least raise warning flags about the security of the modes in question.

This subject is not yet exhausted. See Section 6 for some simple counter-measures to resist chosen-IV attacks, some counter-countermeasures, and their implications for the interpretation of our results.

3.2 Adaptive attacks

Biham's model also differs from ours in another respect: we allow adaptive chosen-text attacks, whereas Biham did not consider adaptive attacks. Moreover, Biham generally required only one encrypted stream for his analyses. In contrast, all of our attacks are cast in the language of adaptive attacks.

We view this distinction as relatively minor. All of our adaptive attacks can be easily converted to non-adaptive attacks with negligible increase in complexity (and, occasionally, a substantial increase in the number of messy details); in short, the adaptivity is merely convenient, not fundamental.

3.3 The cost of chosen texts

One philosophical point is that we try to be explicit about the resource requirements of our attacks, listing separately the number of chosen texts, offline decryptions, and memory words needed. The reader is then free to assign appropriate costs to each resource, according to his or her security environment.

It would be simpler to label each attack with a simple complexity measure that equates the cost of one chosen text with the cost of one trial decryption. Indeed, such a measure has great benefits for simplifying analysis, summarizing results, and comparing modes; and it is a very useful first approximation. The drawback is that highly theoretical attacks needing 2^{56} chosen texts may be equated with more serious attacks needing only 2^{56} trial decryptions. In practice, that distinction can be critical. Therefore, where possible, we aim to improve the quality of the approximation by using more explicit complexity measures.

4   Divide-and-conquer attacks

First, we list some fairly elementary attacks on several modes. These all have
the flavor of "divide-and-conquer" algorithms: namely, we isolate the effect of
each subkey with a chosen-ciphertext probe, and then recover each subkey with
a 2^{56} exhaustive keysearch.

By the end of this section, we will see how to attack the block modes
OFB[M1,M2,...,Mn] and M0|M1|M2|...|Mn, for any n, in the special case where
each mode Mj is either CFB or CFB^{-1}. The intuition is that, in such modes,
we have the relation

$$C_0 = P_0 \oplus E_{K0}(IV0) \oplus \ldots \oplus E_{Kn}(IVn) \tag{1}$$

on the first block; this is highly linear and, therefore, highly suspicious.

We will also see how to attack stream modes of the form OFB$\rightarrow$M1$\rightarrow$
$\ldots \rightarrow$Mn, if each mode Mj is one of OFB, CBC, CBC^{-1}, CFB, CFB^{-1}. The
idea is that we can apply a divide-and-conquer attack that isolates the effect of
the last key Kn (with a single chosen-ciphertext query that probes IVn); we
then strip off the last mode and continue iteratively.

4.1   OFB[CFB,CFB]

Our attack on the OFB[CFB,CFB] mode is composed of three phases; each
phase isolates the effect of one DES key Kj. First, we recover the key $K1$ used
in the first CFB mode by using one chosen-IV chosen-ciphertext query: namely,
we isolate the effect of $K1$ by probing $IV1$. In the second phase, we recover the
key $K2$ by probing $IV2$ with a similar chosen-ciphertext query. Finally, $K0$ is
recovered by exhaustive keysearch.

In the first phase, we probe $IV1$ to isolate the effect of $K1$, and recover $K1$
with a 2^{56} exhaustive keysearch. Let $P[0], C[0]$ be a known plaintext/ciphertext
pair with known IVs. We construct a chosen ciphertext query $C[1]$ as follows.
Pick $IV1[1] \neq IV1[0]$, set $IV0[1] = IV0[0], IV2[1] = IV2[0]$, take $C[1] = C[0]$,
and obtain the decryption $P[1]$ of the new ciphertext. Note that, by Equation 1,

$$P_0[0] \oplus P_0[1] = E_{K1}(IV1[0]) \oplus E_{K1}(IV1[1]).$$

Therefore we may find $K1$ by a 2^{56} exhaustive keysearch, recognizing the right
key value when the above equation holds; with high probability, we expect no
wrong key value to survive the check.

The second phase recovers $K2$ in an entirely analogous fashion, this time
probing $IV2$ instead of $IV1$.

Finally, in the third phase we perform a 2^{56} exhaustive search over $K0$ (the
only remaining unknown key value). Therefore, the total complexity of the attack
is two chosen-ciphertexts and $5 \cdot 2^{56}$ off-line trial encryptions.

4.2   OFB[CFB,CFB^{-1}]

The OFB[CFB,CFB^{-1}] mode can be broken in a way entirely analogous to the cryptanalysis of OFB[CFB,CFB]: probe $IV1$ in one chosen-ciphertext query to recover $K1$, then probe $IV2$ to learn $K2$, and exhaustively search over $K0$. So the OFB[CFB,CFB^{-1}] mode, too, can be broken with two chosen-ciphertexts and $5 \cdot 2^{56}$ off-line trial encryptions.

4.3   OFB[CFB,CFB,CFB]

The OFB[CFB,CFB,CFB] mode can also be broken with the same technique. For this mode, we need three chosen-ciphertexts and $7 \cdot 2^{56}$ off-line trial encryptions.

4.4   CFB|CFB|CFB^{-1} |CFB^{-1}

This mode is also easy to break using the same techniques. (Note that the CFB|CFB|CFB^{-1} |CFB^{-1} mode is illustrated in Figure 1.) As before, in the first phase we can probe $IV0$ to isolate the effect of $K0$ and recover $K0$ by exhaustive search; continue to recover the rest of the keys. In this way, we can break the CFB|CFB|CFB^{-1} |CFB^{-1} mode with a total of three chosen-ciphertexts and $7 \cdot 2^{56}$ off-line trial encryptions.

4.5   OFB→CBC→CBC

OFB→CBC→CBC mode is characterized by the relation

$$C_0 = P_0 \oplus E_{K2}(IV2 \oplus E_{K1}(IV1 \oplus E_{K0}(IV0))).$$

In the first phase of our attack, we probe $IV2$ to isolate the effect of $K2$. More precisely, let $P[0], C[0]$ be a known plaintext/ciphertext pair with IVs $IV0[0], IV1[0], IV2[0]$, and construct a chosen-ciphertext query as follows. Set $C[1] = C[0]$, pick $IV2[1] \neq IV2[0]$, and set $IV0[1] = IV0[0], IV1[1] = IV1[0]$. Next issue a chosen-ciphertext query for the $C[1], IVj[1]$ to get $P[1]$. Finally note that

$$IV2[0] \oplus IV2[1] = D_{K2}(P_0[0] \oplus C_0[0]) \oplus D_{K2}(P_0[1] \oplus C_0[1]);$$

this relation lets us recover $K2$ with a 2^{56} exhaustive search.

The second phase of the attack probes $IV1$ in a similar way to recover $K1$. Finally, $K0$ can be obtained in a third phase by brute force. In sum, this cryptanalysis requires $5 \cdot 2^{56}$ off-line trial encryptions and two chosen-ciphertexts.

4.6   OFB→CFB→CFB

Our attack on OFB→CFB→CFB mode proceeds in a very similar way to that described in the previous paragraph. We probe $IV2$ in a chosen-ciphertext attack, which allows us to isolate the effect of $K2$ by the following relation:

$$C_0[0] \oplus C_0[1] \oplus P_0[0] \oplus P_1[1] = E_{K2}(IV2[0]) \oplus E_{K2}(IV2[1]).$$

Then $K1$ is recovered analogously, and $K0$ by exhaustive keysearch. The total complexity of this attack is two chosen-ciphertext queries and $5 \cdot 2^{56}$ off-line trial encryptions.

5   Narrow-pipe attacks

In this section, we describe a number of narrow-pipe attacks. (By a "narrow pipe", we mean a data channel that is relatively narrow—only 64 bits wide, for instance.) The basic technique is to identify some narrow pipe through which all diffusion is channeled; then you generate a bunch of texts and look for a collision in that narrow pipe. The birthday paradox assures us that we will find a collision in the narrow pipe relatively quickly (within $2^{n/2}$ texts, for a n-bit pipe). Then we hope (1) that we can recognize the collision by looking only at the plaintext and ciphertext, and (2) that we can use that knowledge to deduce some relation which isolates the effect of just one DES key. When the attack is designed correctly, we will be able to find recognizable collisions in the narrow pipe that let us deduce important information about some key Ki standing alone. After recovering Ki with a 2^{56} exhaustive keysearch, we remove the effect of that key and attempt to solve the reduced mode by iterating the attack.

In this section, we show how to break the $CBC|CBC|CBC^{-1}|CBC^{-1}$ block mode, as well as the $OFB[CBC,CBC]$, $OFB[CBC,CBC,CBC]$, and $OFB[CBC, CBC,CBC^{-1}]$ modes.

5.1 $CBC|CBC|CBC^{-1}|CBC^{-1}$

To break $CBC|CBC|CBC^{-1}|CBC^{-1}$ (see Figure 2), we first recover $K0$ by probing $IV1$. Let $P[0], C[0]$ be a known plaintext/ciphertext pair with known IVs, and build a chosen ciphertext query as follows. Pick $IV1[1] \neq IV1[0]$, set $IVj[1] = IVj[0]$ for $j \neq 1$, take $C[1] = C[0]$, and obtain the decryption $P[1]$ of the new ciphertext. Note that

$$IV1[0] \oplus IV1[1] = E_{K0}(IV0[0] \oplus P_0[0]) \oplus E_{K0}(IV0[1] \oplus P_0[1]).$$

Therefore we may find $K0$ by a 2^{56} exhaustive keysearch, recognizing the right key value when the above equation holds; with high probability, the check will eliminate all incorrect guesses at the key.

Once we've learned $K0$ with 2^{56} work and one chosen-ciphertext query, we can peel off the effect of $K0$ and reduce the problem to that of breaking the $CBC|CBC^{-1}|CBC^{-1}$ mode.

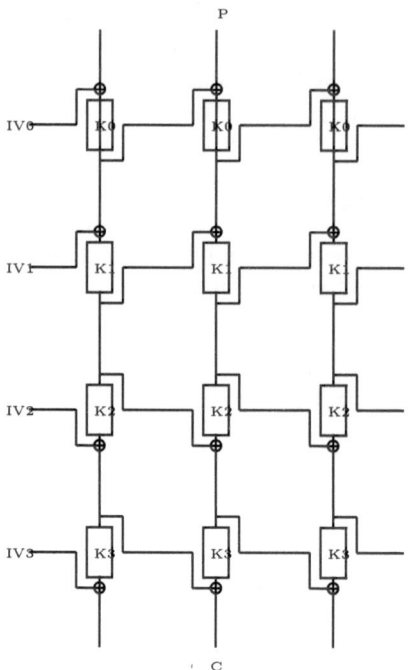

Fig. 2. The $CBC|CBC|CBC^{-1}|CBC^{-1}$ mode.

Biham conjectured that the quad mode $CBC|CBC|CBC^{-1}|CBC^{-1}$ is more secure than any triple mode. Note that our present attack does not immediately disprove Biham's conjecture, since our reduction relies on mounting a chosen-IV chosen-ciphertext query, which is not allowed in Biham's security model. In Section 6.2 we extend it to work with only known-IV queries, which brings us a step closer to Biham's model.

Finishing the attack. We now describe how to finish the attack on $CBC|$ $CBC|CBC^{-1}|CBC^{-1}$. All that remains is to analyze the triple mode $CBC|$ $CBC^{-1}|CBC^{-1}$. Biham has showed how to break this triple mode with 2^{68} chosen plaintexts and 2^{66} work [Bih96]. Nonetheless, in our security model, 2^{68} texts are quite a high barrier, and one might wonder whether there are more efficient attacks.

The answer is yes. We present next a new attack on $CBC|CBC^{-1}|CBC^{-1}$ which requires only 2^{32} chosen-ciphertext chosen-IV queries and $6 \cdot 2^{56}$ trial encryptions. This can be used as a subroutine to develop a full attack on the $CBC|CBC|CBC^{-1}|CBC^{-1}$ quad mode with roughly equivalent complexity.

Breaking $CBC|CBC^{-1}|CBC^{-1}$. We recover $K2$ by probing $IV1$. Fix arbitrary IVs $IV0, IV2$. We construct 2^{32} chosen-ciphertext queries, as follows. For each i, pick $C_0[i]$ and $IV1[i]$ randomly, and let $C_1[i] = IV1[i]$. Now we obtain the decryptions $P[i]$ of those 2^{32} ciphertexts. We search for i, j such that

$P[i] = P[j]$ (using a hash table, so as to avoid increasing the complexity of the attack). Note that two plaintexts will certainly match in the first block (i.e. $P_0[i] = P_0[j]$) if

$$IV1[i] \oplus IV1[j] = E_{K2}(C_0[i] \oplus IV2) \oplus E_{K2}(C_0[j] \oplus IV2) \qquad (2)$$

holds, because then the resulting collision between the second and third DES layers will necessarily propagate up to the plaintext. Moreover, if Equation 2 holds, then in fact the two plaintexts $P[i], P[j]$ will match in their entirety. (This is because the value at the bottom of the third DES layer at the second block is $C_1 \oplus E_{K2}(C_0 \oplus IV2)$; now the choice of $C_1[i], C_1[j]$ ensures that $C_1[i] \oplus E_{K2}(C_0[i] \oplus IV2) = IV1[i] \oplus (E_{K2}(C_0[j] \oplus IV2) \oplus IV1[i] \oplus IV1[j]) = IV1[j] \oplus E_{K2}(C_0[j] \oplus IV2) = C_2[j] \oplus E_{K2}(C_0[j] \oplus IV2)$, so we will get a collision at the bottom of the third layer, and this will necessarily propagate up to the plaintext.) Since we generated 2^{32} ciphertexts and the block size is 64 bits, by the birthday paradox with high probability we will find one pair i, j satisfying Equation 2, and so with high probability we will see $P[i] = P[j]$ for some i, j. On the other hand, because P is two blocks (128 bits) long, the chances of seeing a chance match $P[i'] = P[j']$ by accident is very low. Therefore, we expect to see one match $P[i] = P[j]$, and we can conclude that for such i, j Equation 2 must hold.

Once we've found a pair i, j where Equation 2 holds, we can use it to isolate the effect of $K2$. This lets us recover $K2$ using a 2^{56} exhaustive keysearch. Finally, knowing $K2$ lets us reduce the problem to that of breaking CBC|CBC^{-1} mode, which can be done by standard techniques without any increase in complexity. This lets us break the triple mode CBC|CBC^{-1} |CBC^{-1} with $6 \cdot 2^{56}$ trial encryptions and 2^{32} chosen-ciphertext chosen-IV queries. (In fact, the attack can be extended to work just as efficiently with known-IV chosen-ciphertext queries instead of chosen-IV queries; it just becomes a bit messier to describe.)

Pulling it all together. To summarize, we can apply these techniques to break the CBC|CBC|CBC^{-1} |CBC^{-1} quad block mode with 2^{32} chosen-ciphertext queries and $7 \cdot 2^{56}$ trial encryptions.

Note that we could dramatically reduce the number of chosen-ciphertext queries needed if there were a better way to break the triple mode CBC|CBC^{-1} | CBC^{-1}.

5.2   OFB[CBC,CBC]

For the OFB[CBC,CBC] mode, we use another narrow-pipe attack combined with a birthday argument to find a collision in the OFB streams generated by two different messages. Generate 2^{32} chosen-ciphertext queries, as follows. Fix a 64-bit constant c, fix $IV1, IV2$, and let $C[i] = (c, c, c, c, c)$ for all i. The only value that varies will be $IV0[i]$, which we pick randomly. Obtain the decryptions $P[i]$ of those chosen ciphertexts. Now we search for i, j such that $E_{K0}(IV0[i]) = IV0[j]$; this relation ensures that the two OFB streams for $P[i], P[j]$ will match up (except that they will be out of phase by one block). Of course, given such an i, j, we can recover $K0$ with a 2^{56} exhaustive keysearch.

How can we recognize such a fortunate event? Note that peeling off the second CBC mode during decryption of $C[i]$ leaves $(?, ?, d, e, f)$, while for $C[j]$ we get $(?, d, e, f, ?)$. Furthermore, peeling off the first CBC mode we get $(?, ?, ?, g, h)$ for $C[i]$ and $(?, ?, g, h, ?)$ for $C[j]$. (A word on notation: the question marks "?" just represent arbitrary unknown values; two entries both marked with a "?" need not be equal.) In other words, we can recognize i, j such that $E_{K0}(IV0[i]) = IV0[j]$ by the requirement that $P_3[i] = P_2[j]$ and $P_4[i] = P_3[j]$; false alarms should be very rare, and with 2^{32} chosen ciphertexts, the birthday paradox reassures us that we expect to find at least one such i, j.

Once we've found i, j such that $E_{K0}(IV0[i]) = IV0[j]$, we can recover $K0$ with a 2^{56} exhaustive search. This reduces the problem of breaking OFB[CBC, CBC] to that of breaking (a mode very similar to) CBC|CBC. The latter mode can be broken efficiently with standard techniques, and in fact we can recover $K1, K2$ with one chosen ciphertext and $3 \cdot 2^{56}$ trial encryptions. In total, we can break OFB[CBC,CBC] with 2^{32} chosen ciphertexts and $4 \cdot 2^{56}$ trial encryptions.

5.3 OFB[CBC,CBC,CBC]

OFB[CBC,CBC,CBC] can be broken similarly. First, we recover $K0$ using the same technique as described above for OFB[CBC,CBC]. Then all we need to do is break (a mode very similar to) CBC|CBC|CBC, and that can be done with two chosen ciphertext queries and $5 \cdot 2^{56}$ trial encryptions. In total, our cryptanalysis of OFB[CBC,CBC,CBC] needs 2^{32} chosen ciphertexts and $6 \cdot 2^{56}$ trial encryptions.

5.4 OFB[CBC,CBC^{-1}]

We are not aware of any strong attacks on the OFB[CBC,CBC^{-1}] mode.

We have a narrow-pipe attack that recovers the key with 2^{112} offline trial decryptions, 2^{32} chosen ciphertexts, and no memory; but because this result is so weak, we will refrain from describing it here.

We also have an attack that requires 2^{66} known-IV chosen-ciphertext queries, 2^{56} offline trial decryptions, and 2^{00} memory. This too is highly unrealistic, but we will sketch the attack here for completeness. We proceed much as in Section 5.2, with the additional complication that we must also force an internal feedback channel to match. Choose 2^{64} ciphertexts $C[i] = (c, c, c, c, c)$ for all i. We seek i, j such that $E_{K0}(IV0[i]) = IV0[j]$, which will ensure that the two OFB streams match up (out of phase by one block); we also require that $E_{K2}(C_0[i] \oplus E_{K0}(IV0[i]) \oplus IV2[i]) = IV2[j]$, which yields a collision (out of phase by one block) in the CBC^{-1} layer's internal feedback channel. These two conditions ensure that we can recognize such a pair i, j by the condition $P_{2...4}[i] = P_{1...3}[j]$. Furthermore, the birthday paradox predicts that we will encounter one such i, j; once we've recognized it, we can use the known IVs to recover $K0$ with a 2^{56} exhaustive keysearch. Then the rest of the key material can be obtained with a meet-in-the-middle search.

In short, we are unable to make much progress on the analysis of OFB[CBC, CBC^{-1}], and so we leave it as an open question for others to examine.

5.5 OFB[CBC,CBC,CBC^{-1}]

We present an attack that breaks the OFB[CBC,CBC,CBC^{-1}] mode with one chosen ciphertext query and 2^{112} work. This is an unrealistic attack, but it shows that this quadruple mode does not attain the strength one might ideally hope for in a quad mode, when chosen-IV queries are a threat.

First, we probe $IV2$ to isolate the effect of $K0, K1$. Let $P[0], C[0]$ be a known plaintext/ciphertext pair with known IVs $IVi[0]$. We construct a chosen ciphertext query $C[1], IVi[1]$ by taking $C[1] = C[0]$, letting $IVi[1] = IVi[0]$ for $i = 0, 1, 3$, and picking arbitrary $IV2[1]$ different from $IV2[0]$. Then we have the relation

$$IV2[0] \oplus IV2[1] = E_{K1}(P_0[0] \oplus IV1 \oplus E_{K0}(IV0)) \oplus E_{K1}(P_0[1] \oplus IV1 \oplus E_{K0}(IV0)),$$

which allows us to isolate the effect of $K0, K1$. Now we recover $K0, K1$ with a 2^{112} exhaustive keysearch.

Finally, once we've learned $K0, K1$, we can recover $K2, K3$ with a second exhaustive keysearch. (In fact, we could use the meet-in-the-middle attack on double-DES to recover $K2, K3$, but this will not reduce the total complexity of the full attack significantly.)

The total complexity of the attack is $3 \cdot 2^{112}$ offline trial encryptions and one chosen-ciphertext query. This shows that the quadruple mode OFB[CBC,CBC, CBC^{-1}] is no stronger than triple-DES-ECB (with outer chaining) against chosen-IV chosen-text key-recovery attacks, and so the fourth DES layer seems wasted.

It is interesting to note that the present attack does not apply to the triple mode OFB[CBC,CBC^{-1}], even though this might seem like a paradox at first glance. This leaves open the counter-intuitive possibility that the OFB[CBC, CBC^{-1}] triple mode might well be stronger than the OFB[CBC,CBC,CBC^{-1}] quadruple mode.

6 Discussion

All of our attacks have relied on the ability to control the IV afforded by our attack model. This raises the issue of whether it is possible to prevent these attacks with simple countermeasures. The answer seems to be mixed—yes, there are some simple countermeasures, but they have limitations. We survey some possible approaches here.

6.1 Countermeasures

Encrypt the IVs. One suggestion is to encrypt the IVs before transmission (with, say, triple-DES or quadruple-DES), rather than sending the IVs in the clear. This thwarts the attackers ability to choose the exact values of the IVs.

On the other hand, this approach has a weakness, when one encrypts each IV independently: the attacker can still reuse old IV values in new chosen ciphertexts. Some of our attacks (suitably modified) can be converted to work against this protocol, when (1) they only rely on the ability to force IVj (for some choices of j) to be the same for all chosen ciphertexts and (2) the actual value of IVj is irrelevant. As an illustration, we show that OFB$\rightarrow$CBC$\rightarrow$CBC is no safer with this protocol than before: take

$$IV2[0] = IV2[2] = a, \ IV2[1] = IV2[3] = b,$$
$$IV1[0] = IV1[1] = c, \ IV1[2] = IV1[3] = d$$

for some unknown a, b, c, d, and force IV0 to be constant; then we have the identity

$$D_{K2}(P_0[0]\oplus C_0[0])\oplus D_{K2}(P_0[1]\oplus C_0[1]) = D_{K2}(P_0[2]\oplus C_0[2])\oplus D_{K2}(P_0[3]\oplus C_0[3]),$$

which lets us recover $K2$ with 2^{56} work and four chosen-ciphertext queries, and $K1$ will fall soon thereafter. So encrypting the IVs is no guarantee of safety.

MAC the IVs. Another natural reaction is to simply insist that senders apply a MAC to the IVs which receivers must verify before decrypting. By protecting the integrity of the IVs, this stops chosen-IV attacks.

This approach still leaves the users open to known-IV attacks, when they exist. Some modes are susceptible to known-IV attacks; others may not be. See below for a few illustrations of this danger. In general, the known-IV attacks that we know of usually require more texts than their chosen-IV counterparts, so MACing the IVs may reduce the threat level.

Another argument against this approach is based on engineering considerations. Now we have a new protocol, which is more complicated, and which introduces an whole new primitive to the mix. We end up replacing one failure mode with two: if either the encryption algorithm or the MAC is compromised, then the message keys must be recovered. It is perhaps imprudent to rely on the security of the MAC to protect confidentiality: just as conservative cryptographic design calls for independent session keys for authentication and confidentiality algorithms (to limit the impact of the compromise of any one algorithm), we would do well to avoid linking the security of our MAC with the security of our encryption algorithm.

We could perhaps first encrypt and then authenticate the IVs, to stop both known-IV attacks and attacks which attempt to replay old IVs. However, adding this much complexity to the system may begin to test the limits of one's comfort zone; at the least, more analysis seems needed.

Add redundancy to IVs. Coppersmith et al. [CJM97] have applied a novel countermeasure to stop a chosen-IV attack we discovered on their original CBCM proposal. They limit the possible values for each IV to a small subset: one IV is fixed at 0, and the other 64-bit IV has 44 of its bits fixed at 0. This redundancy limits the ability of an attacker to control the IV, and counters the attack we found.

This is a very clever trick, but it only seems useful in certain cases. Adding redundancy to IVs will not stop most of the attacks listed in this paper. Fixing certain IVs at 0 would deter many of the attacks, but it seems that such a measure could adversely affect security in other ways for a number of the modes proposed by Biham. It seems possible that this countermeasure may introduce as many problems as it solves, and so we are wary of depending upon it for security.

General notes. This is by no means an exhaustive list of available remedies. Nonetheless, we can make some comments that seem broadly applicable.

Many of the obvious countermeasures have not yet been subjected to concerted analysis, and we have attempted to show that there are some pitfalls to watch out for.

Still, it seems likely that techniques can be developed to protect the IVs quite thoroughly, for some (if not all) of Biham's modes. Of course, one has to use them, and use them with extreme care; it is details like this that plague real implementations. The central question is this: will such countermeasures prove cost-effective? or will these advanced modes suffocate under the weight of the extra precautions they require? More research is needed.

6.2   Extending our results to other security models

Stopping chosen-IV attacks is not enough, if the basic ideas behind those attacks can be leveraged into a sharper attack. To illustrate the point, we note that a number of our chosen-IV attacks can be converted to known-IV attacks. Usually, this increases the number of texts needed to mount the attack. These known-IV attacks are invariably more difficult to describe, and—perhaps—more difficult to discover, than their chosen-IV counterparts.

For example, all of Section 4's divide-and-conquer attacks on triple modes can be modified to work with 2^{64} known-IV chosen plaintexts. Use a birthday attack to find two texts with matching values of $IV0, IV1$; then that pair lets you probe $IV2$ and thus recover $K2$, and $K1$ is recovered similarly.

The number of known-IV queries can be reduced by using meet-in-the-middle techniques. For instance, one can break OFB[CFB,CFB] with 2^{32} known-IV chosen plaintexts. Use a birthday attack to find two texts $C[i], C[j]$ with $IV0[i] = IV0[j]$. We learn that

$$E_{K1}(IV1[i]) \oplus E_{K1}(IV1[j]) = E_{K2}(IV2[i]) \oplus E_{K2}(IV2[j]),$$

which lets us recover $K1, K2$ with complexity 2^{56} by a standard meet-in-the-middle attack. (The straightforward implementation of that attack also requires 2^{56} space, though the space requirements can be dramatically reduced by using parallel collision search algorithms [OW96].)

Applying these techniques, we can convert our chosen-IV attacks to attacks which need 2^{32} known-IV chosen texts and $O(2^{56})$ work, for all the triple modes in Section 4, as well as for OFB[CBC,CBC]. OFB[CFB,CFB,CFB] also falls with 2^{32} known-IV chosen texts, due to a piece of blind luck: $K0$ cannot affect

$P_0 \oplus C_0$. Similarly, we can obtain attacks requiring 2^{64} known-IV chosen texts and $O(2^{56})$ work against $CBF|CFB|CFB^{-1}|CFB^{-1}$ and $OFB[CBC,CBC,CBC]$. Our original attack on $OFB[CBC,CBC,CBC^{-1}]$, needing one chosen-IV query and 2^{112} work, can be extended to work with 2^{64} known-IV chosen texts and 2^{112} work. Finally, we can get an attack on $CBC|CBC|CBC^{-1}|CBC^{-1}$ that recovers $K0$ with 2^{64} known-IV chosen texts and 2^{56} trial encryptions, and from there can break the whole quad mode with another $6 \cdot 2^{56}$ trial encryptions and 2^{32} known-IV chosen texts. (These estimates are rough, and the details of the analysis are unchecked.)

Incidentally, these results would disprove Biham's conjectures for several of his modes, *if* we make the major concession of accepting the validity of known-IV attacks. (Additional mild concessions are required in some cases.) For instance, it would show that $CBC|CBC|CBC^{-1}|CBC^{-1}$, $CFB|CFB|CFB^{-1}|CFB^{-1}$, and $OFB[CBC,CBC,CBC]$ are not more secure than all triple modes, if we also assume that there is some triple mode which resists all attacks of complexities less than 2^{64}; the latter assumption is quite reasonable, as triple-ECB-DES with outer chaining is an excellent candidate for one such mode. $OFB[CFB,CFB,CFB]$ fall with complexity 2^{56}, which would disprove (if we accept known-IV attacks) Biham's conjecture that it has a security factor of at least 2^{128}. The triple modes (except $OFB[CBC,CBC^{-1}]$) fall to attacks with 2^{56} complexity (if we accept known-IV attacks), which is less than the conjectured 2^{112} security factor.

These results do not actually refute Biham's conjectured security factors. However, the existence of known-IV attacks of lower-than-expected complexity brings us a step closer to understanding the true security level of these modes.

7   Conclusions

This paper has presented new attacks on all but one of Biham's proposed modes. These attacks rely on the ability to control the IVs, and therefore require quite powerful adversaries which may or may not be a concern in practice. Of all the proposed modes, $OFB[CBC,CBC^{-1}]$ seems to have the best resistance to the chosen-IV attacks which we know of.

These results illustrate the difficulty of building secure modes that contain inner chaining. The danger of internal feedback mechanisms is that the cryptanalyst may be able to probe the internals of the multiple mode of operation by using chosen-text queries; in many cases, this allows the cryptanalyst to isolate the effect of part of the keying material.

This work describes a new failure mode for such systems when the adversary can gain control of IV values. This presents additional evidence for the fragility of constructions based on internal feedback.

We believe that it would be prudent for conservative cryptographic engineers to avoid multiple modes with inner chaining until they are better-understood by researchers. For now, triple-DES-ECB seems to provide more robust—or, at least, better-understood—security.

8   Acknowledgements

The descriptive term "narrow pipe" is due to John Kelsey. The author is deeply grateful to Eli Biham for his comments, which have greatly improved the quality of this work.

References

Bih94a.   E. Biham, "On Modes of Operation," *Fast Software Encryption '93*, LNCS 809, Springer-Verlag, 1994.

Bih94b.   E. Biham, "Cryptanalysis of Multiple Modes of Operation," *ASIA-CRYPT'94*, LNCS 917, Springer-Verlag, 1994.

Bih96.   E. Biham, "Cryptanalysis of Triple-Modes of Operation," Technion technical report CS 885, 1996.

CJM97.   D. Coppersmith, D.B. Johnson, and S.M. Matyas, "Triple DES Cipher Block Chaining with Output Feedback Masking," *IBM Journal of Research and Development*, vol 40, no 2, 1996.

OW96.   P.C. van Oorschot and M.J. Wiener, "Improving implementable meet-in-the-middle attacks by orders of magnitude," *CRYPTO'96*, pages 228-236, Springer-Verlag, 1996.

Differential Cryptanalysis
of the ICE Encryption Algorithm

Bart Van Rompay[1*], Lars R. Knudsen[2**], and Vincent Rijmen[1***]

[1] K.U. Leuven, ESAT-COSIC, K. Mercierlaan 94, B-3001 Heverlee, Belgium
{bart.vanrompay,vincent.rijmen}@esat.kuleuven.ac.be
[2] Dept. of Informatics, University of Bergen, Hi-techcenter, N-5020 Bergen, Norway
larsr@ii.uib.no

Abstract. ICE is a 64-bit block cipher presented at the Fast Software Encryption Workshop in January 1997. It introduced the concept of a keyed permutation to improve the resistance against differential and linear cryptanalysis. In this paper we will show however that we can use low Hamming weighted differences to perform a practical, key dependent, differential attack on ICE. The main conclusion is that the keyed permutation is not as effective as it was conjectured to be.

1 The ICE Algorithm

ICE [7], which stands for *Information Concealment Engine*, is a 64-bit Feistel block cipher with a structure similar to DES, the *Data Encryption Standard* [5]. The standard ICE algorithm takes a 64-bit key and uses 16 subkeys in 16 rounds. There is a fast variant, Thin-ICE, which uses 8 rounds with a 64-bit key, and there are open-ended variants ICE-n which use $16n$ rounds and $64n$-bit keys.

Description of the round function The ICE round function F maps 32-bit inputs to 32-bit outputs, using a 60-bit subkey. First the 32-bit input is expanded to a 40-bit value. A 20-bit subkey performs a keyed permutation and a 40-bit subkey is exored to the resulting value. Finally it uses four 10 to 8-bit S-boxes and a permutation to obtain the 32-bit result of the round function.

Notation In this paper bits are numbered from right to left, starting at bit zero. So the rightmost bit of an n-bit value V is V_0, while the leftmost bit is V_{n-1}. The four S-boxes used in the round function are labeled $S0$, $S1$, $S2$ and $S3$.

---

[*] Sponsored by the Timesec project of the Federal Office for Scientific, Technical and Cultural Affairs (OSTC), Belgium.
[**] This author's work was done during his stay in Leuven as a postdoctoral fellow of the Research Council of the K.U. Leuven.
[***] F.W.O. research assistant, sponsored by the Fund for Scientific Research, Flanders — Belgium.

S. Vaudenay (Ed.): Fast Software Encryption – FSE'98, LNCS 1372, pp. 270–283, 1998.

The expansion function The 32-bit input I to the F function is expanded to four 10-bit values $E0, E1, E2, E3$.

The keyed permutation A 20-bit subkey performs a keyed permutation on the expanded 40-bit text, swapping bits between $E0$ and $E2$, and between $E1$ and $E3$. If permutation key bit $10 + i(i < 10)$ is set, bit i of $E0$ and $E2$ will be swapped. If permutation key bit $i(i < 10)$ is set, bit i of $E1$ and $E3$ will be swapped.

The exor operation The 40-bit result from the keyed permutation is exored with a 40-bit subkey.

The S-boxes ICE uses four 10 to 8-bit S-boxes to map the 40-bit value to a 32-bit value. These S-boxes are similar in structure to those used in LOKI [4,3] in their use of Galois Field exponentiation. From the 10-bit input X, we concatenate X_9 and X_0 to form the row selector R. Bits $X_8...X_1$ form the column selector C. For each row there is a XOR offset value O_R, and a Galois Field prime (irreducible polynomial) P_R. The 8-bit output of an S-box for an input X is given by $(C \oplus O_R)^7 \mod P_R$, under Galois Field arithmetic.

The permutation function Finally the four 8-bit S-box outputs are combined via a P-box into the 32-bit output of the F function.

The key schedule The ICE key scheduling algorithm maps a 64-bit key to 16 60-bit subkeys. Each subkey bit is dependent on only one key bit. The Thin-ICE key schedule is simply the first eight rounds of the standard ICE key schedule. The ICE-n key schedules build on the ICE key schedule and map a 64n-bit key to 16n 60-bit subkeys.

2 Differential Cryptanalysis

Differential cryptanalysis was introduced by Biham and Shamir [2], and can be used to perform chosen plaintext attacks. The basic idea is that two chosen plaintexts with a certain difference $P' = P_1 \oplus P_2$ can encipher to two ciphertexts such that $C' = C_1 \oplus C_2$ has a specific value with non-negligible probability, and such a characteristic (P', C') is useful in deriving certain bits of the key. The heart of differential attacks is the finding and the use of characteristics with high probabilities.

The analysis of ICE in [7] considers only symmetric differences, which have equal left and right 16-bit halves of the 32-bit input to the F function. This is claimed to be the best strategy since they are the only differences not affected by the keyed permutation. As a consequence the attacker has to target at least two S-boxes at a time and the probabilities are too low to be used in a realistic attack.

The approach used in our attack is to use differences with a low Hamming weight (as low as possible). Whether they will be affected by the keyed permutation depends on the values of only a few key bits. The differences used address only one S-box in the round function. In this way we can find characteristics with a probability high enough to (theoretically) recover ICE keys for the algorithm reduced to 15 rounds, in time less than the expected cost for exhaustive search. The complication is that the attack becomes key dependent.

3   Differential Characteristics for ICE

As explained in the previous section we will focus on low Hamming weighted differences that address only one S-box in the round function. It is not possible to build a 2-round iterative characteristic like the one used for the analysis of DES [2], with a difference that addresses only one S-box (using only the middle 6 bits out of the 10 input bits to that S-box so that it is not affected by the expansion in the round function). We can however build 3-round iterative characteristics of the form specified in Figure 1.

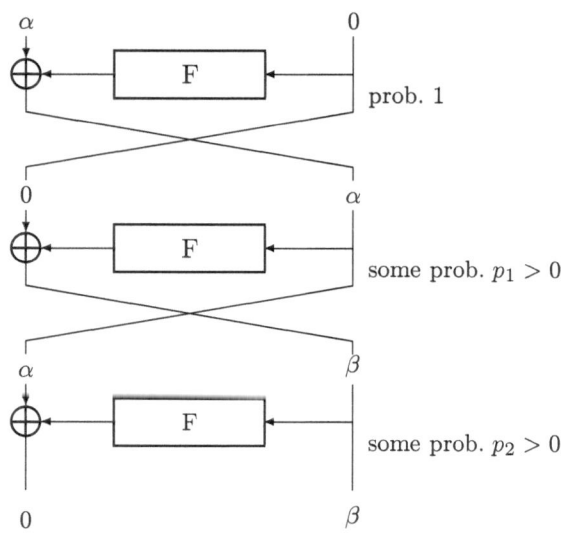

Fig. 1. 3-round iterative characteristic.

Because we restrict the differences α and β to one S-box, they can have a Hamming weight of no more than 4 each, since the 8-bit output of an S-box delivers, after the permutation and the key dependent permutation in the next round, up to 4 bits to an S-box in the next application of the round function.

The characteristic will be valid if α and β are not affected by the keyed permutation in the corresponding rounds. This happens if the permutation key bits used in the bit positions that are set in α and β, are equal to zero (so the difference will not be permuted from the left 20-bit half of the expanded text to the right or vice versa).

There are also characteristics that are valid only if certain permutation key bits are equal to one (corresponding to the bits set in α or β or both). In general we call these *conditional* characteristics (cf. the attack on Lucifer [1]), which have a certain probability with respect to a subset of the key space. Their usage is advisable when they improve the probability over the best probability of a non-conditional characteristic by a factor higher than the inverse of the *key fraction* (the ratio between the size of the subset and the size of the key space), especially if several such characteristics can efficiently share the same structure of chosen plaintexts.

If we consider only differences with Hamming weight one there is a total of 105 conditional characteristics with

$$2^{-13} \geq p_1 \cdot p_2 \geq 2^{-18}.$$

Table 1 lists some of the differences of Hamming weight 1, which can be used to construct a 3-round characteristic with probability $p_1 \cdot p_2 \geq 2^{-15}$, together with the corresponding probabilities. By interchanging the values of α and β we get twice the number of characteristics (except for the fourth entry in the table which has $\alpha = \beta$).

α	β	p_1 (-$\log_2$)	p_2 (-$\log_2$)	round 2	round 3
18	28	6	7	0	0
26	29	6	7.4	1	1
31	26	8	6	1	0
7	7	7	7	0	0
3	10	7.4	7.4	0	0
27	3	7.4	7.4	1	1
4	22	7.4	7.4	1	0
18	30	6	9	1	0
15	23	7	8	0	0
22	7	7	8	1	1

Table 1. Differential characteristics with Hamming weight 1 and $p_1 \cdot p_2 \geq 2^{-15}$. The differences α and β are noted by the bit position in the 32-bit value that is set at one. We also list the required value for the permutation key bits corresponding to α and β in the second and third round of the characteristic.

4   Differential Attack on ICE

4.1   Attack on ICE with 6 rounds

We use the best 3-round characteristic, with $\alpha = 28$ and $\beta = 18$, followed by a trivial round with probability 1. The probability for this 4-round characteristic is $p_1 \cdot p_2 = 2^{-7} \cdot 2^{-6} = 2^{-13}$. It is valid if the bits set in the differences α and β are not permuted in the respective rounds (rounds 2 and 3). This can be translated to the following conditions:

- bit $14 = 0$ for the permutation subkey in round 2.
- bit $2 = 0$ for the permutation subkey in round 3.

Examination of the key scheduling algorithm shows the corresponding condition for the 64-bit user key:

- bit $20 = 1$ and bit $12 = 0$.

Figure 2 shows the 6-round algorithm and the 4-round characteristic. The expected input difference to the round function in round 5 is β, the expected output difference equals the difference in the right half of the ciphertext. This allows us to check if an arbitrary encrypted pair (with the right difference in the plaintext) is a right pair for the characteristic. The difference $\beta = 18$ delivers an input difference to S-box $S1$ or $S3$, depending on the value of the corresponding permutation key bit. So the output differences from S-boxes $S0$ and $S2$ have to be zero, as well as the output difference from either $S1$ or $S3$.

This corresponds to checking the values of $24 - 1 = 23$ bits. So a wrong pair has a probability 2^{-23} of surviving this filtering process. The probability of generating a right pair is much higher (2^{-13}), so when a pair survives the filtering, with a high probability it is a right pair.

For such a right pair we know the inputs and the difference at the output ($\Delta C_L \oplus \beta$) of the last round, and for all possible subkeys we can check whether they correspond. Repeat this for about four right pairs (we need to generate about $4 \cdot 2^{13} = 2^{15}$ pairs of plaintexts), the correct subkey will be suggested each time and can be distinguished from other suggested subkeys.

The signal-to-noise-ratio (the ratio of the number of times the correct key is suggested and the number of times an arbitrary key is suggested) for this attack can be calculated with the method described in [2]. It depends on the number of plaintext pairs m, the probability of the characteristic p, the number k of simultaneous key bits that we count on, the average count a per analysed pair, and the fraction b of the analysed pairs among all the pairs.

$$S/N = \frac{m \cdot p}{m \cdot a \cdot b / 2^k}.$$

In this case we have $m = 2^{15}$ and $p = 2^{-13}$. When concentrating on one S-box we are counting on $k = 20$ key bits (10 used for the permutation and 10 for the exor operation). The average count a equals 2^{12}, since we count on 2^{20}

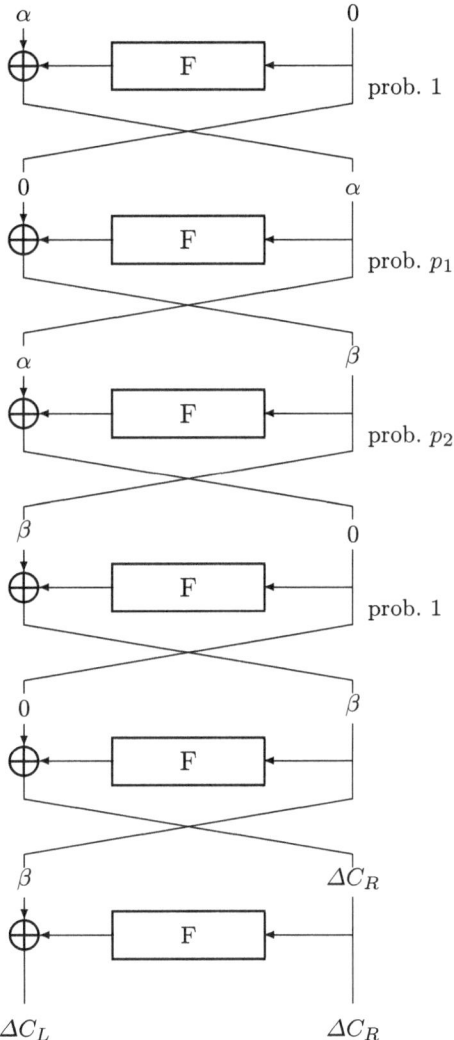

Fig. 2. The characteristic for an attack on 6 rounds.

subkeys and check an 8-bit value (difference at the output of the S-box). The fraction b (filtering) equals 2^{-23}. Hence the signal-to-noise-ratio is:

$$S/N = \frac{2^{15} \cdot 2^{-13}}{2^{15} \cdot 2^{12} \cdot 2^{-23}/2^{20}} = \frac{2^2}{2^{-16}} = 2^{18},$$

and similarly for the other three S-boxes. However $S0$ and $S2$, as well as $S1$ and $S3$, use the same permutation key bits, which we have to determine only once. For the second S-box which uses these permutation key bits we can count on just the 10 exor key bits. In this way we determine all 60 bits of the subkey. The remaining 4 bits of the user key can easily be found by exhaustive search.

4.2 Attack on ICE with 8 rounds (Thin-ICE)

We can extend the previous attack in a straightforward manner, using a 6-round characteristic with a probability $p_1 \cdot p_2 \cdot p_2 \cdot p_1 = 2^{-7} \cdot 2^{-6} \cdot 2^{-6} \cdot 2^{-7} = 2^{-26}$. The attack can be improved however by inserting a round before the first round of the characteristic without reducing the probability, like in the attack on DES [2]. The assumed evolution of differences (during the encryption of a right pair) is shown in Figure 3. In the first round the difference α at the input of the round function is an input difference to S-box $S0$ or $S2$, depending on the value of the corresponding permutation key bit. We guess this bit and repeat the attack if we have guessed wrong. We compensate the difference at the output of the round function of round 1 by using a structure of 2^9 plaintexts:

$$P_i = P \oplus (v_i, 0), \bar{P}_i = P \oplus (v_i, 0) \oplus (0, \alpha) \qquad \text{for } 0 \le i < 2^8,$$

with v_i denoting all the possibilities for the 8 bits that are exored with the output bits from $S0$ or $S2$; (l, r) denotes the left and right 32-bit halves of a 64-bit text.

The probability for the characteristic is $p_1 \cdot p_2 \cdot p_2 = 2^{-7} \cdot 2^{-6} \cdot 2^{-6} = 2^{-19}$. The conditions for it to be valid are:

- bit 14 = 0 for the permutation subkey in round 3.
- bit 2 = 0 for the permutation subkey in round 4.
- bit 2 = 0 for the permutation subkey in round 6.

The corresponding condition for the 64-bit user key is:

- bit 3 = 0, bit 59 = 1 and bit 25 = 1.

In the defined structure there are 2^{16} pairs, of which 2^8 satisfy the first round. These can be isolated in 2^8 time as follows. Since the expected output differences from S-boxes $S1$ and $S3$ in round 7 are zero, we sort the texts according to the values of the corresponding bits in the right half of the ciphertext and find the matching values. We filter these further by S-box $S0$ or $S2$ (like in the 6 round attack), and can expect $2^8 \cdot 2^{-19} = 2^{-11}$ right pairs in a structure. By using 2^{13} structures 4 right pairs are expected. In total however there are $2^{16} \cdot 2^{13} = 2^{29}$ pairs. After filtering for 23 bits we expect there will remain $2^6 = 64$ wrong

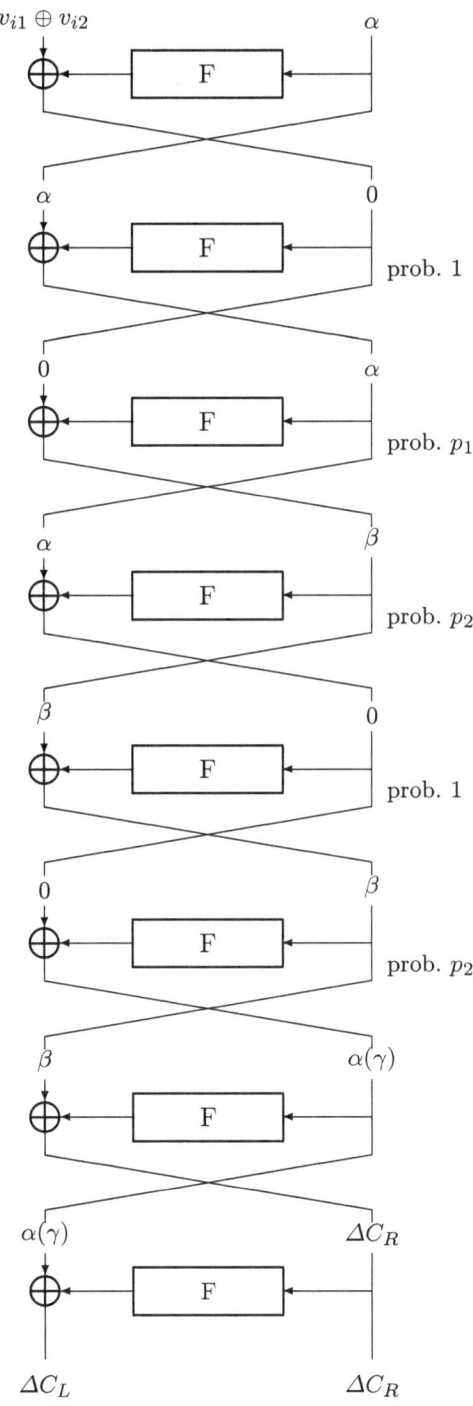

Fig. 3. The characteristic for an attack on 8 rounds.

pairs. For this mixture of right and wrong pairs we try all possible subkeys, concentrating on one S-box at a time.

In the calculation of the signal-to-noise-ratio for this attack there is an extra factor 2^{-8}, imposed by the first round structure ($m = 2^{13} \cdot 2^{16}$, but only $2^{13} \cdot 2^8$ pairs satisfy the first round):

$$S/N = \frac{2^{13} \cdot 2^8 \cdot 2^{-19}}{2^{13} \cdot 2^{16} \cdot 2^{12} \cdot 2^{-23}/2^{20}} = \frac{2^2}{2^{-2}} = 2^4.$$

We can easily extend this attack to make it valid for twice as many keys. Just guess the value of bit 2 of the permutation subkey in round 6. If it equals 1 instead of 0, the round function in that round delivers an output exor different from $\alpha = 28$. With probability 2^{-6} this output exor will be $\gamma = 30$ and we can perform the attack in a similar way. The condition for the 64-bit user key is:

− bit 3 = 0 and bit 59 = 1.

Alternatively we can do an eight round attack using the characteristic with $\alpha = 31$ and $\beta = 26$. According to Table 1 the probability of this characteristic is $p_1 \cdot p_2 \cdot p_2 = 2^{-8} \cdot 2^{-6} \cdot 2^{-6} = 2^{-20}$. The conditions for the permutation key bits translate to the following condition for the user key:

− bit 48 = 1, bit 18 = 1 and bit 48 = 1.

So the permutation key bit in round 6 doesn't impose an extra condition on the user key, and we don't have to guess this bit when using the characteristic with $\alpha = 31$ and $\beta = 26$.

4.3   Practical aspects of the analysis

The attacks for the 6-round version and the 8-round version (Thin-ICE) have been implemented and, on the average, work as predicted. However, using low Hamming weighted differences causes some complications. The input difference to the last round is caused by the output difference from the previous round. That output difference is caused by just one S-box and has a Hamming weight of no more than 8, with an average of 4.

Each S-box in the last round receives 2 bits from these 8 bits. Because the keyed permutation swaps bits between the 'partner' S-boxes $S0 - S2$ and $S1 - S3$, an S-box will finally receive between 0 and 4 from these bits at its input, depending on the value of the permutation subkey. Only these bits can cause an input difference. If S-box $S0$ or $S1$ gets k bits, then respectively $S2$ or $S3$ will get $4 - k$ bits. If a particular S-box gets k bits with a possible difference, the probability to get input difference zero is approximately 2^{-k}. In Table 2 we list the possible values for that probability, and the fraction of subkeys for which it holds.

If the input difference to an S-box is zero, all of the guesses for the permutation subkey that cause a zero difference will be counted, as will all possibilities

probability	fraction of subkeys
1	$1/2^4$
2^{-1}	$4/2^4$
2^{-2}	$6/2^4$
2^{-3}	$4/2^4$
2^{-4}	$1/2^4$

Table 2. Probabilities to get a zero input difference to an S-box.

for the exor subkey. Therefore the attack is less efficient and we have to look for some more right pairs for the characteristic (in practice between 4 and 8), hence use more plaintexts.

For a fraction 2^{-4} of the keys the input difference to the S-box will always be zero, so we can determine only some of the permutation key bits and none of the exor key bits. But then the partner S-box has a probability for zero input difference of only 2^{-4}. We determine the permutation key bits via this S-box, and the 10 exor key bits that we cannot determine can be looked for exhaustively after the differential attack (together with the 4 bits of the user key that are not used in the 60-bit subkey of the last round).

It is possible to exploit the occasions of zero input differences to improve our attack. If the input difference to an S-box in the last round is zero, the output difference is zero as well. In that case we know the corresponding difference at the input to the round function in the second to last round and we can check if its value corresponds to the value that is required for the characteristic. In this way we can do some extra filtering, which is important for the 8 round attack where we expect to get 64 wrong pairs. It will increase the signal-to-noise-ratio and reduce the number of required plaintexts.

4.4 Extending the attack

The 3-round iterative characteristic can be extended in a straightforward way to attack the ICE algorithm with an arbitrary number of rounds. But if the number of rounds exceeds 9, the signal-to-noise-ratio will drop below one, making the attack impossible (the last remark of previous section allows only a slight improvement by extra filtering). There are however several ways to improve the signal-to-noise-ratio.

Counting on more key bits When a pair survives the filtering (and is assumed to be a right pair, following the characteristic), we know the inputs and the difference at the output of the last round and check whether they correspond. In the basic attack we concentrate on one S-box and count on 20 subkey bits (10 used for the permutation and 10 for exoring).

Instead we can consider two partner S-boxes ($S0$ and $S2$, or $S1$ and $S3$) at the same time. They share 10 permutation key bits and both use 10 exor key bits. This allows us to count on 30 key bits, and results in an improvement of the signal-to-noise-ratio by a factor of 2^8 because we check the values of 8 more bits at the output of the second S-box (in the calculation of S/N we have $2^k = 2^{30}$ and $a = 2^{30}/2^{16} = 2^{14}$). In theory further improvements (by a factor of 2^{16} or 2^{24}) are possible by considering respectively three or four S-boxes (50 or 60 key bits).

Checking differences in the first round When a pair is assumed to follow the characteristic we can also check subkey bits in the first round of the algorithm. In this first round we use a special structure (cf. the attack on 8 rounds) and guess the value of the permutation key bit corresponding with the difference of Hamming weight one. Hence we can count on the 10 exor key bits of the S-box where the difference of Hamming weight 1 is located.

Moreover, due to the key schedule some of these subkey bits in the first round represent the same user key bits as some of the subkey bits in the last round of the algorithm. This allows us to improve the signal-to-noise-ratio by a factor of 2^8 by counting on just a few more key bits. Note also that some of the key bits that we count on are already known, because of the condition on the user key for the characteristic.

Filtering in the last round The most important improvement can be made by adapting the characteristic. In the previous attacks we used the characteristics with the highest probabilities. The resulting attacks are called 2R-attacks (cf. Biham and Shamir [2]), because they don't make assumptions for the last two rounds of the algorithm.

Instead we can perform 1R-attacks, using a characteristic up to the last to one round. An example of the last rounds for such a characteristic is shown in Figure 4.

Although the probability of such a characteristic is generally lower than for a 2R-attack, it is useful because it allows much more filtering and an overall reduction of the signal-to-noise-ratio. In the last round we can check if the difference at the output of the round function (ΔC_L in Figure 1) is possible, like we did in the last to one round in the previous attacks. This corresponds to checking the values of 23 bits. But we can also check the difference in the right half of the ciphertext ($\Delta C_R = \alpha$ in Figure 4), thus filter for 32 more bits.

This results in an improvement of the signal-to-noise-ratio by a factor of $2^{32} \cdot p_c$, where p_c represents the factor by which the probability of the characteristic is reduced when we perform a 1R-attack instead of a 2R-attack.

Results for an arbitrary number of rounds Table 3 lists for each number of rounds: the probability of the characteristic, the required number of chosen plaintexts (assuming 4 right pairs for the characteristic are sufficient, and that we need both guesses for the permutation key bit in the first round structure), the number of subkey bits counted on (excluding key bits in the first round

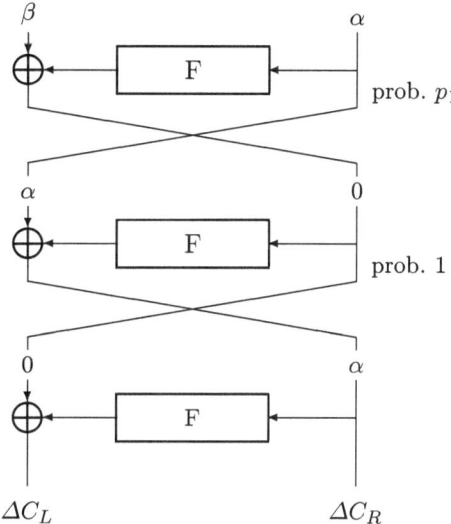

Fig. 4. Characteristic (last rounds) for a 1R-attack.

because of the overlap), the signal-to-noise-ratio and the fraction of keys that can be found with the attack.

When the number of rounds exceeds 9 we list two different attacks: a 2R-attack (where the signal-to-noise-ratio is improved by counting on more key bits and checking the difference in the first round), and a 1R-attack (which has a lower probability and requires more plaintexts). When the number of rounds is a multiple of 3, we have listed only the 1R-attack because it's probability is the same as for the 2R-attack ($p_c = 1$). In the other cases we have $p_c = p_1 = 2^{-7}$ or $p_c = p_2 = 2^{-6}$.

Note that the key fraction is lower for a 1R-attack, because the characteristic imposes more conditions on the user key (except when the number of rounds is a multiple of 3).

The table shows that the differential analysis works for up to 15 rounds of ICE: for this attack 8 key bits are fixed and an exhaustive search would have 2^{56} possibilities, our attack requires at most 2^{56} plaintexts.

4.5   Key dependency of the attacks

We described the previous attacks using the best conditional characteristic. In Table 3 we have listed for how many keys this works. For Thin-ICE the attack works for a fraction 2^{-2} of the keys. For other keys however we can use a different characteristic with a lower probability. We can use several characteristics with the same set of plaintexts, if we use a special structure for these plaintexts. For two characteristics this is a quartet structure (like the one used for the analysis of DES in [2]), for three an octet structure and so on. The number of plaintexts

# rounds	probability †	# plaintexts †	# counters †	S/N †	key fraction †
4	0	4	20	23	all
5	-6	10	20	17	all
6	-13	16	20	18	-2
7	-13	17	20	10	-2
8	-19	23	20	4	-2
9	-26	29	20	5	-4
10	-26	30	20	5	-4
10	-32	36	20	23	-5
11	-32	36	20	-1	-4
11	-39	42	20	24	-6
12	-39	43	20	16	-6
13	-39	43	30	0	-6
13	-45	49	20	10	-7
14	-45	49	50	2	-6
14	-52	55	20	11	-8
15	-52	56	20	3	-8
16	-52	56	60	3	-8
16	-58	62	30	5	-9

Table 3. Differential analysis for an arbitrary number of rounds. († $\log_2$)

we need depends on the characteristic with the lowest probability. If we want to be able to determine as many possible keys with as few possible plaintexts we use the characteristics with the highest probabilities. Table 4 shows the evolution of these numbers for the Thin-ICE algorithm.

# characteristics	# plaintexts ($\log_2$)	key fraction
1	23	25%
3	24	63%
5	25	81%
6	26	88%
8	27	95%

Table 4. For Thin-ICE (8 rounds) : the number of characteristics and plaintexts required versus the fraction of keys that can be found.

5   Conclusion

We have described attacks on the ICE algorithm using differential analysis. The main conclusion of this paper is that keyed permutation does not prevent differential cryptanalysis. Although the analysis is more complicated and becomes key dependent, in our opinion the intention of the design has not been reached.

The best 3-round iterative characteristic that can be used in our attack has a probability of 2^{-13}, which is higher than the probability of 2^{-16} of the best 3-round characteristic for LOKI'91 [6] (a similar block cipher that makes use of four identical 12 to 8-bit S-boxes).

We have also demonstrated a practical attack on the lightweight version Thin-ICE. In its basic form it finds the secret key in 25% of the cases using 2^{23} chosen plaintexts, and in 95% of the cases using 2^{27} plaintexts. The optimal characteristic for our attack on Thin-ICE has a probability of 2^{-19} which is much higher than the probability of the optimal characteristic based on a symmetric input difference, which was shown to be 2^{-56} in [7].

References

1. I. Ben-Aroya and E. Biham, "Differential Cryptanalysis of Lucifer," *Advances in Cryptology – Crypto '93 Proceedings, LNCS 773*, D. Stinson, Ed., Springer-Verlag, 1994, pp. 187–199.
2. E. Biham and A. Shamir, *Differential Cryptanalysis of the Data Encryption Standard*, Springer-Verlag, 1993.
3. L. Brown, M. Kwan, J. Pieprzyk and J. Seberry, "Improving resistance to differential cryptanalysis and the redesign of LOKI," *Advances in Cryptology – AsiaCrypt '91 Proceedings, LNCS 739*, H. Imai, R. Rivest, and T. Matsumoto, Eds., Springer-Verlag, 1993, pp. 36–50.
4. L. Brown, J. Pieprzyk and J. Seberry, "LOKI: A Cryptographic Primitive for Authentication and Secrecy Applications," *Advances in Cryptology – AusCrypt '90 Proceedings, LNCS 453*, J. Seberry and J. Pieprzyk, Eds., Springer-Verlag, 1990, pp. 229–236.
5. FIPS 46, *Data Encryption Standard*, Federal Information Processing Standard (FIPS), Publication 46, National Bureau of Standards, U.S. Department of Commerce, Washington D.C., January 1977.
6. L. Knudsen, "Cryptanalysis of LOKI'91," *Advances in Cryptology – AusCrypt'92 Proceedings, LNCS 718*, J. Seberry and Y. Zheng, Eds., Springer-Verlag, 1993, pp. 196–208.
7. M. Kwan, "The Design of the ICE Encryption Algorithm," *Proceedings of the 4th Workshop on Fast Software Encryption, Haifa, Israel, LNCS 1267*, E. Biham, Ed., Springer-Verlag, 1997, pp. 69–82.

The First Two Rounds of MD4 are Not One-Way

Extended Abstract

Hans Dobbertin

German Information Security Agency
P. O. Box 20 03 63
D-53133 Bonn

dobbertin@skom.rhein.de

In [1] it was shown that there are very effective attacks leading to collisions for the hash function MD4 designed by R. Rivest [3]. A summary of the status of hash functions of the MD4-family with respect to collision-resistance can be found in [2] and [4]. However, attacking the one-wayness of a hash function is a much more demanding challenge, and in case of success it has much more devastating consequences. No result along this line is known for MD4 and its successors. Therefore it is worth to explore how the recently developed new analytic methods for finding collisions can be applied to construct preimages or second preimages. As a first step, we state here the following partial result:

Denote by $MD4^{[12]}$ the reduced version of MD4, where the third round of its underlying three-round compression function is cancelled, but everything else of its specification is kept (e.g. initial value, padding rule).

$$\boxed{MD4^{[12]} \ \textit{is not one-way.}}$$

It takes less than one hour to find preimages on a PC. Second preimages take a few minutes or even less than a millisecond if the initial value is free.

Example. Assume we want to find a preimage of

$$V = 0x00000000 \ \ 0x00000000 \ \ 0x00000000 \ \ 0x00000000.$$

We have constructed the following message $M = M_0||M_1||...||M_{28}$, which is hashed to this value V by $MD4^{[12]}$:

$M_0 = 0xB5B6AC17$	$M_8 = 0xFF9405C3$	$M_{16} = 0x814F4825$	$M_{24} = 0x847064AD$
$M_1 = 0xB5B6AC17$	$M_9 = 0xFF9405C3$	$M_{17} = 0x814F4825$	$M_{25} = 0x05DDD0F5$
$M_2 = 0xB5B6AC17$	$M_{10} = 0xFF9405C3$	$M_{18} = 0x814F4825$	$M_{26} = 0xD462FA71$
$M_3 = 0x85574A58$	$M_{11} = 0xC26EA1D5$	$M_{19} = 0x814F4825$	$M_{27} = 0x56A79DEC$
$M_4 = 0x4353212D$	$M_{12} = 0x015CB5D0$	$M_{20} = 0x9919C508$	$M_{28} = 0x00000080$
$M_5 = 0x4353212D$	$M_{13} = 0x81BBD193$	$M_{21} = 0x9919C508$	
$M_6 = 0x4353212D$	$M_{14} = 0x1DEF9763$	$M_{22} = 0x9919C508$	
$M_7 = 0x3E30333E$	$M_{15} = 0xADE9028B$	$M_{23} = 0x2FD7B0F9$	

We anticipate that a similar attack works for the last two rounds of MD4.

S. Vaudenay (Ed.): Fast Software Encryption – FSE'98, LNCS 1372, pp. 284–292, 1998.
© Springer-Verlag Berlin Heidelberg 1998

Technical Details for Checking the Example. According to the padding rule, before processing, a message has to be extended by a bit string

$$P = 100...0 \text{ (bin)} \, \| \, \ell,$$

where ℓ is the 64-bit representation of the bit-length of the (unextended) message. In P, between 1 on the left and ℓ on the right side, the minimal number of zeros is placed such that the bit-length of the extended message becomes a multiple of $512 = 16 \times 32$ (to allow an iterative application of the compression function, which takes 16 words as input). The above M has bit-length 29×32. This means that

$$\ell = 00000000000003A0 \text{ (hex)},$$

and P is a 96-bit string with the little-endian representation $P = P_0 \, \| \, P_1 \, \| \, P_2$:

$$P_0 = 0x00000080,$$
$$P_1 = 0x000003A0,$$
$$P_2 = 0x00000000.$$

Denote by MD4$^{[12]}$-Compress the compression function of MD4$^{[12]}$, i.e. the first two rounds of the MD4 compression function. In order to compute the MD4$^{[12]}$ hash value of M, first MD4$^{[12]}$-Compress is applied to $M_0, ..., M_{15}$ with the following fixed initial value, which is a part of the specification of MD4:

$$IV = 0x67452310 \ \ 0xEFCDAB89 \ \ 0x98BADCFE \ \ 0x10325476.$$

This gives the output

$$C = \text{MD4}^{[12]}\text{-Compress}(IV; M_0, ..., M_{15})$$
$$= 0xA86FDECC \ \ 0x25BF84C9 \ \ 0xDB95C842 \ \ 0xD0B260B9.$$

C is then the initial value for the second application of MD4$^{[12]}$-Compress with $M_{16}, ..., M_{28}, P_0, P_1, P_2$ as input. The output is the hash value of M:

$$\text{MD4}^{[12]}(M) = \text{MD4}^{[12]}\text{-Compress}(C; M_{16}, ..., M_{28}, P_0, P_1, P_2)$$
$$= 0x00000000 \ \ 0x00000000 \ \ 0x00000000 \ \ 0x00000000.$$

How to invert the first two rounds of MD4 compression

Suppose $IV = (IV_0, IV_1, IV_2, IV_3)$ and the compress value $C = (C_0, C_1, C_2, C_3)$ are given. Set $H = (H_0, H_1, H_2, H_3) := C - IV$. Our approach is described in the following tables (the underlined intries are those which are up-dated in the repsective steps):

ROUND ONE OF MD4 COMPRESSION

input	register A	register B	register C	register D	step
	IV_0	IV_1	IV_2	IV_3	
X_0	$*$	IV_1	IV_2	IV_3	step 0
X_1	$*$	IV_1	IV_2	$*$	step 1
X_2	$*$	IV_1	$*$	$*$	step 2
X_3	$*$	$*$	$*$	$*$	step 3
X_4	$*$	$*$	$*$	$*$	step 4
X_5	$*$	$*$	$*$	$*$	step 5
X_6	$*$	$*$	$*$	$*$	step 6
X_7	$*$	$*$	$*$	$*$	step 7
X_8	P_0	$*$	$*$	$*$	step 8
X_9	P_0	$*$	$*$	P_3	step 9
X_{10}	P_0	$*$	P_2	P_3	step 10
X_{11}	P_0	P_1	P_2	P_3	step 11
X_{12}	K	P_1	P_2	P_3	step 12
X_{13}	K	P_1	P_2	K	step 13
X_{14}	K	P_1	K	K	step 14
X_{15}	K	B_3	K	K	step 15

ROUND TWO OF MD4 COMPRESSION

input	register A	register B	register C	register D	step
X_0	K	B_3	K	K	step 16
X_4	K	B_3	K	K	step 17
X_8	K	B_3	K	K	step 18
X_{12}	K	B_2	K	K	step 19
X_1	K	B_2	K	K	step 20
X_5	K	B_2	K	K	step 21
X_9	K	B_2	K	K	step 22
X_{13}	K	B_1	K	K	step 23
X_2	K	B_1	K	K	step 24
X_6	K	B_1	K	K	step 25
X_{10}	K	B_1	K	K	step 26
X_{14}	K	B_0	K	K	step 27
X_3	H_0	B_0	K	K	step 28
X_7	H_0	B_0	K	H_3	step 29
X_{11}	H_0	B_0	H_2	H_3	step 30
X_{15}	H_0	H_1	H_2	H_3	step 31

That is, we assume that the contents of the A-, D-, and C-registers equals a constant K in steps 12,16,20,24, steps 13,17,21,25, and steps 14,18,22,26, respectively.

Idea. In round two the majority function g is applied, and by the described approach we can separate the contents in the B-registers from those in the other registers in round two.

Algorithm. Choose K and B_0 randomly. Now $X_0, ..., X_{11}$, and X_{15} are fixed by the steps 16,20, 24,28,17,21,25,29,18,22,26,30, and 31, respectively: for instance in step 16 we have the equation

$$K = (K + g(B3, K, K) + X_0 + K_1) \ll 3 = (K + K + X_0 + K_1) \ll 3$$

with $K_1 = \mathtt{0x5A827999}$. Thus $X_0 = (K \ll 29) - 2K - K_1$, and so on.

Now compute P_0, P_1, P_2, P_3 by applying $X_0, ..., X_{11}$ in step 0,...,11. The values P_0, P_1, P_2, P_3 allow next to determine X_{12}, X_{13}, X_{14} from the steps 12,13,14.

Use X_{15} in step 15 to compute B_3, X_{12} in step 19 to compute B_2, X_{13} in step 23 to compute B_1, and finally X_{14} in step 27 to compute another value for B_0 which is "derived from above". If the latter B_0-value matches the chosen one, then we have found a preimage, otherwise try again.

Thus we need about 2^{32} trials to be successful. However, the algorithm can be speeded up by a factor of 100 or more if we use "continuous approximation" for the computation of B_0 (where K is fixed respectively, in order to have sufficient continuity; see second part of the below C-progam).

References

1. H. Dobbertin, *Cryptanalysis of MD4*, Fast Software Encryption (Third Workshop on Cryptographic Algorithms, Cambridge 1996), Lecture Notes in Computer Science, Springer-Verlag 1996, pp. 55-72.
2. H. Dobbertin, *The status of MD5 after a recent attack*, CryptoBytes, The technical newsletter of RSA Laboratories, vol. 2/2, Sommer 1996, pp. 1-6.
3. R. Rivest, *The MD4 message-digest algorithm*, Request for Comments (RFC) 1320, Internet Activities Board, Internet Privacy Task Force, April 1992.
4. M.J.B. Robshaw, *On recent results for MD2, MD4 and MD5*, Bulletin 4, RSA Laboratories, November 1996 (see http://www.rsa.com/PUBS/).

Appendix

C-program inverting the reduced MD4 with cancelled third compression round

The first part of the program below is a modifiction of the above algorithm, which allows to match the redundancy in the input required by the padding rule. On a Pentium PC the program finds a hash-preimage in about 15-20 minutes on the average.

```
#define UL unsigned long
#define shift(x,i) (UL)(((x)<<(i))^((x)>>(32-(i))))
#define f(x,y,z) ((x)&(y) & (~(x))&(z))
#define g(x,y,z) (UL)((x)&(y) | (x)&(z) | (y)&(z))
#include <stdio.h>

main(int ac,char *av[]){
  int i,k,sh,trials,Zeros,Ones,record,weight;
  UL KK,KKK,KKKK,K1,diff,test;
  UL AX,BX,CX,DX,P0,P1,P2,P3,AY,BY,CY,DY;
```

```
UL X0,X1,X2,X3,X4,X5,X6,X7,X8,X9,X10,X11,X12,X13,X14,X15;
UL Delta_A,Delta_B0,B0_basic,P2_basic;
UL AA,B0,B1,B2,B3,C0,C1,C2,C3;
UL H0,H1,H2,H3,HH0,HH1,HH2,HH3,HHH0,HHH1,HHH2,HHH3;
UL Q0,Q1,Q2,Q3,IV0,IV1,IV2,IV3;
UL M0,M1,M2,M3,M4,M5,M6,M7,M8,M9,M10;
UL M11,M12,M13,M14,M15,M16,M17,M18,M19;
UL M20,M21,M22,M23,M24,M25,M26,M27,M28;
UL PP0,PP1,PP2;

if(ac!=2){
  fprintf(stdout,"Usage: %s seed\n",av[0]);
  exit(1);
}

srand(atoi(av[1]));

K1 = 0x5A827999;
KK = 0x57902134;
KKK= 0x57902134;

/* We have here a special case of a more general algorithm. In
general KK and KKK are different, but have only a small Hamming
difference. How to choose these constants will be explained in
the complete paper about this attack.*/

Zeros=0;
Ones=0;
trials=0;

/* Here you can specify the hash value  (HH0,HH1,HH2,HH3) */
HH0=0x0;
HH1=0x0;
HH2=0x0;
HH3=0x0;
/********************************************************/

HHH0=HH0;
HHH1=HH1;
HHH2=HH2;
HHH3=HH3;

/* Here starts the first part: searching M16,...,M28 */

START_I:

  record = 33;

  AA = KK;
  H2 = rand();
  H3 = rand()&0xffffff7f;
  H0 = KK;
  X15 = 0x0;
  P1 = shift(KK,13)-KK-X15;
  X14= 0x3A0;
  KKKK = shift(KKK+KK+X14+K1,13);
  H1 = shift(KKKK+g(H2,H3,KK)+X15+K1,13);

  IV0=HH0-H0;
  IV1=HH1-H1;
  IV2=HH2-H2;
  IV3=HH3-H3;

  X0 = shift(KK,29)-AA-KK-K1;
  X1 = shift(KK,29)-KK-KK-K1;
  X2 = X1;
  X3 = X1;
  X4 = shift(KK,27)-KK-KK-K1;
```

```
X5 = X4;
X6 = X4;
X7 = shift(H3,27)-KK-KK-K1;
X12= shift(KK,19)-KK-KK-K1;
X13= shift(KKK,19)-KK-KK-K1;
X14= shift(KKKK,19)-KKK-KK-K1;

P2_basic = rand();

AX=IV0;BX=IV1;CX=IV2;DX=IV3;
AX = shift(AX+f(BX,CX,DX)+ X0, 3);
DX = shift(DX+f(AX,BX,CX)+ X1, 7);
CX = shift(CX+f(DX,AX,BX)+ X2,11);
BX = shift(BX+f(CX,DX,AX)+ X3,19);
AX = shift(AX+f(BX,CX,DX)+ X4, 3);
DX = shift(DX+f(AX,BX,CX)+ X5, 7);
CX = shift(CX+f(DX,AX,BX)+ X6,11);
BX = shift(BX+f(CX,DX,AX)+ X7,19);
Q0=AX;Q1=BX;Q2=CX;Q3=DX;

for(i=0; i<250; i++){

  trials=trials+1;

  sh=i&0x1f;
  diff=shift(1,sh);
  P2 = P2_basic^diff;
  C3 = shift(P2+f(KK,AA,P1)+X14,11);

  P3  = shift(KK,25)-f(AA,P1,P2)-X13;
  P0  = shift(AA,29)-f(P1,P2,P3)-X12;

  X8  = shift(P0,29)-f(Q1,Q2,Q3)-Q0;
  X9  = shift(P3,25)-f(P0,Q1,Q2)-Q3;
  X10 = shift(P2,21)-f(P3,P0,Q1)-Q2;
  X11 = shift(P1,13)-f(P2,P3,P0)-Q1;

  C2  = shift(C3+KK+X8+K1,9);
  C1  = shift(C2+KK+X9+K1,9);
  C0  = shift(C1+KK+X10+K1,9);

  Delta_A = shift(H2,23)-g(H3,KK,KKKK)-K1-C0;
  Delta_A = Delta_A^X11;

  weight=0;
  for(k=0; k<32; k++){Delta_A=shift(Delta_A,1);weight=weight+(Delta_A&1);}

  if(weight<record+2){
    P2_basic = P2;
  }
  if(weight<record){
    record=weight;
    if(record<2){
if(record==1){Ones=Ones+1;}
      fprintf(stdout,"Part I:   Hamming dist. %i ",record);
      fprintf(stdout,"Trials %i ",trials);
      fprintf(stdout,"Ones %i  Zeros %i\n",Ones,Zeros);
      fprintf(stdout,"%8.8X %i\n\n",Delta_A,i);
    }
  }

  if(weight==0){

    Zeros = Zeros+1;
    test = g(KKKK,KK,C0)^KK;
    if(test!=0){goto START_I;}
    test = g(KKK,KK,C1)^KK;
    if(test!=0){goto START_I;}
```

```
        M16=X0;
        M17=X1;
        M18=X2;
        M19=X3;
        M20=X4;
        M21=X5;
        M22=X6;
        M23=X7;
        M24=X8;
        M25=X9;
        M26=X10;
        M27=X11;
        M28=X12;
        PP0=X13;
        PP1=X14;
        PP2=X15;
        trials=0;

        H0=IV0;
        H1=IV1;
        H2=IV2;
        H3=IV3;
        IV0=0x67452301;
        IV1=0xefcdab89;
        IV2=0x98badcfe;
        IV3=0x10325476;
        H0=-IV0+H0;
        H1=-IV1+H1;
        H2=-IV2+H2;
        H3=-IV3+H3;
      goto START_II;
    }
  }
  goto START_I;


  /* Here starts the second part: searching M0,...,M15 */

START_II:
  record=33;
  KK = rand();
  B0_basic=rand();

  X0 = shift(KK,29)-KK-KK-K1;
  X1 = shift(KK,29)-KK-KK-K1;
  X2 = shift(KK,29)-KK-KK-K1;
  X3 = shift(KK,29)-KK-KK-K1;
  X4 = shift(KK,27)-KK-KK-K1;
  X5 = shift(KK,27)-KK-KK-K1;
  X6 = shift(KK,27)-KK-KK-K1;
  X8 = shift(KK,23)-KK-KK-K1;
  X9 = shift(KK,23)-KK-KK-K1;
  X10= shift(KK,23)-KK-KK-K1;

  for(i=0; i<250; i++){

    trials=trials+1;

    sh=i&0x1f;
    diff=shift(1,sh);
    B0=B0_basic^diff;

    X7 = shift(H3,27)-g(H0,B0,KK)-KK-K1;
    X11= shift(H2,23)-g(H3,H0,B0)-KK-K1;
    X15= shift(H1,19)-g(H2,H3,H0)-K1-B0;
```

```
AX=IV0;BX=IV1;CX=IV2;DX=IV3;
AX = shift(AX+f(BX,CX,DX)+ X0, 3);
DX = shift(DX+f(AX,BX,CX)+ X1, 7);
CX = shift(CX+f(DX,AX,BX)+ X2,11);
BX = shift(BX+f(CX,DX,AX)+ X3,19);
AX = shift(AX+f(BX,CX,DX)+ X4, 3);
DX = shift(DX+f(AX,BX,CX)+ X5, 7);
CX = shift(CX+f(DX,AX,BX)+ X6,11);
BX = shift(BX+f(CX,DX,AX)+ X7,19);
AX = shift(AX+f(BX,CX,DX)+ X8, 3);
DX = shift(DX+f(AX,BX,CX)+ X9, 7);
CX = shift(CX+f(DX,AX,BX)+X10,11);
BX = shift(BX+f(CX,DX,AX)+X11,19);
P0=AX;P1=BX;P2=CX;P3=DX;

X12 = shift(KK,29)-P0-f(P1,P2,P3);
X13 = shift(KK,25)-P3-f(KK,P1,P2);
X14 = shift(KK,21)-P2-f(KK,KK,P1);

B3 = shift(P1+KK+X15,19);
B2 = shift(B3+X12+KK+K1,13);
B1 = shift(B2+X13+KK+K1,13);
Delta_B0 = shift(B1+X14+KK+K1,13)-B0;

weight=0;
for(k=0; k<32; k++){Delta_B0=shift(Delta_B0,1);weight=weight+(Delta_B0&1);}

if(weight<record+2){
  B0_basic = B0;
}
if(weight<record){
  record=weight;
  if(record<2){
   if(record==1){Ones=Ones+1;}
    fprintf(stdout,"Part II:   Hamming dist. 1  ");
    fprintf(stdout,"Trials %i  ",trials);
     fprintf(stdout,"Ones %i\n",Ones);
    fprintf(stdout,"%8.8X %i\n\n",Delta_B0,i);
   }
}

if(weight==0){
  fprintf(stdout,"Cancel the third round ");
  fprintf(stdout,"of the MD4 compression function,\n");
  fprintf(stdout,"then the following message M=M0,...,M28 has ");
  fprintf(stdout,"the hash value\n");
  fprintf(stdout,"H  = 0x%8.8X ",HHH0);
  fprintf(stdout,"0x%8.8X ",HHH1);
  fprintf(stdout,"0x%8.8X ",HHH2);
  fprintf(stdout,"0x%8.8X:\n\n",HHH3);
  fprintf(stdout,"M0 = 0x%8.8X;  ",X0);
  fprintf(stdout,"M1 = 0x%8.8X;\n",X1);
  fprintf(stdout,"M2 = 0x%8.8X;  ",X2);
  fprintf(stdout,"M3 = 0x%8.8X;\n",X3);
  fprintf(stdout,"M4 = 0x%8.8X;  ",X4);
  fprintf(stdout,"M5 = 0x%8.8X;\n",X5);
  fprintf(stdout,"M6 = 0x%8.8X;  ",X6);
  fprintf(stdout,"M7 = 0x%8.8X;\n",X7);
  fprintf(stdout,"M8 = 0x%8.8X;  ",X8);
  fprintf(stdout,"M9 = 0x%8.8X;\n",X9);
  fprintf(stdout,"M10= 0x%8.8X;  ",X10);
  fprintf(stdout,"M11= 0x%8.8X;\n",X11);
  fprintf(stdout,"M12= 0x%8.8X;  ",X12);
  fprintf(stdout,"M13= 0x%8.8X;\n",X13);
  fprintf(stdout,"M14= 0x%8.8X;  ",X14);
  fprintf(stdout,"M15= 0x%8.8X;\n",X15);
  fprintf(stdout,"M16= 0x%8.8X;  ",M16);
  fprintf(stdout,"M17= 0x%8.8X;\n",M17);
```

```
        fprintf(stdout,"M18= 0x%8.8X;  ",M18);
        fprintf(stdout,"M19= 0x%8.8X;\n",M19);
        fprintf(stdout,"M20= 0x%8.8X;  ",M20);
        fprintf(stdout,"M21= 0x%8.8X;\n",M21);
        fprintf(stdout,"M22= 0x%8.8X;  ",M22);
        fprintf(stdout,"M23= 0x%8.8X;\n",M23);
        fprintf(stdout,"M24= 0x%8.8X;  ",M24);
        fprintf(stdout,"M25= 0x%8.8X;\n",M25);
        fprintf(stdout,"M26= 0x%8.8X;  ",M26);
        fprintf(stdout,"M27= 0x%8.8X;\n",M27);
        fprintf(stdout,"M28= 0x%8.8X;\n\n",M28);
        fprintf(stdout,"The corresponding padding string is P0,P1,P2:\n");
        fprintf(stdout,"P0 = 0x%8.8X;  ",PP0);
        fprintf(stdout,"P1 = 0x%8.8X;  ",PP1);
        fprintf(stdout,"P2 = 0x%8.8X;\n\n",PP2);
        exit(1);
    }
  }
  goto START_II;
}
```

Differential Cryptanalysis of KHF

David Wagner

University of California, Berkeley
daw@cs.berkeley.edu

Abstract. Bakhtiari et al recently proposed a fast message authentication primitive called KHF. This paper shows that KHF is highly vulnerable to differential cryptanalysis: it can be broken with about 37 chosen message queries. This suggests that the KHF design should be reconsidered.

1 Introduction

Recent applications in secure networking have spurred research in cryptographic primitives for message authentication. In particular, there is great demand for a high-speed MAC that can be implemented in software. In 1996, Bakhtiari, Safavi-Naini, and Pieprzyk proposed a new MAC primitive, called KHF, for fast software message authentication [BSF95]. See Section 2 for a brief description of their primitive.

One of the main contributions of the KHF work is its careful attention to using Boolean functions which can be shown to have very good non-linearity, as well as other desirable theoretical properties. Nonetheless, this paper shows how to break KHF efficiently, despite its solid theoretical foundation in cryptographic Boolean function theory.

In this paper we show how to break KHF with differential cryptanalysis [BS93]. The attack requires just 37 chosen messages, and is described in Section 3.

2 Description of KHF

We give a brief overview of the KHF message authentication algorithm here, omitting those details that are irrelevant to our attack.

Before processing, padding is prepended and appended to the message, and the result is split into 32-bit blocks denoted by $M_1, M_2, \ldots, M_n$. Two 128-bit state buffers ($X_{1\ldots4}$ and $Y_{1\ldots4}$) and a 512-bit redundancy buffer ($B_{1\ldots16}$) are used internally. The key is fed into the initial values of the X, Y buffers and optionally also into the padding, with the exact details depending on the mode of operation.

We perform n rounds, one for each message block. The i-th round uses M_i, B_i, and Y to update the three buffers: it calculates $T = f_i(Y_1, Y_2, Y_3, Y_4, M_i)$, modifies X using T, xors M_i into B_{a_i} (where a_i is derived from T), and then

S. Vaudenay (Ed.): Fast Software Encryption – FSE'98, LNCS 1372, pp. 293–296, 1998.
© Springer-Verlag Berlin Heidelberg 1998

modifies Y using T and B_i. (These modifications do not depend implicitly on any unmentioned variables.)

The round functions f_i are defined as

$$
f_i(A,B,C,D,E) = \begin{cases} (A\&E) \oplus (B\&C) \oplus ((B \oplus C)\&D) & i = 1 \bmod 5 \\ A \oplus (B\&(A \oplus D)) \oplus (((A\&D) \oplus C)\&E) & i = 2 \bmod 5 \\ A \oplus (C\&D\&E) \oplus ((A\&C)|(B\&D)) & i = 3 \bmod 5 \\ B \oplus ((D\&E)|(A\&C)) & i = 4 \bmod 5 \\ D \oplus E \oplus (((D\&E) \oplus A)\& \sim (B\&C)) & i = 0 \bmod 5 \end{cases}
$$

where the logical operations $\oplus, \&, |, \sim$ are performed on each bit of the 32-bit words independently. Here $\oplus$ represents bitwise exclusive-or, $\&$ represents bitwise and, $|$ represents bitwise or, and $\sim$ represents bitwise logical negation.

3   Analysis

First, we exhibit a useful high-probability differential characteristic [BS93] for KHF.

The main observation is that a one-bit difference into the input of the round function f is very likely to yield a zero output difference (i.e. a collision). This arises from the fact that the f_i operate on each bit position independently; in other words, the KHF round functions have very poor diffusion properties across bit positions. More precisely, flipping one bit in E leaves $f_i(A,B,C,D,E)$ unchanged with the probabilities given below:

$i \bmod 5$	1	2	3	4	0
Prob	1/2	1/2	3/4	5/8	3/8

This corresponds to a differential characteristic for f_i of the form $(0,0,0,0,2^j) \mapsto 0$. Such characteristics hold with very high probabilities despite the focus on the construction of f from highly non-linear Boolean function theory.

Next, we show how to extend the high-probability differential characteristic for the f function into a characteristic for the whole algorithm. Forcing such a difference into one M_i leaves $T = f_i(Y_1, Y_2, Y_3, Y_4, M_i)$ unchanged, and thus leaves the X and Y buffers unchanged; however, the B_{a_i} word does change. Therefore, to obtain a collision for the entire KHF calculation, we introduce the same one-bit difference into two message words, M_i and M_j where $i < j$. While this introduces a difference into the B buffer at the i-th round, we hope that it gets canceled out later in the j-th round before the difference has a chance to propagate. This gives us two conditions for success: first, we must have $a_i = a_j$; second, we require that B_{a_i} is not present in the list $B_i, B_{i+1}, \ldots, B_{j-1}$. (Subscripts on B are taken modulo 16.)

To calculate the characteristic's probability, we can model the a_i as effectively random integers selected from the set $\{1, 2, \ldots, 16\}$, since we do not know the value of T or Y. For best chance of success we suggest taking $j = i + 1$ and

$i = 3 \bmod 5$; then the probability that both conditions hold is

$$\frac{3}{4} \cdot \frac{5}{8} \cdot \frac{15}{16} \cdot \frac{1}{16} = \frac{225}{8192} \approx \frac{1}{36}.$$

To recap, introducing the same one-bit difference into M_i and M_{i+1} (for some $i = 3 \bmod 5$) will yield a collision in the final output of KHF with this $1/36$ probability. We have implemented the attack and empirically confirmed this analysis.

We now describe how to break KHF using differential cryptanalysis. We have given a class of differences which, when xored into the message input, yield a collision in the KHF output with very high probability. This makes breaking the KHF message authentication scheme easy. For example, to forge the MAC digest $\mathrm{KHF}(K, M)$ on a message M under a chosen-plaintext threat model, we could obtain the MAC digests $D_i = \mathrm{KHF}(K, M \oplus \Delta_i)$ (where each Δ_i is one of the xor differences of Hamming weight two described above) and look for repetitions in the list of D_i. With high probability the repeated value will be the desired MAC digest $\mathrm{KHF}(K, M)$. By using the high-probability characteristic described above, this attack will need only about 72 chosen plaintext queries[1] to break a system using KHF for message authentication.

Other attacks are also possible. For example, to learn $\mathrm{KHF}(K, M\|Y)$ (where $\|$ denotes concatenation on 32-bit boundaries), we offer the following differential attack needing 37 chosen plaintext queries. Find M' (of the same length as M) so that $\mathrm{KHF}(K, M\|X) = \mathrm{KHF}(K, M'\|X)$ (where X has the same length as Y but is otherwise arbitrary), by using the differential attack developed above. The collision $\mathrm{KHF}(K, M\|X) = \mathrm{KHF}(K, M'\|X)$ typically arises because the internal states after processing M and M' match. Now use a single chosen plaintext query to learn $\mathrm{KHF}(K, M'\|Y)$. The relation $\mathrm{KHF}(K, M\|Y) = \mathrm{KHF}(K, M'\|Y)$ will hold with very high probability; this lets us deduce $\mathrm{KHF}(K, M\|Y)$, as desired. Note that we have never used $M\|Y$ in a chosen-message query, but we obtain the MAC digest corresponding to $M\|Y$, so we have broken the MAC primitive. The attack lets us break the KHF message authentication scheme with about 37 chosen plaintexts, on average.

This latter attack is based on the observation that "internal collisions" can lead to MAC forgery. Of course, this property is hardly novel; it was one of the central techniques used to break a number of MAC primitives in [PO95].

4 Conclusion

KHF was designed around a set of Boolean functions with excellent non-linearity properties. The round function f on 32-bit input words was constructed by applying a good Boolean function to each bit position independently. This ensures

---

[1] One caveat: this assumes that the message is sufficiently long (60 bytes) so that we can easily find 72 two-bit Δ_i values of the right form. Shorter messages can also be attacked with similar differential characteristics that have a somewhat lower probability; the disadvantage is that we will need slightly more chosen plaintext queries.

that we achieve the best possible diffusion within any one bit position, but its fatal flaw is that it gives very poor diffusion across bit positions.

Our cryptanalysis of KHF takes advantage of this weakness in the round function. We note that this design process—start with a good Boolean function and extend it to a n-bit function by bitwise parallel evaluation—has fundamental flaws: no matter how many good properties the Boolean function has, the full n-bit round function will have very poor diffusion across bit positions. Paying too much attention to the theoretical properties of the underlying Boolean function can be dangerous, if that causes the rest of the algorithm structure to receive less attention. It is important to start from a solid foundation, but that alone is not enough.

To summarize, KHF suffers from serious vulnerabilities to differential attack, and should be considered insecure. The flaws that we have found do not inspire confidence in the design process, nor is it clear whether there is any simple way to fix these flaws without radically modifying the structure of KHF. Therefore, we recommend that the KHF design be abandoned.

References

BSF95.   S. Bakhtiari, R. Safavi-Naini, and J. Pieprzyk, "Keyed hash functions," *Cryptography: Policy and Algorithms*, E. Dawson and Jovan Golic (Eds), Lecture Notes in Computer Science, vol. 1029, Springer-Verlag, 1996, pp.201-214.

BS93.   E. Biham and A. Shamir, *Differential Cryptanalysis of the Data Encryption Standard*, Springer-Verlag, 1993.

PO95.   B. Preneel, P.C. van Oorschot, "MDx-MAC and building fast MACs from hash functions," *Advances in Cryptology—CRYPTO '95*, Springer-Verlag, 1995, pp. 1–14.

Author Index